MOLECULAR
ASSOCIATION

MOLECULAR ASSOCIATION

VOLUME 1

Edited by

R. FOSTER

Chemistry Department
University of Dundee

1975

ACADEMIC PRESS
LONDON   NEW YORK   SAN FRANCISCO
A Subsidiary of Harcourt Brace Jovanovich, Publishers

CHEMISTRY

ACADEMIC PRESS INC. (LONDON) LTD.
24/28 Oval Road,
London NW1 7DX

United States Edition published by
ACADEMIC PRESS INC.
111 Fifth Avenue
New York, New York 10003

Copyright © 1975 by
ACADEMIC PRESS INC. (LONDON) LTD.

Library of Congress Catalog Card Number: 75 15221
ISBN: 0 12 262701 6

Printed in Great Britain by
Page Bros (Norwich) Ltd, Norwich

Contributors

R. S. DAVIDSON, *Chemistry Department, University of Leicester, Leicester LE1 7RH, England.*

K. M. C. DAVIS, *Chemistry Department, University of Leicester, Leicester LE1 7RH, England.*

D. J. KLEIN, *Department of Physics, University of Texas, Austin, Texas, U.S.A.*

N. KULEVSKY, *Chemistry Department, University of North Dakota, Grand Forks, North Dakota, 58201, U.S.A.*

Z. G. SOOS, *Department of Chemistry, Princeton University, Princeton, N.J., U.S.A.*

Acknowledgements

Grateful acknowledgements are due to the following for permission to use published data. References to the authors concerned are made in the text.

The American Chemical Society (*Journal of the American Chemical Society*): Ch. 4, Fig. 6; (*Journal of Physical Chemistry*): Ch. 3, Fig. 4.

The American Institute of Physics (*Journal of Chemical Physics*): Ch. 1, Figs 5, 16, 17, 18, 20, 24; (*Physical Review*): Ch. 1, Figs 10, 11, 14, 15.

The American Physical Society (*Physical Review Letters*): Ch. 1, Fig. 9.

The Chemical Society (*Journal of the Chemical Society*): Ch. 3, Table 1.

The Chemical Society of Japan (*Bulletin of the Chemical Society of Japan*): Ch. 1, Fig. 1.

Elek Science, Ltd. (Molecular Complexes, Volume 1, ed. R. Foster) Ch. 3, Tables 2, 4.

International Union of Crystallography (*Acta Crystallographica*): Ch. 1, Fig. 7.

Koninklijke Nederlandse Chemische Vereniging (*Recueil des Travaux chimiques des Pays-bas*): Ch. 3, Fig. 5.

Pergamon Press, Ltd., (*Spectrochimica Acta*): Ch. 3, Fig. 3; (*Tetrahedron Letters*): Ch. 4, Fig. 7.

Preface

This series of volumes is concerned with the interactions between various chemical species which, although giving rise to new properties, nevertheless normally cause relatively little perturbation of the component species. As has been pointed out previously[1], there is finally no sharp demarcation between the weakest of such interactions, for example, the interatomic attraction in liquid helium, and other interactions, which, though classed as complexes, involve binding energies which are comparable with those of many classical chemical bonds. Between these extremes are many interactions which are conventionally termed "complexes" such as the weak electron–donor–acceptor (EDA) complexes and hydrogen-bonded complexes.

Indeed the term "complex" has assumed a variety of meanings in the language of chemistry. It has been used to describe the products of certain reactions in which, beyond argument, electrons have been redistributed in such a way that we should really describe the product as a new compound: for example, sigma complexes formed by the attack of nucleophiles on aromatic species; also inorganic coordination complexes. For the chemist the term "complex" also has the suggestion of a definite stoichiometry. However, we may well wish to discuss systems where we shall not want to presume such a model, but rather to suppose that we are dealing with a cluster of molecules which may in themselves only represent a time-averaged model of an aggregation of varying geometry and varying stoichiometry.

For these reasons, and also for bibliological convenience of distinguishing this present series from a two-volume collection of essays recently published[2], the particular title "Molecular Association" has been chosen. However, it is not intended to exclude interactions which may involve *non*-molecular species (e.g. ions, atoms) although the majority of the systems discussed will doubtless be concerned with interactions between molecules.

I should also emphasise that I do not wish to restrict the subject matter to electron–donor–acceptor interactions, although again there is no doubt that many contributions will involve discussion of these interactions, since there has been such a remarkable acceleration of interest in this area. It is perhaps relevant to note that it is just twenty-five years since this fertile field was effectively brought to the notice of a wide spectrum of scientists through the incisive work of Robert Mulliken.[3] There continues to be considerable debate concerning the nature of the forces involved in the ground state of these complexes.[4] However, there is little doubt as to the essential nature of the new absorption band as assigned by Mulliken. Interest in EDA interactions ranges from the so-called "contact charge-transfer" phenomena, through the still relatively weak interactions in the solid liquid and vapour phases, to systems where electron transfer can readily occur. Practical applications of these latter systems are rapidly increasing in electrically conducting devices including photoconduction. These often involve polymeric substrates particularly in the field of photo-image devices. Some of these systems are effectively ion–ion interactions following the complete transfer of an electron from the donor to the acceptor.

The nature of molecular interactions in biological systems, both in "normal" systems and in cases where a foreign (drug) species has been introduced, is one of the fundamental questions in molecular biology and the point at which the biophysicist, biochemist, biologist and chemist meet.

As the mechanisms of chemical reactions are probed more carefully, the nature of the intermolecular forces before, during, and at the end of the reaction become of importance. In this area we require information, not only of the ground-states of molecules but also in systems where one or more of the molecular species is in an electronically excited state.

It is intended in this and subsequent volumes to provide contributions which are critical reviews rather than exhaustive literature surveys with little or no comment.

The present volume contains a chapter by Soos and Klein on a theoretical approach to charge-transfer in solid state complexes. There is a chapter by Davidson on the involvement of EDA complexes in photochemical reactions. This essay is in many ways complementary to the recent critical reviews by Colter and Dack[5] of the implication of EDA complexes in non-photo-chemical reactions. The chapter by Davis on the effect of solvent on the properties of EDA complexes collates information from many sources and provides an obvious follow-on to the recent review of such complexes in the vapour phase by Tamres.[6]

The chapter on dielectric properties of molecular complexes in solution by Kulevsky has been purposely restricted in its scope. This decision was made when it became known that a wider-ranging review was being written by Price for a monograph edited by Yarwood.[7]

I should like to thank the contributors and the publishers for their patient cooperation in the various stages of production of this volume.

February 1975

                                                                    Roy Foster

REFERENCES

1. R. Foster, "Organic Charge-transfer Complexes", Academic Press, London and New York (1969) p. v.
2. "Molecular Complexes", Vols 1 and 2 (ed. R. Foster), Elek Science, London; Crane Russak, New York (1973 and 1974).
3. See reference 1. *Also*: R. S. Mulliken and W. B. Person, "A Lecture and Reprint Volume", Wiley-Interscience, New York (1969).
4. M. W. Hanna and J. L. Lippert. In "Molecular Complexes", Vol. 1, (ed. R. Foster), Elek Science, London; Crane Russak, New York (1973).
5. A. K. Colter and M. R. J. Dack. In "Molecular Complexes". Vols 1 and 2 (ed. R. Foster), Elek Science, London; Crane Russak, New York (1973 and 1974).
6. M. Tamres. In "Molecular Complexes", Vol. 1, (ed. R. Foster), Elek Science, London; Crane Russak, New York (1973).
7. A. H. Price. In "Spectroscopy and Structure of Molecular Complexes," (ed. J. Yarwood), Plenum Press, London and New York (1973), Ch. 7.

CONTENTS

Chapter 1. Charge-Transfer in Solid State Complexes
Z. G. Soos and D. J. Klein

Chapter 2. Dielectric Properties of Molecular Complexes in Solution
N. KULEVSKY

Chapter 3. Solvent Effects on Charge–Transfer Complexes
K. M. C. DAVIS

Chapter 4. Photochemical Reactions Involving Charge-Transfer Complexes

R. S. DAVIDSON

1 Charge-Transfer in Solid-State Complexes

Zoltán G. Soos

Department of Chemistry, Princeton University, Princeton, N.J., U.S.A.

Douglas J. Klein

Department of Physics, University of Texas, Austin, Texas, U.S.A.

1

I. INTRODUCTION

A. Contrast of Solution and Solid-state complexes

Solutions of electron donors (D) and acceptors (A) exhibit absorptions not associated with either individual molecule.[1] Mulliken[2] proposed that the new absorptions involve an electron transfer from D to A, as indicated by

$$\psi_{\text{complex}}(DA) \xrightarrow{h\nu_{\text{CT}}} \psi_{\text{complex}}(D^+A^-). \tag{1}$$

Here $h\nu_{\text{CT}}$ is the charge–transfer (CT) excitation energy, and the CT band is explicitly associated with the molecular complex formed between a D and an A molecule. The nature of such DA complexes has been the subject of extensive experimental and theoretical studies.[3–5] The energetics of dimer formation, the geometry of the complex, the systematics of varying D, A, and solvents, and the nature of the CT absorption have been explored. The CT excitation is greatly enhanced,[3, 4, 6] as suggested by Mulliken,[2] *via* small admixtures of the excited singlet configuration $^1|D^+A^-\rangle$ into the ground state, $|DA\rangle$. However, the additional configuration-interaction stabilization due to this small admixture does not[6, 7] usually dominate the binding energy or the geometry of the complex. As discussed in Section II, the full quantum-mechanical treatment of a DA dimer thus requires explicit consideration of a variety of intermolecular forces.

There have also been extensive experimental studies of crystalline complexes of π-molecular donors and acceptors.[8] It is now evident that aromatic donors and acceptors usually crystallize in molecular stacks.[9–11] All intermolecular separations are normal van der Waals distances, except for a slight ($\lesssim 0.4$ Å) shortening of the interplanar separation. The π overlap of D and A molecules along the stacks then leads to a CT band in addition to the molecular (Frenkel) excitations expected in molecular crystals. In fact, the observation[12] of solid-state CT bands polarized perpendicular to the molecular plane, and thus parallel to the stack axis, was an early confirmation of Mulliken's proposed electron-transfer mechanism.[2, 3, 4]

In spite of the early recognition of CT bands in π-molecular solids, it has proved very difficult to apply the Mulliken theory to crystalline complexes. The reason is quite simple. Solid-state π-molecular complexes contain infinite molecular stacks,[9, 10] for example[13] of the type ... DADA ... shown in Fig. 1 for D = N,N,N',N'-tetramethyl-p-phenylenediamine (TMPD) and A = tetracyanobenzene (TCNB). Even the most highly idealized application of configuration interaction, with each donor restricted to D or D^+ and each acceptor to A or A^-, leads to an infinite secular determinant. The theoretical description of solid-state complexes thus lags far behind the analysis of DA dimers, however difficult and even controversial the dimer problem has been.[3, 4, 6] It must be clearly recognized at the outset that many results for dimers cannot be simply transferred to DA solids.

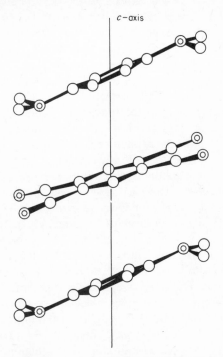

FIG. 1. The infinite donor–acceptor stack in TMPD–TCNB viewed along [1$\bar{1}$0]. [Reproduced with permission from Y. Ohashi, H. Iwasaki and Y. Saito, *Bull. Chem. Soc. Japan*, **40**, 1789 (1967)].

The thermodynamics of complex formation for dimers can readily be extended to DA crystals[3, 8, 14] but it is then difficult to determine the various polarization, CT, and other processes relating the energy of the solid to that of isolated D and A molecules. The importance of the stability of the dimer, which then controls the observed equilibrium in solution, is greatly diminished in the solid, which provides an inherently static problem. By contrast, the equilibrium geometry is readily available in the solids, but must usually be inferred from rather incomplete data in solution. Thus the experimental focus also shifts on going from dimers in solution to DA solids. The fundamental CT excitation indicated in (1) is, in fact, the principal feature of interest in both dimers and solids.

B. Molecular–Exciton Approach

We develop in Sections II and III a general analysis of charge transfer in π-molecular crystals. The underlying assumptions of molecular-exciton theory[15–18] are used throughout. The observation that the solid-state spectra of typical closed-shell molecular crystals, for example of benzene, naphthalene, or anthracene, are closely related to gas-phase spectra makes it natural to describe such molecular crystals in terms of weakly overlapping, unperturbed molecules. Molecular-exciton theory[15–18] then provides, often in great detail, a phenomenological approach for predicting the small crystal shifts and splittings relative to gas-phase transitions. In most cases, degenerate first-order perturbation theory suffices, since the crystal interactions are small compared to the spacing of the molecular energy levels. Vibrational, triplet, and singlet excitons all reflect shifts and splittings induced by the lower symmetry of the crystalline environment.

The crystal structures[9–11] of π-molecular CT crystals support the occurrence of weakly overlapping, essentially molecular D and A constituents. Their infrared and ultraviolet spectra also show absorptions characteristic of D and A molecules.[19, 20] Even the strongest π donors and acceptors, which crystallize[9] in *ionic* lattices of D^+ and A^- radicals, show spectra characteristic of the molecular ion-radicals. These spectroscopic observations provide[9, 19, 20] the experimental basis for the sharp separation of π-molecular CT crystals into those with largely neutral and those with largely ionic ground-states. The sharp separation has been understood theoretically[21, 22] for weakly overlapping molecular crystals with the Madelung (electrostatic) energy providing the stability of the ionic lattice.[22, 23] Thus π-molecular CT solids contain largely unperturbed molecular constituents, and it is natural to use isolated-molecule wave functions in a phenomenological approach to solid-state complexes.[24]

Molecular-exciton theory[15–18] is especially useful when the crystal interactions are weak enough to permit focusing on a *single* molecular excitation, which becomes N-fold degenerate in a solid of N noninteracting molecules. In most cases, the *lowest* molecular triplet or singlet state provides the basis for the Frenkel excitons of typical closed-shell molecular solids. In sharp contrast, the restriction to the *lowest* electronic excitation in

π-molecular CT crystals leads to a theory[25] in which only CT states are included. The occurrence of a lowest CT, rather than Frenkel, excitation results in a fundamentally different form of molecular-exciton theory.[24, 25] The additional low-energy electron transfer in π-molecular CT solids is the decisive electronic feature of the solid-state complexes reviewed here. The two-state configuration-interaction picture of the idealized DA dimer[3, 4] can then be extended to the infinite molecular stacks encountered in the solid state.

The most striking difference between CT and and the usual molecular solids occurs in ion-radical solids based on the strongest π-donors and acceptors. Crystals of these *open-shell* molecular ions, D^+ and/or A^-, are contrasted in Table 1 with crystals of nonoverlapping closed-shell molecules.

TABLE I. Comparison of ionic charge–transfer and free-radical crystals with molecular solids in the limit of non-overlapping sites

Property	Ionic CT or FR crystal	Molecular crystal
Constituents	Ion-radicals D^+ or A^- Open shell ($S = \frac{1}{2}$)	Neutral molecules Closed shell
Crystal ground State	2^N spin degeneracy Paramagnetic	Nondegenerate Diamagnetic
Lowest excitation	CT ($\parallel$ to Stack)	Frenkel (molecular)
Crystal binding	Madelung and intermolecular	Intermolecular

The enormous spin degeneracy in either ionic CT or free-radical (FR) solids should especially be noted. The magnetic properties of ion-radical solids[23, 24] provide important information about the ground-state charge distribution and complement optical studies of CT absorptions. The molecular-exciton approach to solid-state complexes is applicable to any molecular solid with low-lying CT excitation.[25] We nevertheless will focus primarily on the relatively rare *ionic* CT and FR solids, rather than on the more common *neutral* complexes of weak π-donors and acceptors, because the ion-radical systems exhibit unusual magnetic, electric, and optical properties that have been central to solid-state studies.[24, 26, 27]

C. π-Molecular Ion–Radical Solids

The strong π-acceptors 7,7,8,8-tetracyanoquinodimethane (TCNQ) and chloranil are shown in Fig. 2, together with the strong π-donors p-phenylenediamine (PD), TMPD, and tetrathiofulvalene (TTF). While these molecules

Representative Molecules

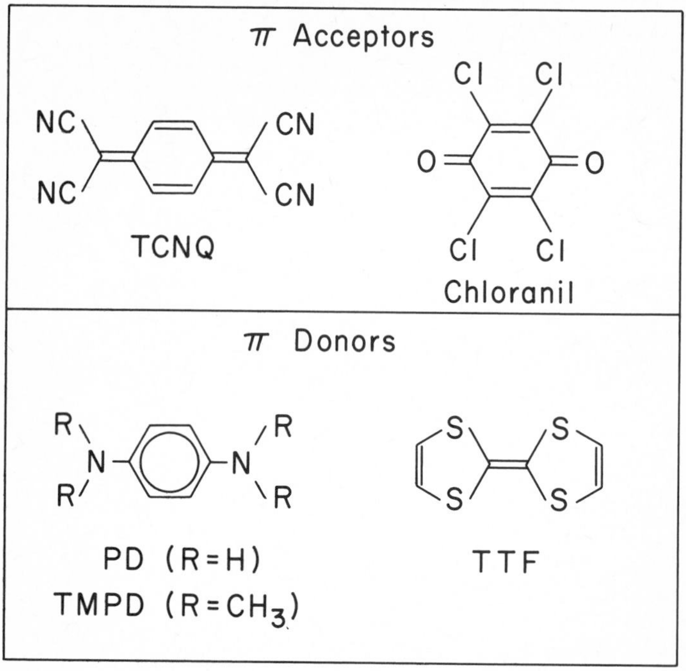

FIG. 2. Representative examples of strong π-donors and strong π-acceptors.

are by no means the only donors and acceptors that form ionic CT or FR crystals, they are certainly among the most versatile and thoroughly studied.[24] TMPD–TCNQ and TMPD–chloranil form mixed[9] stacks of ion-radicals, as shown by $\ldots D^+A^-D^+A^-\ldots$. The low-lying CT excitation is now a "back charge-transfer"[21]

$$|\ldots D^+A^-D^+A^-\ldots\rangle \xrightarrow{h\nu_{CT}} |\ldots D^+ADA^-\ldots\rangle. \qquad (2)$$

and involves returning the electron to the donor. TCNQ forms extensive series of FR salts,[28] with diamagnetic metallic or organic cations located between different $\ldots A^-A^-A^-A^-\ldots$ stacks providing electrical neutrality, as discussed in greater detail in Section III. Chloranil also forms[29] FR salts with alkali metals and with organic or organometallic bases.[9] In these FR crystals, A^- acts as both the donor and the acceptor, as indicated by the CT excitation

$$|\ldots A^-A^-A^-A^-\ldots\rangle \xrightarrow{h\nu_{CT}} |\ldots A^-AA^{--}A^-\ldots\rangle. \qquad (3)$$

Ion-radical salts are therefore solid-state analogues of self-complexes.[30] TMPD and TTF form analogous self-complexes, for example with various halides, in which the cation radicals D^+ form stacks and the fundamental CT excitation is

$$|\dots D^+ D^+ D^+ D^+ \dots\rangle \xrightarrow{hv_{CT}} |\dots D^+ DD^{++} D^+ \dots\rangle. \qquad (4)$$

CT processes in self-complexes based on open-shell molecular ion-radicals do not involve an *orbital* excitation, since the highest-occupied molecular orbital is half-filled and can accommodate another electron; hv_{CT} is then related to the correlation energy for placing two electrons on the same molecular site.

As indicated in Section III, there are many FR salts with complex stoichio-metry, as shown by $Cs_2 TCNQ_3$. Now formally neutral TCNQ sites also occur in the ion-radical stack.[9, 31] When the sites are crystallographically non-equivalent, a mixed-valence organic system is achieved. Iida[32] has suggested that the additional, even lower-energy, CT bands in complex salts involve transfers between radical and neutral sites, as shown by

$$|\dots AA^- A^- AA^- A^- AA^- A^- \dots\rangle \xrightarrow{hv_{CT}} |\dots AA^- AA^- A^- A^- AA^- A^- \dots\rangle. \qquad (5)$$

for a 2:3 salt like $Cs_2 TCNQ_3$. Now there is no requirement that a TCNQ site become doubly charged and the lower energy of (5) is readily understood. The complex salt $TMPD(TCNQ)_2$ contains segregated stacks[33] in which all TCNQ sites are equivalent; several other 1:2 salts have served to establish[9] the geometry of $TCNQ^{-\frac{1}{2}}$. It is evident that complex salts permit additional, potentially very facile,[34, 35] mechanisms for charge conduction along the FR stack.

Ion-radical crystals provide a small but versatile subclass of π-molecular solids with a lowest CT excitation. Their potentially very high conductivity, as shown by TTF–TCNQ[36–38] or NMP–TCNQ[39–41] (NMP = N-methylphenazine), make them organic "metals", while other TCNQ systems span the entire spectrum from semiconductors to insulators.[24] A number of recent reviews[26, 27, 42, 43] have examined these stacked one-dimensional conductors, and they remain an area of active interest to both solid-state physicists and chemists. In the present context of charge-transfer in solid-state complexes, the very unusual and varied physical properties of ion-radical systems provide a natural test for the phenomenological theory. Indeed, the occurrence of a lowest CT excitation will be combined with structural information to account, at least qualitatively, for the unusual magnetic, optical, and electric properties.

D. Hubbard Models and CT Crystals

The Mulliken theory of CT in dimers and the related optical studies in solid-state complexes have been of primary interest to chemists. One of our major goals is to demonstrate that π-molecular CT solids have important common features with the widely used, idealized Hubbard model[44] of solid-state physics, the properties of which are reviewed in Section IV. The physical origins of Hubbard models are illustrated by Van Vleck[45] for a hypothetical chain of H atoms which, like FR crystals, have one unpaired electron per site. The correlation energy $U > 0$ is defined to be the energy for putting two spin-paired electrons on the same site, as might occur in a CT excitation. The Mulliken CT integral t is defined in Section II.A and allows electron transfer among adjacent sites while conserving the spin orientation. The resulting Hamiltonian for a one-dimensional lattice is

$$H_{Hu} = t \sum_{n\sigma} (a_{n\sigma}^+ a_{n+1\sigma} + a_{n+1\sigma}^+ a_{n\sigma}) + U \sum_n a_{n\alpha}^+ a_{n\beta}^+ a_{n\beta} a_{n\alpha} . \tag{6}$$

where the electron creation and annihilation operators $a_{n\sigma}^+$, $a_{n\sigma}$ are defined for the odd electron at each site. The derivation of Hamiltonians related to (6) for CT crystals is given in Section II, while a summary of the properties of (6) is presented in Section IV.

The Hubbard Hamiltonian provides a simple, idealized model for studying both electron correlation effects, as represented by the U term in (6), and electron delocalization effects, as represented by the t term. The $t = 0$ and $U = 0$ limits correspond, respectively, to isolated sites and to a one-band metal. When t/U is small and there is one electron per site, (6) represents a Mott insulator;[46] Hubbard models have consequently been applied to metal–insulator transitions in various transition-metal oxides.[47, 48] The strong correlations for $U \gg t$ are further manifested by the effective Heisenberg spin Hamiltonian,[45, 49] a limiting case of (6) discussed in Section IV

$$H_H = \frac{4t^2}{U} \sum_n (S_n \cdot S_{n+1} - \tfrac{1}{4}). \tag{7}$$

H_H describes the low-temperature magnetic properties of a free-radical chain with $S_n = \tfrac{1}{2}$ at the nth sites. The antiferromagnetic exchange $2J = 4t^2/U > 0$ provides a direct connection[23] between the magnetic and optical properties of CT and FR crystals, in which $h\nu_{CT} = \Delta E_{CT}$ is a measure of the energy for producing a doubly-occupied site. The identification[23, 50]

$$\Delta E_p \approx t^2/\Delta E_{CT} . \tag{8}$$

relates the activation energy for paramagnetism, ΔE_p, to the CT excitation, ΔE_{CT}, and to the Mulliken integral, t, which also enters directly in electric conduction. The magnetic properties of CT crystals summarized in Section V illustrate some of the consequences of antiferromagnetic exchange.

These general considerations apply to any solid containing open-shell atomic or molecular sites. Le Blanc[34, 51] noted that a large U could qualitatively explain the semiconductivity of most $TCNQ^-$ salts, which would be metallic on the basis of simple band theory ($U = 0$). The same conclusion holds for Wurster's blue perchlorate, $TMPD–ClO_4$, which contains a regular $TMPD^+$ stack above 186 K. Strebel and Soos give[22] a systematic analysis π-molecular CT complexes in terms of modified Hubbard models. The restriction of π-electron overlap to the ion-radical or molecular stack produces a one-dimensional Hubbard model, with small t, while the occurrence of a low-lying CT excitation is associated with U. The theoretical development in Section II shows that π-molecular CT solids are, in fact, possibly the best available realizations of Hubbard models.

The longstanding theoretical interest in both the linear Heisenberg antiferromagnetic chain (7) and the linear Hubbard model (6) thus has direct bearing on π-molecular CT solids. The Mulliken problem for CT in solid-state complexes has, in effect, already been investigated as an idealized problem in electron correlations. Indeed, the same formal problem occurs in the analysis of π-electrons in polyenes[52] or in aromatic molecules.[53]

E. Relation to Other Work: Scope of Review

The occurrence of a CT excitation below any Frenkel states is basic to our present formulation of a consistent phenomenological treatment of the electronic states of π-molecular CT complexes. The extension of the Mulliken treatment[2, 3, 4] for DA dimers is accomplished by a novel application[25] of molecular-exciton theory. Subsequent mathematical analysis is facilitated by solid-state results for Hubbard models. Thus crystalline π-molecular complexes lead to problems which overlap in part with experimental and theoretical studies in both chemistry and physics.[24] The one-dimensionality of the ion-radical stacks is a common structural feature with either inorganic conductors[27] based on stacked transition-metal complexes or inorganic insulators[54, 55] based on the exchange-coupled transition-metal ions. In addition to a structural focus, the fact that π-molecular CT crystals are organic solids, and even prototypes for polymers, naturally suggests a common approach to physical properties like conductivity. Before outlining our development, we list previous reviews which overlap in part the present topic of CT in solid-state complexes. These reviews also provide more comprehensive compilations of the voluminous original literature.

The books by Foster,[3] Mulliken and Person,[4] and Rose[5] focus primarily on DA dimers and review both the experimental and theoretical situations. Hanna and Lippert[6] provide a critical review of intermolecular and CT contributions in dimers. These problems are common to solid-state complexes, as discussed in Section II.

Herbstein's[9] extensive compilation of π-molecular crystal structures (up to 1971) shows many of the ion-radical salts based on the molecules in Fig. 2, with special emphasis on the overlap patterns of TCNQ and TMPD crystals. The crystal structure is a basic input in our phenomenological treatment, as discussed in Section III. Prout and Kamenar[10] review the structures of other, primarily neutral, complexes.

Briegleb's important book[8] still provides the most comprehensive optical and thermodynamic data on neutral π-molecular CT crystals. The extensive optical studies[32, 56–63] of Matsunaga, Akamatu, Iida, Nagakura and their colleagues have elucidated the CT transition and its polarization in both neutral and ionic stacked complexes. Offen[64] summarizes pressure effects on the optical spectra, while Prout and Wright's brief review[11] is an early attempt to relate structural and electronic features in solid-state complexes.

The conductivity of organic solids has been discussed by LeBlanc,[34] by Gutmann and Lyons,[14] and by Kommandeur.[65] The comprehensive (to 1967) conductivity data collected by Gutmann and Lyons[14] may be subject to reinterpretation, since it is largely based on compaction studies. The conductivity work by Siemons, Bierstedt, and Kepler[35] demonstrates the importance of single-crystal measurements in TCNQ salts.

The magnetic properties of insulating or semiconducting ion-radical salts have been reviewed by Nordio, Soos, and McConnell[23] and extended to ionic CT crystals by Soos.[66] The purely magnetic properties included in Section V require a collective treatment, and pose problems that at first are quite distinct from those associated with CT excitations and with conductivity. More extensive theoretical treatments of exchange in one-dimensional spin systems are reviewed by Hone[67] and by Richards and Hone.[68] Their principal focus is the unusual one-dimensional EPR lineshape observed[69] in $N(CH_3)_4MnCl_3$ (TMMC), with a linear chain of $S = \frac{5}{2} Mn^{++}$ ions, and subsequently in several linear chains of $S = \frac{1}{2} Cu^{++}$ ions.

The vast inorganic literature of metal–metal interactions provides one-dimensional structures for halides of most of the first-row transition metals.[70] Hatfield and Whyman[71] survey the many possibilities afforded by Cu(II) systems, as was already noted by Kato, Jonassen, and Fanning.[72] Kokoszka and Gordon,[73] Kokoszka and Duerst,[54] Martin,[74] Sinn,[75] and Ginsberg[76] have all reviewed exchange interactions, primarily in small transition-metal clusters. The similarity of exchange-coupled inorganic and organic systems will certainly be exploited further, even though the mechanism for exchange is probably different.[24] Thus exchange in inorganic magnetic chains is probably not dominated by CT interactions and localized transitions (d–d bands) usually occur below CT excitations.

Shchegolev[26] has reviewed Soviet work on the highly conducting TCNQ salts, while Zeller[27] has focused on the partly oxidized Krogmann[77] salts, like $K_2Pt(CN)_4$ $Br_{0.3} \cdot xH_2O$, based on tetracyanoplatinate ions. There is little doubt that electron–phonon interactions will prove important not only for the good conductors, but also for charge-carriers in either insulating or semiconducting organic solids. However, we do not consider the role of polarons[14, 78, 79] here. Indeed, we only mention current theories, for example the disorder model,[43, 80, 81] for highly conducting TCNQ salts in passing. These often controversial models for conduction presuppose an understanding of the electronic states of π-molecular CT crystals, and are consequently areas for potential extensions of our phenomenological model.

The underlying assumptions of molecular-exciton theory have been extensively discussed.[15–18] Rice and Jortner's[17] comprehensive analysis of theoretical difficulties with quantum-mechanical computations in even the simplest closed-shell molecular crystals emphasizes the need for a phenomenological approach and provides a background for the discussion of molecular parameters in Section VI.

We present in Section II a phenomenological theory for the electronic states of a molecular crystal and show that the restriction to a lowest CT excitation permits a convenient fermion representation. The structural variety of ion-radical crystals based on strong π-donors and acceptors is shown in Section III to provide several different applications of the phenomenological theory. A qualitative analysis of optical and electric properties follows at once, even without solving completely the various Hubbard models. The exact, numerical, perturbative and self-consistent analyses of Hubbard models and of one-dimensional spin systems are summarized in Section IV. The magnetic properties of insulating and semiconducting CT crystals are reviewed in Section V, with emphasis on uncorrelated spin excitations in $\ldots D^+A^-D^+A^- \ldots$ stacks. The identification and approximate computation of the parameters appearing in the phenomenological theory are discussed in Section VI. A general phenomenological approach to CT in solid-state complexes leads to far too many possible applications for a survey of reasonable length. Even the topics selected here span both theoretical and experimental work, as well as both optical and magnetic properties. We have consequently endeavoured to make each Section reasonably self-contained.

II. DEVELOPMENT OF PHENOMENOLOGICAL THEORY

A. The Ion–Radical Dimer

Instead of the more familiar[3–5] DA dimer, we discuss here a hypothetical ion-radical dimer, a complex whose ground state is primarily ionic, $|D^+A^-\rangle$. A CT excitation like (1) then produces a largely neutral $|DA\rangle$ pair. The occurrence of ion-radicals in both ionic CT and FR crystals reflects the Madelung stabilization of the ionic lattice.[22, 23] Such stabilization simply shifts the relative energies of the ionic and neutral configurations, and we consider the situation shown in Fig. 3, with $|D^+A^-\rangle$ the ground state. We follow Mulliken's[2] original postulate that $|D^+A^-\rangle$ and $|DA\rangle$ are interacting, but strongly orthogonal, wave functions. Such an approach is natural for solid-state problems, since it avoids the "nonorthogonality catastrophe"[82] for an infinite basis and is further indicated for a phenomenological theory in which explicit (*ab initio*, for example) computations are not attempted.

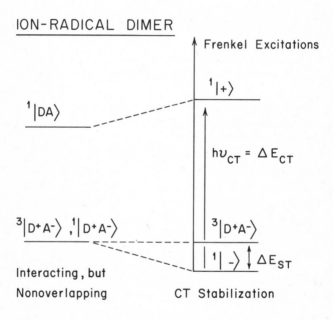

FIG. 3. Schematic energy-level diagram for an ion–radical dimer.

The highest occupied molecular orbital of D and the lowest unoccupied molecular orbital of A are both taken to be non-degenerate. Then the ion-radical dimer D^+A^- reduces to degenerate singlet and triplet configurations in the limit of zero differential overlap. The charge distributions of $^1|D^+A^-\rangle$ and $^3|D^+A^-\rangle$ are very nearly identical, and any multipole interactions are very nearly the same, as indicated in Fig. 3. The Coulomb Hamiltonian, H connects only states with the same spin multiplicity, since it is an excellent approximation to neglect spin–orbit interactions for aromatic molecules containing primarily such light atoms as C, H, or N. Configuration interaction then connects only $^1|D^+A^-\rangle$ and $|DA\rangle$, which is necessarily a singlet, if higher (Frenkel) states are neglected. Taking $E(D^+A^-)$ as the zero of energy, we obtain

$$H = \begin{pmatrix} 0 & t \\ t & \Delta E \end{pmatrix}. \tag{9}$$

where $\Delta E = E(DA) - E(D^+A^-) > 0$, and t is the Mulliken CT integral

$$t = \langle D^+A^- | H | DA \rangle \tag{10}$$

The Mulliken CT integral t is formally identified[22] with the Hubbard "hopping" integral in π-molecular CT complexes and plays a central role in correlating the physical properties of ion-radical crystals.[24] As indicated above, it represents the mixing of the singlet configurations $^1|D^+A^-\rangle$ and $|DA\rangle$ by the electrostatic Hamiltonian for the rigid dimer.

The solution of (9) leads to the eigenvalues

$$\lambda_\pm = \tfrac{1}{2}\Delta E \pm \tfrac{1}{2}[(\Delta E)^2 + 4t^2]^{\frac{1}{2}} \tag{11}$$

with λ_- corresponding to the ground state and λ_+ to the excited state. The corresponding eigenfunctions are

$$^1|\pm\rangle = a_\pm\, ^1|D^+A^-\rangle + b_\pm |DA\rangle, \tag{12}$$

where the expansion coefficients are

$$a_\pm = \frac{\lambda_\pm - \Delta E}{[t^2 + (\Delta E - \lambda_\pm)^2]^{\frac{1}{2}}} \tag{13}$$

$$b_\pm = \frac{t}{[t^2 + (\Delta E - \lambda_\pm)^2]^{\frac{1}{2}}}$$

provided that $^1|D^+A^-\rangle$ and $|DA\rangle$ are normalized functions. The CT excitation in Fig. 3 is given by

$$h\nu_{CT} = (\Delta E^2 + 4t^2)^{\frac{1}{2}} \tag{14}$$

while the stabilization of $^1|D^+A^-\rangle$ relative to $^3|D^+A^-\rangle$ provides a singlet-triplet splitting of

$$\Delta E_{ST} = -\lambda_- \approx t^2/\Delta E \tag{15}$$

for $t \ll \Delta E$. Thus the CT stabilization of $^1|D^+A^-\rangle$ relative to $|DA\rangle$ does not alter the usual results[3–5] for the CT transition, and in addition provides an important contribution[23] to the singlet–triplet splitting of the ion-radical dimer. The magnetic properties are, of course, strongly dependent on ΔE_{ST}.

An analogous treatment may be used for self-complex dimers, A^-A^- or

D^+D^+, with the additional feature that both singlet excited states $|AA^{--}\rangle$ and $|AA^{--}A\rangle$ or $|DD^{++}\rangle$ and $|D^{++}D\rangle$ now occur. If the dimer has a centre of inversion, the allowed CT transition[30] is to the state $2^{-\frac{1}{2}}(|AA^{--}\rangle - |A^{--}A\rangle)$, whose energy is not shifted by configuration interaction. The CT excitation then remains (14), but the stabilization of $^1|A^-A^-\rangle$, or of $^1|D^+D^+\rangle$, has increased to

$$\Delta E_{ST} = \tfrac{1}{2}(\Delta E^2 + 8t^2)^{\frac{1}{2}} - \tfrac{1}{2}\Delta E \approx 2t^2/\Delta E. \qquad (16)$$

Self-complexes without centres of inversion, or at crystallographically non-equivalent sites, could show two CT transitions. Both $TCNQ^-$ anion radicals[83] and PD^+ or $TMPD^+$ cation radicals[84] form self-complex dimers in solution. A more complete theoretical analysis is of course possible for such dimers.

The CT stabilization of the singlet state, as indicated in (15) and (16), corresponds to the kinetic exchange in Anderson's theory[49] of superexchange between transition-metal ions. The direct, or Heisenberg, contribution to ΔE_{ST} is[85, 49]

$$J_d = \langle D^+ \otimes A^- | HP | D^+ \otimes A^- \rangle \qquad (17)$$

where P permutes electrons between the two ion-radicals and $|D^+ \otimes A^-\rangle$ is a product function. (17) reduces to $\langle D^+(1)A^-(2)|e^2/r_{12}|D^+(2)A^-(1)\rangle$ when only the two electrons in the half-filled molecular orbitals are considered.[23] Since π-electron overlap is small even along the stack, we assume that J_d is a small contribution to ΔE_{ST}. As discussed in detail by Herring[85] and summarized below in Section II.C, it is by no means straightforward to estimate the relative contributions to ΔE_{ST} even in the simplest case of two H atoms. The strong antiferromagnetic ($\Delta E_{ST} > 0$) exchange in ionic CT and FR solids suggests that (15) or (16), which is necessarily antiferromagnetic, is more important than J_d, which is generally ferromagnetic in sign.

B. Intermolecular Forces

The phenomenological treatment of the ion-radical dimer closely parallels the zeroth-order approach to the DA dimer. In particular, the connection between the isolated-molecule wave functions and the dimer states $|D^+A^-\rangle$ and $|DA\rangle$ has been studied in some detail by Murrell, Randic, and Williams[86] and, for π-molecular dimers, by Hanna and coworkers.[6, 87] It is evident that a crystal of D and A molecules will present an even more difficult problem in intermolecular forces. The cataloguing of various contributions to the ground-state energy and geometry of a rigid dimer, while not unique, nevertheless focuses attention on the different types of matrix elements of the

Schrödinger Hamiltonian. These considerations make it premature, in our opinion, to identify quantitatively various contributions to the stability of π-molecular CT crystals. As emphasized in Section II.D, the general phenomenological theory is developed in terms of molecular wave functions that already contain many classical electrostatic contributions. The present subsection may be skipped by readers not interested in the details[88, 89] of "exchange perturbation" theories.

The wavefunction for an isolated, rigid DA complex has the general form

$$|DA\rangle = \sum_{rs} a_{rs} \mathscr{A}(|Dr\rangle \otimes |As\rangle) + \sum_{tu} b_{tu} \mathscr{A}(|D^+t\rangle \otimes |A^-u\rangle) \qquad (18)$$

Here $|Dr\rangle|As\rangle$, $|D^+t\rangle$, $|A^-u\rangle$ are, respectively, the normalized molecular wavefunctions of the isolated donor, isolated acceptor, and their isolated ion-radicals. The coefficients a_{rs} and b_{tu} describe the configuration-interaction mixing for the dimer, and $\mathscr{A}$ is the antisymmetrizer. The largest coefficient for a DA dimer is expected to be a_{00}, while the largest coefficient for an ion-radical dimer is b_{00}. It will clearly be necessary in practice to limit sharply the expansion (18) by selecting the physically most important terms.

The quantum theory of intermolecular forces[88, 90–92] analyses the contributions to (18) by comparing various types of matrix elements of the electrostatic Hamiltonian, H, of the dimer:

(1) Crystal-field, or classical Coulomb, interactions represent the contribution of the *permanent* multipoles of the isolated molecular ground state, for example $|D^+0\rangle$ and $|A^-0\rangle$ in an ion-radical solid. If only these interactions were important, the neutral dimer $|DA\rangle$ would be accurately represented by the non-antisymmetrized product $|D0\rangle \otimes |A0\rangle$.

(2) Induction interactions represent the polarization, or distortion, of the charge distributions for $|D0\rangle$ and $|A0\rangle$ and are discussed in terms of *induced* multipoles. There is still no overlap of the molecular charge distributions. The ground states $|DA\rangle$ would then reduce to a product function of, for instance, the form $(\sum_r a_r |Dr\rangle)(\sum_s a_s |As\rangle)$. It is likely that induction is less important in the solid than in a dimer, since the solid generally contains D and A sites with considerable symmetry. There are often, for example, inversion centres at D^+ or A^- sites for ion-radical crystals based on TMPD, TTF, TCNQ and chloranil.

(3) Resonance interactions involve off-diagonal Coulomb matrix elements which become important owing to a degeneracy of the isolated molecular states. Such contributions arise primarily for identical molecules, one of which is excited, and are consequently important in crystals, as found in molecular-exciton theory.[15–18]

(4) Dispersion, or van der Waals, interactions also involve off-diagonal matrix elements and represent instantaneous electrostatic interactions between oscillating multipoles. If dispersion dominates, then a neutral $|DA\rangle$ state could be represented by a superposition of product functions $|Dr\rangle \otimes |As\rangle$, with all $b_{tu} = 0$ in (18), but $|DA\rangle$ would no longer separate into a simple product as was the case for induction interactions.

(5) Resonant charge-transfer interactions involve off-diagonal matrix elements like (10) that become important on account of a degeneracy of the isolated-molecular states. Now specific electrons cannot be assigned to a given molecule and the charge distributions overlap. Such interactions arise when an extra (odd) electron occurs alternately with one molecule and then another, as can happen in complex FR salts.

(6) Exchange interactions also involve differential overlap and are most simply associated with the kinetic exchange contributions (15) or (16) or with the diagonal Heisenberg exchange (18). Both types of exchange are second-order in the intermolecular differential overlap.

It is interesting that CT stabilization associated with the Mulliken CT integral t may well be one of the smallest contributions to the interaction energy, but leads directly to the additional absorption characteristic of CT solids or dimers. Furthermore, all these contributions arise from a *single* interaction, the quantum-mechanical electrostatic potential. (Additional coupling with the electron spins or the nuclear motions are usually small and are neglected.) Thus it is somewhat artificial to specify the relative importance of various intermolecular contributions, since there is no rigorous, unique theoretical decomposition. However, it is evident that once a finite basis is adopted in (18), then the matrix elements of H with respect to the approximate molecular orbitals can be classified as indicated above.

The crystal–field, induction, dispersion, and exchange interactions are all comparable[87, 93, 94] in largely neutral dimers, such as tetracyanoethylene (TCNE) with benzene and p-xylene or benzene and I_2. Exchange contributions are also important. Thus the accurate representation of the dimer ground-state wavefunction and geometry remain quantum-mechanical problems of some complexity. The isolated-molecular multipole moments and polarizabilities may be used to estimate crystal–field or induction contributions even in the solid state. However, as already noted, the binding energy of the solid has not been of primary concern. Gutmann and Lyons[14] use, instead, such multipole and polarization arguments to estimate the stabilization of excited states, and in particular of charge carriers, in organic semiconductors.

As a final remark on the difficulty of applying quantum theory systematically to the weak, but complex, intermolecular forces[88–92] in molecular dimers or crystals, we note there is even some difficulty in identifying the relative magnitudes of the neutral and ionic components of a dimer.[86] Thus care must be taken in speaking about the coefficients $a_\pm$ and $b_\pm$ in (13). The difficulty arises because both the neutral configurations $\{|Dr\rangle \otimes |As\rangle\}$ and the ionic configurations $\{|D^+t\rangle \otimes |A^-u\rangle\}$ are complete, and hence either would suffice for the expansion (18). Once again, the restriction to a finite (incomplete) set of neutral and ionic configurations restores linear independence. The use of neutral and ionic configurations is thus physically motivated and is dependent on the nearness to the isolated-molecular limit. An ionic configuration like $|D^+0\rangle|A^-0\rangle$ is certainly more sensible than the alternative of including many high-energy (continuum) neutral configurations. No real difficulty arises for only a few configurations, and our discussion will necessarily be restricted to a few molecular states in any case.

C. Orthonormal Basis for Dimer

The neutral and ionic configurations in (9) need not be orthogonal in general, but may have an overlap matrix

$$S = \begin{pmatrix} 1 & S' \\ S' & 1 \end{pmatrix} \tag{19}$$

with $S' = \langle D^+A^- | DA \rangle$. The triplet state $^3|D^+A^-\rangle$ is automatically orthogonal to $|DA\rangle$. The overlap was neglected in the introductory analysis of Section II.A and does not create difficulties. Indeed, we show below that it is advantageous to define the phenomenological states $^1|D^+A^-\rangle$ and $|DA\rangle$ to be orthonormal. Such an approach is natural in our molecular-exciton treatment, which is in large part based on the realization that current ab initio quantum-theoretical computations are not nearly accurate enough to deal with the small energy shifts arising from the interactions of molecules containing around 100 electrons.

Since our overlap matrix (19) has only positive eigenvalues, the square root matrix is defined.[95] The new orthonormal dimer basis

$$\{S^{-\frac{1}{2}}|DA\rangle,\ S^{-\frac{1}{2}1}|D^+A^-\rangle\} \tag{20}$$

could have been defined in the first place in Section II.A. Contributions to intermolecular interactions which do not depend on a finite intersite differential overlap are not greatly changed on transforming to (20), at least when the molecules are weakly interacting. In fact, the usual estimates of the crystal–field, induction, resonance and dispersion interactions are predicated on vanishing S'. The charge–transfer and exchange matrix element depend critically on S' and can be significantly changed on using an orthonormal basis.

The general result[85, 96] is that an orthonormal basis tends to increase the kinetic exchange and to decrease the direct exchange J_d. To see this, we observe that orthogonalization will primarily require phase changes in the wave functions in the intermolecular region. Such oscillations in the orthogonal basis give large kinetic-energy derivatives and hence also large CT matrix elements t. Since J_d in (17) may be considered to be the classical self-energy of the exchange charge distribution between A^- and D^+, orthogonalization leads to a small oscillatory exchange charge distribution with a vanishing net charge, and hence to a reduced J_d. These qualitative conclusions have been verified by quantitative examples in very simple systems, and orthogonalization[96] can indeed make an order-of-magnitude difference in J_d. The importance of kinetic, or CT, exchange in Anderson's theory of superexchange[49] between transition-metal ions can be traced to his use of orthonormal (or Wannier) orbitals.

D. Site Representation for Molecular Crystals

It is evident that any molecular solid or metal may be described if all the localized site functions, including those for ionized sites, are known. The strength of the interactions between the sites does not matter, since the basis of localized states is complete. On the other hand, the general crystal Hamiltonian remains as insoluble in such a site representation as it was in the position representation. Molecular-exciton theory[15–18] is a familiar site representation in which the weakness of the intermolecular forces between neutral, closed-shell molecules is invoked to allow a low-order perturbation expansion about the isolated-molecule states. We develop a similar restricted description for π-molecular CT crystals in Section II.E. Since the generality of site representations for molecular solids has not been widely appreciated, we first discuss a general site representation, following Hubbard's analysis[97] of d electrons in transition metals and Klein and Soos's treatment[25] of ionic CT and FR solids.

The basis functions for a molecular crystal are taken to be antisymmetrized products of the form

$$|1r_1, 2r_2, \ldots Nr_N\rangle \equiv \mathscr{A}|1r_1\rangle \otimes |2r_2\rangle \ldots \otimes |Nr_N\rangle \qquad (21)$$

where $|ir_i\rangle$ is a normalized wavefunction for the (r_i)th state of the ith molecule and $\mathscr{A}$ is the antisymmetrizer for all electrons in the crystal. The molecular states $|ir_i\langle$ are taken to be *strongly orthogonalized*, with

$$\langle ir_i \otimes jr_j | \mathsf{P} | ir_i \otimes jr_j \rangle = 0 \qquad (22)$$

if the permutation P interchanges two or more electrons between sites i and j. Strong orthogonality is easily achieved, at least in principle, by constructing the molecular states $|ir_i\rangle$ from the Wannier functions of the solid.[17, 25] Any crystal state can be constructed by superposition of functions of the form of (21), including states with different degrees of ionization.

We now define Hubbard's X-operators[97] to induce transitions between the molecular states $|ir_i\rangle$. The matrix elements of the desired X operators are[25]

$$\langle 1r_1, 2r_2, \ldots Nr_N | X_i^{rs} | 1s_1, 2s_2, \ldots Ns_N \rangle = \delta_{r_i r} \delta_{s_i s} \prod_{j \neq i} \delta_{r_j s_j} \qquad (23)$$

X_i^{rs} changes state s of molecules i to state r, while leaving all other sites unchanged. The full electrostatic crystal Hamiltonian can be expanded in the basis (21) on which the shift operators X_i^{rs} are defined. The general result is

$$H = \sum_n h_n + \tfrac{1}{2} \sum_{mn}' V_{mn} \qquad (24)$$

where we introduce the prime notation to exclude the $m = n$ term in the sum. Here h_n contains all intrasite contributions and is given by

$$h_n = \sum_{rs} \epsilon_n^{rs} X_n^{rs} \qquad (25)$$

where ϵ_n^{rs} is, in the limit of isolated molecules, just the exact Hamiltonian matrix element between molecular states r and s. If we choose the molecular states to diagonalize ϵ_n^{rs}, then the ϵ_n^r are simply interpreted as molecular energies perturbed somewhat by an average crystalline environment.[25] The terms in V_{ij} representing interactions between sites i and j are given by

$$V_{ij} = \sum_{rstu} v_{ij}^{rstu} (-1)^{n_{it}(n_{js} + n_{ju})} X_i^{rt} X_j^{su} \qquad (26)$$

with $i \neq j$. The scalars ϵ_i^{rs} and v_{ij}^{rstu} are simply the one- and two-molecule matrix elements of H in the molecular configuration (21), while n_{it} in (26) gives the number of electrons in state t of the ith site. The coefficients ϵ_i^{rs} and v_{ij}^{rstu} may be chosen,[25] as summarized in Section VI.A, to account for crystal–field and induction interactions and, in an average way, for other inter-molecular interactions as well. Thus we define the canonical crystal-perturbed molecular states, $|nr_n\rangle$, for the nth site to be eigenfunctions of h_n in (25). It is perhaps worth emphasizing that, in a phenomenological treatment, the ϵ_n^{rs} appear as adjustable parameters and may therefore be defined, along with the associated eigenstates, in the most convenient fashion.

Explicit formulas for ϵ_n^{rs} and for v_{mn}^{rstu} are summarized in Section VI.A. There is some freedom (or ambiguity) in their definition, even if we interpret them physically as crystal-perturbed molecular states. Such states, when close to the unambiguous limit of isolated molecules, may contain different types of intermolecular correlations. Since *all* intramolecular correlations are included, the $|nr_n\rangle$ will generally not correspond to simple molecular-orbital states. For completeness, we note that three- and four-site interactions also occur in (24), but that no more than four molecules arise for pair-wise Coulomb interactions. However, these terms involve only three- or four-centre exchange integrals for different molecules and are expected[97] to be negligible in molecular solids, where intersite differential overlap is small. In any case, even (24) will have to be simplified further.

The commutation relations[25] of the X operators follow from their definition in (23)

$$[X_i^{rs}, X_i^{tu}] = \delta_{st}X_i^{ru} - \delta_{ur}X_i^{ts} \tag{27}$$

$$X_i^{rs}X_j^{tu} + (-1)^{(n_{ir} + n_{is})(n_{jt} + n_{ju})}X_j^{tu}X_i^{rs} = 0 \qquad (i \neq j) \tag{28}$$

Although there are techniques for dealing directly with such cumbersome commutation relations, it is almost always more convenient to transform to a second-quantized representation based on either fermion or boson operators. The general crystal Hamiltonian (24) for any molecular solid, whether containing closed- or open-shell sites, becomes quite different once we restrict the crystal states in (21). In the closed-shell systems encountered in molecular-exciton theory, Frenkel excitations lead[98, 99] approximately to a boson description or, for a one-dimensional system with a single Frenkel excited state per site, to either[100] bosons or fermions. The restriction to the lowest CT excitation in the π-molecular solids discussed here also leads to a convenient fermion representation[25] of (24). The difference between the molecular-exciton theory of closed-shell systems and the theory of π-molecular CT crystals thus reflects the choice of the low-lying states retained in the general site representation (24). Since it has proved very difficult to extend molecular-exciton theory to include (higher-energy) ionized states[14, 101] it will also be cumbersome in the present model to include (higher-energy) Frenkel states. In particular, such additional Frenkel states would lead to a complicated mixed representation.

E. Minimum Basis for CT Crystals

Several types of π-molecular crystals were identified in Section I.C. as showing CT transitions below any Frenkel (molecular) state. The mixed stacks encountered in either neutral CT complexes (...DADA...) or ionic complexes (...$D^+A^-D^+A^-$...) require at least the three states $|A\rangle$, $|A^-\alpha\rangle$, and $|A^-\beta\rangle$ for an A site and three states $|D\rangle$, $|D^+\alpha\rangle$, and $|D^+\beta\rangle$ for a D site. Then either the CT excitation (1) or (2) can be described by crystal states (21) in which only three molecular states per site are allowed. The self-complexes encountered in ion-radical stacks require, in addition, the molecular states $|A^{--}\rangle$ for an A site and $|D^{++}\rangle$ for a D site, as seen by inspection of the CT processes (3) and (4). These four states per site also suffice for complex FR stacks illustrated in (5) and are mathematically equivalent[25] to the one-electron orbital description used in Hubbard models.[45]

The minimum basis for CT crystals is therefore

$$|nr\rangle = |Ag\rangle, |A^-\alpha\rangle, |A^-\beta\rangle, \text{ or } |A^{--}\alpha\beta\rangle \qquad (29)$$

for an acceptor at the crystal site n and

$$|nr\rangle = |D^{++}g\rangle, |D^+\alpha\rangle, |D^+\beta\rangle, \text{ or } |D\alpha\beta\rangle \qquad (30)$$

for a donor at site n. The spin degeneracy of the ion-radicals A^- and D^+ is explicitly shown. (29) and (30) represent the *ground states* for molecular species with different numbers of electrons and are conveniently taken to be the exact, orthorgonal, crystal-perturbed site eigenfunctions of h_n in (25). The orbital nondegeneracy of the various states has been implicitly assumed. The restriction of the state index r to the four values g, α, β, and $\alpha\beta$ in (29) or (30) sharply limits the flexibility of basis and excludes all Frenkel excitations. However, these four states per site, which may also be indexed by their charges, are the important ones for CT processes and become exact in the limit of unperturbed isolated-molecule states. In some applications, when electron correlations are large, it is possible to ignore $|A^{--}\rangle$ and $|D^{++}\rangle$, thus further simplifying the basis.

The molecular states (29) and (30) are *exact* many-electron states and include, by definition, all electron correlations at a given molecular site, including core electrons.[25] The basis (21) for a crystal of N molecules now contains 4^N states and, even in the most idealized application of the dimer discussed in Section II.A, leads to a secular determinant of dimensions $4^N \times 4^N$. Some reduction is achieved by considering only states with the same number of electrons, but the number of allowed crystal states will nevertheless remain too large for direct computations. We therefore seek a second-quantized representation for the X operators connecting the states (29) and (30) of the minimum basis.

It is convenient to define the "vacuum" state $|ng\rangle$ to be $|A\rangle$ for an acceptor at site n and $|D^{++}\rangle$ for a donor at n. Now the other states are formally obtained by adding electrons, although the appropriate X operators actually first annihilate a many-electron state and then create another many-electron state with a different number of electrons. The operator $a_{n\sigma}^+$ is defined[25] to create a z component of spin, $\pm\frac{1}{2}\hbar$, by adding an electron at site n, as shown by

$$a_{n\sigma}^+|ng\rangle = |n\sigma\rangle, \qquad \sigma = \alpha, \beta \qquad (31)$$

and $|n\sigma\rangle$ is one of the ion-radical states in (29) or (30). The complete definitions in terms of shift operators are

$$a_{n\sigma}^+ \equiv X_n^{\alpha, g} + X_n^{\alpha\beta, \beta}$$
$$a_{n\beta}^+ \equiv X_n^{\beta, g} - X_n^{\alpha\beta, \alpha} \tag{32}$$

where $\alpha\beta$ denotes either $|A^{--}\rangle$ or $|D\rangle$, depending on whether n is an A or D site. The adjoint operator $a_{n\sigma}$ decreases the total number of electrons by one and changes the z component of spin at site n by $\mp\frac{1}{2}\hbar$ for $\sigma = \alpha, \beta$. It follows from the commutation relations (27) and (28), after some tedious manipulation, that the site operators $a_{n\sigma}^+$ and $a_{n\sigma}$ satisfy fermion anti-commutation relations,

$$[a_{n\sigma}^+, a_{n'\sigma'}^+]_+ = [a_{n\sigma}, a_{n'\sigma'}]_+ = 0 \tag{33}$$
$$[a_{n\sigma}, a_{n'\sigma'}^+]_+ = \delta_{nn'}\delta_{\sigma\sigma'}$$

These fermion operators connect the various states of the minimum basis (29) and (30) and automatically maintain antisymmetrization. Thus we have

$$|n\alpha\beta\rangle = a_{n\alpha}^+ a_{n\beta}^+ |ng\rangle = -a_{n\beta}^+ a_{n\alpha}^+ |ng\rangle \tag{34}$$

for the $|A^{--}\rangle$ or $|D\rangle$ states.

The consequences of applying the operators $a_{n\sigma}^+$ or $a_{n\sigma}$ in the minimum basis are most simply seen by considering them to be one-electron operators for the highest occupied molecular orbital of D and the lowest unoccupied molecular orbital of A, depending on whether n is a D or an A site. This *formal* connection with one-electron creation and annihilation operators, which always satisfy the conventional fermion anticommution relations (33), will permit in Section IV the borrowing of mathematical results for Hubbard models to discuss π-molecular crystals.

F. Fermion Representation of H

In (24) of Section II.D, we expressed the full crystal electrostatic Hamiltonian in terms of one- and two-site contributions. The one-site operator h_n is given in (25) in terms of the shift operators, while the intermolecular contributions V_{mn} are given in (26). The restriction to the minimum basis (29) and (30) limits the state indices r or s to one of four choices for either D or A sites. The shift operators must, finally, be expressed in terms of the Fermion operators $a_{n\sigma}^+$ and $a_{n\sigma}$. The inverse of (32) and of similar expressions are thus needed to obtain a fermion representation for H. We find,[25] for instance, that

$$X_n^{g,g} = a_{n\alpha}a_{n\alpha}^+a_{n\beta}a_{n\beta}^+$$

$$X_n^{\sigma,g} = a_{n\sigma}^+a_{n\bar\sigma}a_{n\bar\sigma}^+ \qquad (35)$$

$$X_n^{\alpha\beta,g} = a_{n\alpha}^+a_{n\beta}^+$$

Such relations may be verified by substituting (32) into the right-hand site of (35) and by using the fact that

$$X_n^{rs}X_n^{tu} = \delta_{st}X_n^{ru} \qquad (36)$$

as follows from (27). The crystal Hamiltonian may therefore be systematically expanded in terms of the fermion operators of the minimum basis. The $v_{mn}^{gg,\,gg}$ contribution in (26) contains octic terms in a operators, as shown by taking $X_n^{g,\,g}X_m^{g,\,g}$ in (35). Thus H remains too cumbersome even after achieving a fermion representation. Instead of writing down the general expression, we examine the coefficients ϵ_n^{rs} and v_{nm}^{rstu} and seek approximations to the matrix elements that will result in no more than quartic terms in the operators $a_{n\sigma}^+, a_{n\sigma}$.

As a final remark about the shift operators X, we note that they are often the generators[102, 103] to a direct product of unitary or symplectic groups. The conversion from conventional fermion operators to such generators is sometimes[102] made to apply group-theoretic techniques to many-body problems.

The site Hamiltonian h_n does not contain products of X operators and is at most quartic in $a_{n\sigma}^+a_{n\sigma}$. The phenomenological treatment of any site with four levels, two of which are degenerate, requires two splitting parameters. We take $\epsilon_{ng} = 0$ as the zero of energy for both $|A\rangle$ and $|D^{++}\rangle$ sites. Then ϵ_n is the energy of the spin-degenerate ion-radical states $|A^-\sigma\rangle$ or $|D^+\sigma\rangle$ and $\epsilon_{n\alpha\beta}$ is the energy of the states $|A^{--}\rangle$ or $|D\rangle$, with two electrons more than the vacuum state $|ng\rangle$. The ϵ_n and $\epsilon_{n\alpha\beta}$ may of course be quite different for D and A sites and are not restricted to be positive.

It is convenient to define the parameter U_n by

$$U_n \equiv \epsilon_{n\alpha\beta} - 2\epsilon_n \qquad (37)$$

U_n represents the intrasite correlation for two spin-paired electrons above the vacuum state $|ng\rangle = |A\rangle$ or $|D^{++}\rangle$. In Hubbard models with frozen cores, U_n is positive and only involves the repulsion between two electrons on site n

$$U_n^{(0)} \simeq e^2\langle r_{12}^{-1}\rangle \qquad (38)$$

where r_{12} is the interelectronic separation. The dimensions of the typical molecules in Fig. 2 indicate that $\langle r_{12}^{-1}\rangle \approx (5\text{Å})^{-1}$, which leads to $U_n^{(0)} \approx 3\,\text{eV}$.

In the present analysis, the interactions with core electrons are also included, and U_n need not be equal to $U_n^{(0)}$. Furthermore, the crystal-perturbed site functions may contain an average polarization contribution, and such polarization has been argued[34, 40, 104] to reduce $U_n^{(0)}$ significantly.

The site Hamiltonian h_n thus reduces to

$$h_n = \epsilon_n(a_{n\alpha}^+ a_{n\alpha} + a_{n\beta}^+ a_{n\beta}) + U_n a_{n\alpha}^+ a_{n\beta}^+ a_{n\beta} a_{n\alpha} \qquad (39)$$

The two parameters $U_n > 0$ and ϵ_n are related to the observed structures of specific π-molecular CT crystals. It is evident that crystal symmetry will usually reduce the number of adjustable parameters in (39), for example by making both ϵ_n and U_n independent of n in any crystal with one site per unit cell.

G. Matrix Elements

The site representation developed in Section II.D is quite general, but needs further simplification to reduce the number of phenomenological parameters. The restriction to the minimum basis (29) and (30) in Section II.E is based on the experimental observation that π-molecular CT crystals are both molecular solids and have a lowest CT excitation. Although a fermion representation was achieved in Section II.F, the resulting Hamiltonian is still complex, principally because the intermolecular coefficients v_{mn}^{rstu} in (26) are complicated expressions of the $a_{n\sigma}^+$, $a_{n\sigma}$ operators. We have thus used the fact that CT crystals show a lowest excitation corresponding to an electron transfer, but have so far not invoked simplifications arising from the one-dimensional nature of π-molecular CT crystals or from the small π-electron overlap even along the stack, as indicated by only a modest shortening the interplanar separations from the van der Waals distance of 3·5–3·6 Å. It is then natural to assume, for example, that all intersite differential overlaps between molecules in different stacks vanish. We are fortunate that π-molecular CT crystals, in addition to the common *electronic* feature of a lowest CT band, have the common *structural* feature of stacks of planar molecules or molecular ions. The absence of magnetic ordering in organic ion-radical crystals is an immediate consequence of their one-dimensionality.[23] The simplification of the intermolecular matrix elements of H can therefore be carried out in general for these crystals.

We consider first the intermolecular Coulomb interactions that provide the stability of ion-radical crystals. These long-range, three-dimensional interactions depend on the charge at the various molecular sites. The charge operators are simply

$$\rho_n = -\sum_\sigma a_{n\sigma}^+ a_{n\sigma} \qquad (n = \text{A site})$$

$$\rho_n = 2 - \sum_\sigma a_{n\sigma}^+ a_{n\sigma} \qquad (n = \text{D site}) \qquad (40)$$

and reflect the electron filling above the vacuum state $|A\rangle$ or $|D^{++}\rangle$. The π-electron charge densities of the ion-radicals D^+ or A^- are delocalized over the molecules. The point-charge approximation, as used in typical inorganic crystals,[105] must therefore be replaced by[106–108]

$$M(m, n) = \sum_{\mu \in m, \, \nu \in n} \frac{\rho_\mu \rho_\nu}{|R_{\mu m} - R_{\nu n}|} \qquad (41)$$

where ρ_μ and ρ_ν are, respectively, the charge densities at the μth atom of the mth singly-charged ion and of the νth atom of the nth singly-charged ion: the sum indicated has μ ranging over the atoms of site m and ν over atoms of site n. Delocalization of π-electron charge densities reduces $|M(n, m)|$ by factors of two or more, but $M(m, n)$ can readily be evaluated.[106] The densities ρ_μ, ρ_ν may be taken from either approximate quantum-molecular computations[109] or, via the McConnell relation, from solution hyperfine data.[110, 111] Variations in different approximations are unimportant in $|M(n, m)|$, and the main point is that a delocalized distribution, however crudely known, should be used.

The crystal Coulomb interaction is then

$$V_C = \tfrac{1}{2} \sum_{mn}{}' |M(m, n)| \rho_m \rho_n \qquad (42)$$

Two approximations should be noted in (42). First, the crystal perturbation of the site functions has been neglected, since (41) was computed from data on isolated ion-radicals. Induction forces (See Section II.B.2) are neglected, with the relatively high symmetry of ion-radical sites in many crystals providing some justification for an inevitable assumption. Clearly, (41) cannot be evaluated unless the charge distribution is known. The second assumption in (42) is that the charge distribution in ion-radicals simply scales with the molecular orbitals of a singly charged radical. This is probably a good approximation for small deviations from unit charge, but not so good for a $TCNQ^{--}$ site. Fortunately, the ground states of ion-radical solids are usually close to singly charged or uncharged, with only complex FR salts providing fractional charges like $\tfrac{1}{2}$ or $\tfrac{2}{3}$. The scaling approximation in (42) is therefore not expected to be seriously in error. An approximation similar to (41) is also used in largely neutral dimers[87] and converges more rapidly than a standard multipole expansion.

In the limit of nonoverlapping, singly-charged sites with $\langle \rho_n \rangle = \pm 1$ in (42), the ground-state expectation value of V_C is just the Madelung energy, $MN/2$. Metzger[106, 107] has computed M for representative ionic CT and FR crystals. The very small Madelung energy [107] of NMP–TCNQ should be noted. In the more general[108] case of π overlap along the stack, the ground-state expectation value of V_C contains a scaled Madelung energy, as indicated by $\langle \rho_n \rangle \langle \rho_m \rangle$ in (42), as well as additional stabilization from $\langle \rho_n \rho_m \rangle - \langle \rho_n \rangle \langle \rho_m \rangle$. The latter may be important in the delicate balance between a neutral and an ionic lattice.[108] Improvements to the self-consistent, ground-state approximation to V_C are discussed in Section VI.B.

The Coulomb interactions (42) are thus simply approximated *via* $M(n, m)$, and contain only terms up to quartic in the $a_{n\sigma}^+$, $a_{n\sigma}$ operators, as may be verified by substituting (40) into (42). The treatment of $M(m, n)$ illustrates how various coefficients in v_{mn}^{rstu} may be identified by looking for the largest multipole contribution. It also emphasizes that the great flexibility of ϵ_n^{rs} or v_{mn}^{rstu} in expressing various crystal perturbations does not preclude estimating various leading terms by considering either isolated sites or nonoverlapping charge distributions.

The π-electron overlap along the molecular or ion-radical stack leads to one-dimensional electron transport. The Mulliken CT integral (10) is regained in the solid, with all additional molecular states simply providing unit overlap in

$$t = \langle \ldots D^+ A^- D^+ A^- \ldots | H | \ldots D^+ A D A^- \ldots \rangle \tag{43}$$

Similar CT integrals occur for stacks of A^- or D^+ ion-radicals in simple FR salts. A variety of other CT processes involving electron transfers between sites with either fewer or more than two electrons also occur. They lead to additional modifications of the model and are discussed further in Section VI.B. Here we assume a single CT integral that is independent of the electron occupancy of the two adjacent sites. The transfer terms in v_{mn}^{rstu} then reduce to[25]

$$H_t = \sum_{n\sigma} t_{nn+1}(a_{n\sigma}^+ a_{n+1\sigma} + a_{n+1\sigma}^+ a_{n\sigma}) \tag{44}$$

where t_{nn+1} is the Mulliken CT integral for adjacent sites n and $n + 1$ in the molecular stack. (44) preserves the restriction to transferring electrons with a fixed spin, which led in Section II.A to configuration stabilization of only the $^1|D^+A^-\rangle$ state of the ion-radical dimer. The first approximation for H_t thus leads to the usual Hubbard-model result in (6), again with the proviso that the site operators (44) connect many-electron states.

We have neglected many additional contributions[25] in v_{mn}^{rstu}: electron-transfer between nonadjacent sites or between sites in different stacks; two-electron transfers; the direct exchange J_d in (17), which can be derived in the site representation and is small for the orthonormal crystal basis; induction and dispersion contributions that are contained in an average way in ϵ_n, U_n, and even in $M(m, n)$ if the molecular charge distributions are perturbed. As more information about π-molecular CT crystals becomes available, their inclusion, as summarized in Section VI, will become necessary. In the present state of refinement, the four crystal parameters afforded by the one-site terms h_n, V_C and H_t are amply sufficient.

H. Charge-Transfer Hamiltonian

The leading terms of the crystal electrostatic Hamiltonian in the minimum basis (29) and (30) are

$$H_{CT} = \sum_{n\sigma} \epsilon_n a_{n\sigma}^+ a_{n\sigma} + \sum_n U_n a_{n\alpha}^+ a_{n\beta}^+ a_{n\beta} a_{n\alpha}$$
$$+ \tfrac{1}{2} \sum_{nn'}' |M(n, n')| \rho_n \rho_{n'} + \sum_{n\sigma} t_{nn+1} (a_{n\sigma}^+ a_{n+1\sigma} + a_{n+1\sigma}^+ a_{n\sigma}) \qquad (45)$$

H_{CT} is the sum of the one-site Hamiltonian h_n in (39), of the crystal Coulomb interaction V_C in (42), and of the transfer term H_t in (44). The sums indicated in (45) are over all the sites in the crystal, with H_t restricted to transfers along the molecular stack. H_{CT} is the molecular-exciton result for π-molecular CT crystals[25] and is the fundamental equation of the present model.

H_{CT} contains at most quartic terms in the site creation and annihilation operators. Thus standard many-fermion techniques[112, 113] can be applied to obtain approximate solutions, as discussed in Section IV. Since H_{CT} commutes with the number operator

$$N_e = \sum_{n\sigma} a_{n\sigma}^+ a_{n\sigma} \qquad (46)$$

for a given stack, the number of electrons is conserved not only for the entire crystal, but also within each stack. Solutions of H_{CT} are therefore to be found in a particular subspace with N_e electrons above the vacuum states $|D^{++}\rangle$ and $|A\rangle$. In most cases, the stoichiometry will dictate $N_e = \langle N_e \rangle$. For example, either neutral or ionic 1:1 CT crystals have $N_e = N$, the number of sites in the stack. *Simple* FR crystals based entirely on A^- or on D^+ stacks also have $N_e = N$ and are $\tfrac{1}{2}$-filled. Complex TCNQ salts, with formally neutral TCNQ sites, illustrate less than $\tfrac{1}{2}$-filled stacks, with $N_e < N$, while complex TMPD salts would illustrate more than $\tfrac{1}{2}$-filled stacks, with $N_e > N$. The second-quantized representation of H_{CT} is natural for discussing molecular states with different numbers of electrons that occur for CT excitations.[24, 25]

We emphasize that H_{CT} is very likely the simplest realistic description for π-molecular CT crystals.[24] For example, single-ion Frenkel excitations in TCNQ salts usually begin around 1·5 eV, while the CT band is around 1·0 eV. Thus it is by no means guaranteed that such Frenkel excitations are always unimportant. It is certain, on the other hand, that including even a single Frenkel excitation per site would considerably increase the mathematical complexity of the resulting site representation. Additional improvements in the intermolecular terms v_{mn}^{rstu} or in the crystal parameters ϵ_n, U_n, t_{nn+1} and $M(m, n)$ *within* the minimum basis can readily be envisaged, and such refinements will undoubtedly be required as experimental studies expand. The simplest molecular-exciton result for CT crystals, H_{CT}, thus requires four states per site and is consequently somewhat more complicated than typical molecular-exciton theory, where two states (ground and excited) suffice.

Two-state models were in fact first proposed to deal with CT crystals. For example, each site in a ... DADA ... stack can be taken as either charged or neutral in an effort to describe the charge distribution in the crystal.[100, 114] However, the spin degeneracy associated with either D^+ or A^- sites is then inevitably lost. In the opposite limit of looking at the magnetic properties of ion-radical solids, the spin degeneracy of D^+ or A^- sites leads to a two-state model.[23] Now, however, the CT excitations and conductivity along the stack cannot be described. Two-level models are therefore quite useful for various aspects of π-molecular CT crystals, but cannot provide a general approach, and a four-state model like H_{CT} is required for these crystals. The conversion to a fermion representation allows us to express H_{CT} in a simple form that is mathematically equivalent to having only one orbital per site, as in the Hubbard model (6). The demonstration of this equivalence is an important result of the site representation.

III. CLASSIFICATION OF π-MOLECULAR CT CRYSTALS

A. Structural Variations of Molecular Stacks

The vast majority of organic molecular solids are closed-shell systems with a lowest Frenkel (molecular) excitation. They may be analysed in terms of molecular-exciton theory and will not be discussed here. We focus instead on molecular crystals with a lowest CT excitation, the vast majority of which have essentially neutral (... DADA ...) ground states of closed-shell D and A sites. Only the strongest π donors and acceptors, such as shown in Fig. 2, form ion-radical crystals.

Mulliken and Person[4] review in some detail the various types of donor and acceptor molecular orbitals that participate in dimer formation. Such classification of dimers again emphasizes the relative primitiveness, and the divergence of interests, in the description of solid-state complexes. The detailed character of the relevant molecular orbitals is suppressed in our phenomenological Hamiltonian H_{CT} in (45). It turns out that all the π-molecular complexes involve the highest-occupied D orbital and the lowest-unoccupied A orbital, as indicated in the minimum basis (29) and (30). However, an infinite DA chain based on π–σ^* overlap would present no difficulties, and any molecular solid with a lowest CT excitation lends itself to a minimum-basis site representation.

The central problem in solid-state complexes is to describe various collective properties. A DA dimer provides a two-level configuration-interaction picture summarized in Section II.A. The corresponding problem for a DA crystal is obviously more complicated. The charge distribution, γ, for a 1:1 CT complex is defined by the ground-state charge distribution, $\ldots D^{+\gamma}A^{-\gamma}D^{+\gamma}A^{-\gamma}\ldots$, along the molecular stack. Long-range Coulomb interactions must be included and provide the sharp experimental separation[9, 23] between π-molecular CT crystals with largely neutral and largely ionic ground states. The magnetic properties of ion-radical crystals also require a collective description, since the antiferromagnetic exchange (7) along the molecular stack is by far the largest term in the spin Hamiltonian. These collective properties, as well as conduction along the stack and the crystal modification of the CT excitation (14), are to be described by H_{CT}.

Detailed crystallographic information[9, 10] is, as already emphasized, a basic input to our phenomenological theory of π-molecular CT crystals. We now use structural information to simplify H_{CT}, without invoking additional approximations.

Fig. 4 shows a partial classification[24] in which the first division is into complexes with neutral and ionic ground states. *Neutral* complexes are typically diamagnetic semiconductors or insulators that contain a lowest

Classification of Ion-Radical Crystals

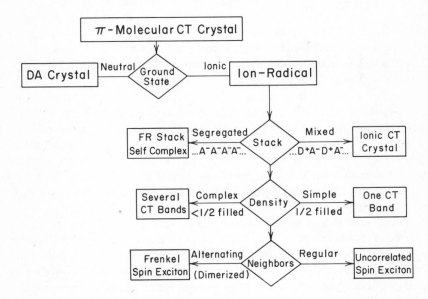

FIG. 4. Schematic classification of π-molecular CT crystals, with emphasis on the structural variations in ion-radical crystals.

CT absorption polarized along the stack. Many such complexes based on weak donors and acceptors have been studied,[3, 8, 9, 10] but will not be emphasized here. Kommandeur and coworkers[115, 116] showed that the weak paramagnetism in small-gap semiconductors is associated with thermally produced charge carriers D^+ or A^-, since the same activation energy for conduction and for paramagnetism was observed in $1:2$ pyrene–iodine and $2:3$ perylene–iodine complexes.

The ion-radical complexes in Fig. 4 are further divided into crystals containing mixed $(\ldots D^+A^-D^+A^-\ldots)$ stacks and FR crystals based on segregated ion-radical stacks. The former are illustrated in Section III.C and are typically paramagnetic semiconductors. The segregated stacking of either A^- or D^+ ion-radicals in self-complexes is best illustrated by the hundred or more TCNQ salts, but analogous series of FR salts based on $D = $ PD, TMPD, or TTF and on $A = $ chloranil or bromanil are also known.[24, 9] FR crystals provide the few known examples of highly conducting, or "metallic", organic systems[26, 36–43, 81] that are currently at the focus of intensive solid-state studies, but the majority of FR systems are paramagnetic semiconductors.

The stoichiometry of ion-radical crystals provides another subdivision indicated in Fig. 4. Ionic $1:1$ complexes have one unpaired electron per site, as indicated by D^+ or A^- ion-radicals, and are simple, or $\frac{1}{2}$-filled, with $N_e = N$ electrons for N sites. Crystals with stoichiometries like DA_2, D_2A, or D_2A_3 are complex and have $N_e \neq N$ in (46). Partial fillings, not equal to $\frac{1}{2}$, for H_{CT} are often more difficult to treat accurately. TCNQ forms complex salts with many metals and, when several stoichiometries are possible, the complex salt is not only the better conductor,[34, 35] but often shows additional low-energy CT absorptions.[32]

The final classification in Fig. 4 is into regular and alternating stacks. In a regular stack, each ion-radical interacts equally strongly with its two neighbours in the stack; in an alternating stack, the interaction with one neighbour is much stronger, and ion-radical dimers (or higher n-mers) are produced.[117, 23] Triplet spin excitons are characteristic of alternating stacks and were among the first features of ion-radical salts studied.[118–122] The regular versus alternating separation is primarily important for elucidating magnetic properties.[66]

The structural and stoichiometric possibilities in Fig. 4 lead to rather different simplifications of H_{CT}. The various physical properties of FR salts, which show the entire spectrum of simple/complex and regular/alternating are thus qualitatively explained, as shown below, without detailed solutions to H_{CT}. Other structural features, for example the effects of crystallographically disordered cations[80, 81] or of doping arising from slightly nonstoichiometric[123] complexes are also included[24] naturally into H_{CT}. It is therefore possible to describe either neutral or ionic CT crystals and all the FR crystals by the same general site Hamiltonian, H_{CT} in (45).

B. Collective Physical Properties

There are thousands of neutral π-molecular CT crystals and several hundred with ion-radical ground states. It is unfortunate that only a handful of the many solids that have been studied have been subjected simultaneously to structural, optical, electric, and magnetic investigations. All provide information about CT in solid-state complexes, and all are needed to estimate various crystal parameters in a theoretical analysis. The wider experimental range afforded by electric and magnetic studies in ionic complexes thus permits a far more detailed application of H_{CT} than is possible in neutral complexes. There are, of course, many additional results available for a few solid-state complexes such as NMP–TCNQ[26, 39–41] and, recently, for TTF–TCNQ,[36–38, 124, 125] both excellent organic "metals". The elementary physical properties discussed here represent, by contrast, the *minimum* requirement for correlating structural, optical, electrical and magnetic properties.

The crystal structure is, of course, essential for reducing the parameters in H_{CT}. The lack of structures, together with the occurrence of both solvent-free and solvent-containing crystal modifications, has hampered static susceptibility studies in neutral FR systems.[24, 126, 127] Structures based on stacks of planar subunits of aromatics or of transition-metal complexes[71] often show several modifications. Rb–TCNQ[128, 129], NMP–TCNQ[130, 131] and K–chloranil[132] are all examples and the two modifications may fall into different subclasses in Fig. 4. Many ion–radical solids also undergo phase transitions,[23, 29, 133–135] as illustrated by the dimerization[136, 137] of the TMPD$^+$ stack in Wurster's blue perchlorate below 186 K. Structural changes may alter the classification, but cannot be described by H_{CT} without additional terms for lattice vibrations and electron–phonon interactions. Herbstein's[9] compilation of π-molecular structures and the many recent TCNQ structures are therefore invaluable for a theoretical discussion of CT effects.[24]

The fundamental CT processes indicated in (1) through (5) have generally been investigated by diffuse reflectance spectroscopy in ion-radical crystals.[56–63, 138] Their intensity is so large that transmission measurements require inconveniently thin, noncrystalline samples. The polarization of the CT band is, as expected, along the ion-radical stack. Only a few temperature-dependent studies have been reported: for TMPD–ClO$_4$,[63] for the neutral CT crystal NEP,[63] for Rb–TCNQ(I),[135] and for K–chloranil.[139] The intensity, but not position, of the CT absorption depends on temperature.

Since H_{CT} does not describe the higher-energy Frenkel excitations, no discussion of these molecular transitions will be included beyond noting that they are beginning to be assigned.

The conductivity of organic semiconductors has been difficult to interpret, in part on account of the potentially large contribution of even trace impurities in very weakly conducting systems.[14, 34, 65] Ion-radical CT crystals are typically rather good conductors and include the most highly conducting organic materials. Thus impurity effects should be relatively less important. Studies on O(TCNQ)$_2$,[140] where Q = quinolinium, and NMP–TCNQ[41, 26] show that the activation energy for conduction along the stack, ΔE_c, is reproducible even though the absolute conductivity varies from sample to sample. Both of these crystals have been discussed in terms of the disorder model,[80, 81] which in one-dimension yields an activation energy for $T^{\frac{1}{4}}$ rather than T. The point here is that reasonably reproducible conductivity results are possible, although several rather different types of theoretical interpretations have been proposed. The single-crystal activation energy ΔE_c along the stack

$$\sigma_\parallel = \sigma_0 \exp\left(-\Delta E_c/kT\right) \qquad (47)$$

will be the characteristic electric property considered. Powder (compaction) data are often similar[14, 35, 141] but may show sufficient differences, especially when ΔE_c is small, to be distinctly less valuable than single-crystal data. Structural changes will of course change ΔE_c. It is perhaps worth pointing out that Magnus Green Salt,[142, 143] a stacked inorganic solid of alternating square-planar $PtCl_4^{--}$ and $Pt(NH_3)_4^{++}$, has recently been shown to be an extrinsic (impurity) semiconductor[144] and that its presumed CT band around 7 kK has been reassigned.[145] The conductivities of organic solids are by no means free of artifacts.

The static susceptibility and EPR spectra of ion-radical CT crystals provide information that, in previous reviews, has not been successfully correlated with optical and electric data. Indeed, the EPR and CT studies remained largely parallel and unrelated areas, even though it was recognized that CT stabilization was an important contribution to the antiferromagnetic exchange along the stack.[23] H_{CT} provides a relation between the magnetic properties associated with the ground-state spin-degeneracy in Table I and the optical and electric properties arising from CT excitations.[24] In many cases, the paramagnetic susceptibility or the EPR intensity satisfies

$$\chi T \approx \exp\left(-\Delta E_p/kT\right) \qquad (48)$$

and defines an activation energy for paramagnetism, ΔE_p, of the order $t^2/\Delta E_{CT}$ in (8) (or (15) for an ion–radical dimer and (16) for a self-complex dimer). ΔE_p, ΔE_c, and ΔE_{CT} thus provide characteristic energies associated, respectively, with magnetic, electric and optical properties.

C. Mixed Simple Stacks: Neutral or Ionic

1:1 complexes based on mixed stacking of D and A sites, as shown in Fig. 1, are the solid-state analogues of DA dimers. The structure and CT bands of neutral complexes are well known.[8–11] Table II contains physical properties for several ionic 1:1 complexes. TMPD–TCNQ[146] and TMPD–chloranil[147] both crystallize in mixed stacks, as does the less completely characterized cosublimed[148] PD–chloranil (PDC). The small[149–151] ΔE_p in TMPD–TCNQ permits considerable variations in the paramagnetism in contrast to the larger[148, 152] ΔE_p of PDC and TMPD–chloranil. The CT bands of TMPD–TCNQ,[61] of TMPD–chloranil[60, 62, 138] and of PDC[60, 62] are polarized parallel to the stack. Single-crystal conductivity data are available for TMPD–TCNQ[151] and for TMPD–chloranil,[152] but only compaction results[153] for PDC. There are many other potentially ionic complexes between TCNQ and aromatic amines,[151] as well as other chloranil, bromanil, or PD complexes[9, 154] and even mixed stacks of TCNQ and square-planar Pt or Pd complexes.[155] Additional structural information for ionic 1:1 complexes would be desirable.

TABLE II. Physical properties of 1:1 ionic CT crystals with mixed, simple, regular stacks

Crystal structure	ΔE_{CT}/eV	ΔE_c/eV	ΔE_p/eV
...D$^+$A$^-$D$^+$A$^-$... TMPD–TCNQ[146]	0.95[61]	0.36[151]	0.068[149] 0.075[150] 0.08[151]
TMPD–chloranil[147]	1.0[60]a 1.2[62] < 1.2[138]	0.45[152]	0.13–0.15[152]
PD–chloranil[148]b	1.4[60]a 1.2[62]	0.43[153]a	0.13[148]

^{a} Powder data.
^{b} Magnetically non-equivalent stacks.

The mixed stacks for the complexes in Table II are crystallographically equivalent. We focus attention on a single stack and treat the long-range Coulomb interactions self-consistently.[22] It is convenient to label the sites in the stack consecutively, with even integers assigned to A and odd integers to D sites. The intrasite Hamiltonian h_n in (39) now contains only two site energies, ϵ_A and ϵ_D, and two correlation terms, U_D and U_A. If any multiple of N_e is added to H_{CT}, all the energies for given N_e are merely shifted. We neglect a shift of $-\frac{1}{2}(\epsilon_A + \epsilon_D)$ and represent the variation of the site energies along the stack by

$$\epsilon_n = (-1)^n \epsilon \tag{49}$$

with $\epsilon > 0$ indicating that the highest-occupied molecular orbital on a D site (odd n) has lower energy than the lowest-unoccupied A site (even n). The intrasite correlation

$$U = \langle D(1,2)|h_n|D(1,2)\rangle \tag{50}$$

is appropriate for a D site and, as indicated in (38), is a simple repulsion in the frozen-core approximation. Thus U may be estimated from quantum-molecular computations for either D or A^{--} sites. No qualitatively new features occur on retaining both U_D and U_A, and (50) will be used for both D and A^{--} sites.

Electron transfer along the stack is given by H_t in (44), and a single transfer integral t suffices in most 1:1 complexes, since each D or A site interacts equally strongly with two neighbours in the stack. The overall electric neutrality of the crystal requires equal ground-state charges on D and A sites, or $\langle \rho_D \rangle = -\langle \rho_A \rangle$ in (40). Thus the *ground-state* charge distribution is found from the Madelung approximation to the crystal Coulomb energy,

$$\langle V_C \rangle^{(0)} = -\tfrac{1}{2}|M|N\langle \rho_A \rangle^2 \tag{51}$$

for a stack containing $N/2$ sites of both D and A. $\langle V_C \rangle^{(0)}$ suppresses charge fluctuations and may therefore fail for excited-state properties. It is the simplest self-consistent expression for describing a solid containing stacks of ionicity $\gamma = -\langle \rho_A \rangle$. The site representation of H_{CT} reduces, with the aid of (49) and the approximation (51), to the modified Hubbard model[22] for a mixed, simple, regular stack

$$H_{MSR} = \sum_{n\sigma}\{(-1)^n \epsilon a_{n\sigma}^+ a_{n\sigma} + t(a_{n\sigma}^+ a_{n+1\sigma} + a_{n+1\sigma}^+ a_{n\sigma})\}$$
$$+ U\sum_n a_{n\alpha}^+ a_{n\beta}^+ a_{n\beta} a_{n\alpha} - \tfrac{1}{2}|M|N\langle \rho_A \rangle^2 \tag{52}$$

The solution of H_{MSR} and of other special cases of H_{CT} are deferred to Section IV. They all illustrate consequences of the intrasite correlations retained in the U term.

It is instructive to examine the $t = 0$ limit[22] of H_{MSR} which then reduces to a sharp neutral-ionic transition.[21, 23] The neutral lattice, with $\langle \rho_A \rangle = 0$, has ground-state energy $(-2\epsilon + U)N/2$, as expected for $N/2$ doubly-occupied neutral D sites. The ionic lattice has $\langle \rho_A \rangle = -1$ and energy $-\tfrac{1}{2}|M|N$ corresponding to singly-occupied D^+ and A^- sites. A superionic lattice, with $\langle \rho_A \rangle = -2$ and energy $(-2M + 2\epsilon + U)N/2$, would correspond to a $\ldots D^{++}A^{--}D^{++}A^{--} \ldots$ stack. The neutral-ionic transition for

$t = 0$ thus simply requires

$$2\epsilon - U \gtrless |M| \tag{53}$$

with $\gtrless$ corresponding to neutral and ionic, respectively. (53) describes the balance between the monopole–monopole (for delocalized π-electrons) stabilization of the ionic lattice relative to purely intrasite considerations. In zeroth-order, we identify

$$2\epsilon - U^{(0)} = I_D - EA_A \tag{54}$$

where I_D is the ionization potential of D, and EA_A is the electron affinity of A. Here we have neglected not only exchange and CT forces for both the neutral and ionic lattice, but also the induction and dispersion interactions that are expected to stabilize the ionic lattice preferentially. The case $t = 0$ thus establishes the important result that, other things being equal, the lattice is completely neutral when $(I_D - EA_A)$ is less than M and completely ionic when $(I_D - EA_A)$ is greater than M. The self-consistent solution[22] of H_{MSR} shows that a sharp neutral-ionic boundary remains even for small finite t, with $\langle \rho_A \rangle$ changing discontinuously from almost zero to almost unity, as shown in Fig. 5 for several values of t/ϵ as a function of U. A small gap against single-particle excitations persists even at the transition.[156]

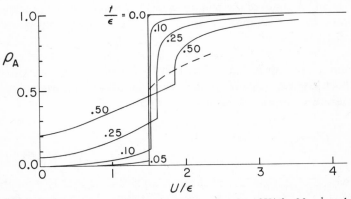

FIG. 5. The ground-state acceptor charge density $\langle \rho_A \rangle$ as a function of U/ϵ for $M = \frac{1}{2}\epsilon$ and various t/ϵ values. The $t = 0$ result is exact. $\langle \rho_A \rangle$ increases discontinuously from its small U value to the dashed line defining the neutral → ionic transition. [Reproduced with permission from P. J. Strebel and Z. G. Soos, *J. Chem. Phys.*, **53**, 4077 (1970).]

Although CT complexes have been studied as a function of hydrostatic pressure[157, 158, 64] and are routinely examined as a function of temperature, it has not been possible to identify a truly borderline complex which can be cycled from neutral to ionic and back. There almost certainly are such CT crystals, but they are necessarily rare. After all, both M and $(I_D - EA_A)$ are of the order of volts. Even if other intermolecular contributions are retained, the required crystals would need two energies of the order of volts that are so nearly balanced that charges in kT or in pressure can reverse the order.

The principal features of MSR stacks in Table II are readily understood. The CT excitation involves (2), and such excitations are also required for conduction. Spin excitations are at lower energy, since no doubly occupied sites are required, and are discussed in greater detail in Section V.

D. Crystals of Dimers and Complex CT Crystals

The regular mixed stacking of aromatic molecules or molecular ions is the most common structure encountered in π-molecular CT crystals. Prout and Kamenar[10] review the structures of n–σ^* complexes, which often form crystals of DA dimers and two-dimensional sheets as well as molecular chains. Here the n-donors are amines, nitrogens in heterocycles, or oxygen; the acceptor, usually a halogen, has an available σ^* level. All the complexes are apparently largely neutral and their structural properties are complicated by twinning, disorder, and instability. We are not aware of physical studies that would warrant a detailed theoretical analysis at this time. The following remarks merely sketch how H_{CT} would appear for a molecular crystal of n–σ^* dimers.

The D and A states of the minimum basis (29) and (30) are now the donor n-orbital and the acceptor σ^*-orbital, respectively. If only pairwise CT between a given D and A occurs, the crystal reduces to a collection of dimers. This is simulated by taking an *alternating* stack with two transfer integrals, a normal t for a dimer and a $t' \approx 0$ for interactions between dimers. Each DA dimer may be analyzed by the Mulliken method, illustrated in Section IIA, whose second-quantized representation is

$$H_{dimer} = \sum_{n=1}^{2} \sum_{\sigma} (-1)^n \epsilon a_{n\sigma}^+ a_{n\sigma} + U \sum_{n=1}^{2} a_{n\alpha}^+ a_{n\beta}^+ a_{n\beta} a_{n\alpha}$$
$$+ \sum_{\sigma} t(a_{1\sigma}^+ a_{2\sigma} + a_{2\sigma}^+ a_{1\sigma}) + \langle V_C \rangle^{(0)} \qquad (55)$$

This two-electron, six-state problem has ϵ, t and U defined as given above for the 1:1 stack, with $n = 1$ for D and $n = 2$ for A. The six two-electron states are $|DA\rangle$, $^1|D^+A^-\rangle$, $^3|D^+A^-\rangle$ and $|D^{++}A^{--}\rangle$. The triplet configuration does not interaction with $|DA\rangle$, and the high-energy state $|D^{++}A^{--}\rangle$ is neglected. It is typical that the second-quantized expression (55) for a dimer, with only a few states, is far more cumbersome than the direct position representation, whereas just the opposite is true for infinite systems. ·

Coulomb interactions couple the dimers to the rest of the crystal. The Madelung approximation (51) is again used, and leads to a crystal shift of the ionic state $^1|D^+A^-\rangle$. Differential overlap between dimers is ignored. The ground state of a crystal of equivalent dimers then reduces to a product function of dimer states

$$^1\psi_{DA} = a|DA\rangle + b^1|D^+A^-\rangle \tag{56}$$

where the mixing coefficients a and b are identical for all dimers. The crystal-perturbed states $|DA\rangle$ and $^1|D^+A^-\rangle$ could, in principle, be different from isolated-dimer states on account of induction, resonance, and dispersion contributions. The assumption of zero differential overlap between dimers nevertheless leads to a simple, two-level Hamiltonian

$$H_{dimer} = \begin{pmatrix} -2\epsilon + U & t \\ t & -|M|b^2 \end{pmatrix} \tag{57}$$

The diagonal term $-|M|b^2$ represents the approximation (51) for the Madelung stabilization of the ionic form. We note that (57) is now a self-consistent problem, with the expansion coefficient b occurring in H_{dimer} also. When $2\epsilon - U - |M| > 0$, the neutral crystal has lower energy even for complete electron transfer. In this case, we always have $B = E(DA) - E(D^+A^-)$ satisfying

$$B = 2\epsilon - U - |M|b^2 > 0 \tag{58}$$

and the expansion coefficients for the λ_- ground state in (11) are given by (13). The value of b in (56) is therefore

$$b^2 = t^2\{t^2 + \tfrac{1}{4}[B + (B^2 + 4t^2)^{\frac{1}{2}}]^2\}^{-1} \tag{59}$$

Since $B > 0$ for any b according to (58), the limit $t \to 0$ leads to $b \to 0$, as in the isolated dimer. On the other hand, B decreases with increasing b, and the crystal of dimers shows a greater degree of configuration interaction than the isolated dimer, with $b = 0$ in H_{dimer}. Finally, the most interesting case of $2\epsilon - U - |M| \approx 0$ again leads to sharp neutral-ionic transitions for small t. (59) provides a hexic self-consistency relation in b^2 for the case $B > 0$, and illustrates how the Coulomb stabilization in a crystal is important even the limit of non-overlapping DA dimers.

Complex CT crystals, for example with a 1:2 molecular stack of the type ...DAADAADAA..., provide another subclass whose physical properties have yet to be studied in detail. Some crystals are paramagnetic,[151] suggesting an ion–radical ground state. The assumption that all A sites are crystallographically equivalent leads to the choice

$$\epsilon_n = -\epsilon \cos(2\pi n/3) \tag{60}$$

for the site energies. The first D site has $n = 0$, the next two A sites have $n = 1,2$ and have the same site energy, etc. Now two transfer integrals t between DA and AA overlaps are required in H_{CT}, but otherwise there are no new features. 2:1 complexes of the type D_2A have $\epsilon < 0$ in (60) and the labelling begins with an A site at $n = 0$.

The site energy ϵ_n in H_{CT} is therefore convenient for distinguishing between D and A sites in any solid-state complex. The zeroth-order approximation of using a single site energy for all D sites and another for all A sites is rigorously valid in many crystals with crystallographically equivalent sites. The structure determines the most important DA overlaps and thus the types of transfer integrals required in the site representation of H_{CT} given in (45). The n–σ^* crystals of dimers reduce to a slight modification of the isolated or solution dimer; the molecular stacks in π–π^* complexes lead to H_{MSR} for 1:1 CT crystals; complex CT crystals provide alternating stacks, with either $\gtrsim \frac{1}{2}$-filling, in Fig. 4; two- or three-dimensional CT crystals, presumably based on nonplanar donors and acceptors, require CT integrals leading to two- and three-dimensional networks.

Our main reason for including briefly these rather less common CT complexes is to emphasize that the crucial approximation leading to the site representation developed in Section II is the occurrence of a CT band below all Frenkel states. The topology of the crystalline D and A lattice is of secondary importance. This is somewhat obscured by the preponderance of one-dimensional stacks in the most extensively studied π-molecular complexes. We now consider the additional structural variations indicated in Fig. 4 for the self-complexes of π-molecular ion radicals.

E. FR Crystals with Simple Regular Stacks

A regular one-dimensional array of H atoms, or of any other free-radical, is potentially a CT system, since no orbital excitation is required for an electron transfer. Neutral organic free radicals[126, 127] are often rather bulky, as illustrated by DPPH (diphenyl picryl hydrazyl), PAC (picryl amino carbazyl) and BDPA (bisdiphenylene phenyl allyl). DPPH forms various solvent-free and solvent-containing structures[159] and is neither stacked nor a CT crystal.[160] NEP is a more compact neutral radical, shown in Fig. 6, whose CT band and paramagnetism point to a CT system,[30, 63] although the crystal structure has not been published. The ion-radicals $TMPD^+$, TTF^+, $TCNQ^-$, and chloranil$^-$ based on the strong donors and acceptors in Fig. 2 provide a variety of one-dimensional arrays of self-complexes with known structures.

The simplest ion–radical stack, the simple regular case in Fig. 4, occurs[136] in TMPD–ClO_4 above 186 K, in the recently reported[129] second form of Rb–TCNQ, in the α-form[132] of K-chloranil and possibly in NMP–TCNQ if the disorder[130] of the diamagnetic NMP^+ sites is neglected. In TMPD–ClO_4 and Rb–TCNQ(II), there is a single type of ion in the stack. Single

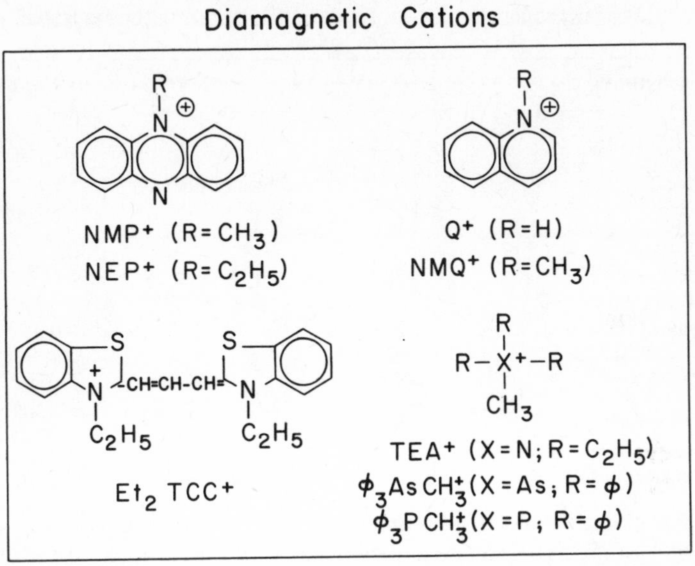

Diamagnetic Cations

FIG. 6. Diamagnetic organic cations that occur as counterions in TCNQ salts.

values of U_n, of ϵ_n, and of t_{nn+1} are therefore required. Since both ClO_4^- and Rb^+ or K^+ are expected to occur as diamagnetic ions that help to isolate the FR stack,[23] there is no doubt that the D^+ stack in TMPD–ClO_4 has one electron per site, as does the A^- stack in Rb–TCNQ(II) or K–chloranil. The site energy H_{CT} then reduces exactly to

$$\left\langle \sum_{n\sigma} \epsilon_n a_{n\sigma}^+ a_{n\sigma} \right\rangle = \langle \epsilon N_e \rangle = \epsilon N \qquad (61)$$

and is an additive constant that may be neglected without loss of generality. It is then also true that $\langle \rho_n \rangle = \pm 1$ not only in the ground state, but also in the 2^N spin states associated with the ground-state charge distribution of a FR stack. H_{CT} thus simplifies to

$$H_{SSR} = \sum_{n\sigma} t(a_{n\sigma}^+ a_{n+1\sigma} + a_{n+1\sigma}^+ a_{n\sigma}) + U \sum_n a_{n\alpha}^+ a_{n\beta}^+ a_{n\beta} a_{n\alpha} \qquad (62)$$

for a segregated, simple, regular stack. H_{SSR} is mathematically equivalent to the one-dimensional Hubbard model (6), although the complete identification holds best for the ground-state charge distribution in view of our treatment of $\langle V_C \rangle^{(0)}$ in (51). The magnetic properties are, in favourable cases like H_{SSR}, given by simple models, but electric or optical excitations which redistribute charge involve additional approximations for the long-range Coulomb interactions.

TABLE III. Physical properties of π-molecular FR crystals with segregated, simple, regular stacks

Crystal structure	$\Delta E_{CT}/eV$	$\Delta E_c/eV$	$\Delta E_p/eV$
$\ldots D^+D^+D^+D^+ \ldots$			
TMPD–ClO$_4$[136]	1·47[63]	Weak	<0·01[168, 167]
($T > 186$ K)	1·5[56]^a	Semiconduction[164]	
TMPD–I[166]	1·1[61]	—	0·009[167]
$\ldots A^-A^-A^-A^- \ldots$			
Rb–TCNQ (II)[129]	0·78[135]^a	0·16–0·19[135]	—
($T > 230$ K)		0·16–0·22[163]	
K–chlornail[132]	1·5[58]	—	~0·05[29, 165]
(α form)	1·43[139]		($T < 150$ K)
NMP–TCNQ[130]^b	0·53[57, 169]^a	0·037[41]	0·0[40, 26]
		($T < 150$ K)	(not activated)

^{a} Powder data.
^{b} Disordered cations.

SSR stacks in Table III are rare, especially at low temperature. The χ_p data[26, 40] for NMP–TCNQ, a proposed[40, 41] quantitative SSR system, are shown in Section V.B to be in qualitative disagreement[161, 162] with a simple Hubbard model, and NMP–TCNQ is perhaps better represented[24] as slightly less than $\frac{1}{2}$-filled, or complex, salt. The Mott insulator expected for H_{SSR} when $t \ll U$ is consistent with the large activation energies[135, 163] ΔE_c in Rb–TCNQ(II); TMPD–ClO$_4$ is also a fairly poor semiconductor,[164] except possibly at the regular-to-dimer phase transition at 186 K; no ΔE_c value has been reported for K–chloranil, but this is likely to be a poor to medium semiconductor and undergoes phase changes,[29, 165] accompanied by large hysteresis effects in χ_p, at low temperature. TMPD–I[166, 167] and TMPD–Br[167] are probably similar to TMPD–ClO$_4$. We note that the structural disorder[130] in NMP–TCNQ leads naturally to random site energies $\{\epsilon_n\}$ in H_{CT}, and thus to various proposed disorder models.[26, 80, 81] No quantitative analysis of disorder is available, however, and interrupted strands[170] (finite chains) may also be important.

Various types of instabilities are associated with one-dimensional metals[171] and magnetic insulators.[172, 173] The absence of π-molecular SSR stacks as $T \to 0$ is suggestive, but it should be emphasized that the driving force for dimerization in specific crystals has not been identified. For example, the measurement[174] of $\Delta H = T\Delta S$ for the TMPD–ClO$_4$ transition at 186 K immediately rules[23] out a purely spin-dependent interpretation, since the entropy of the spin systems is far too small to account for the observed ΔS. Phase transitions inevitably require a much fuller description of the lattice vibrations than is currently available for molecular crystals and pose difficult problems in any system. In some ways, the theorems about the instability of one-dimensional systems are misleading in that there is no guarantee that *all* relevant contributions are one-dimensional. Indeed, it is expected that the lattice dynamics are three- dimensional.

TABLE IV. Physical properties of π-molecular FR crystals with segregated, simple, alternating stacks (triplet exciton crystals)

Crystal structure	$\Delta E_{CT}/eV$	$\Delta E_c/eV$	$\Delta E_p/eV$
...$(D^+D^+)(D^+D^+)$...			
TMPD–ClO$_4$[137,b]	1·44[63]	Weak semi-	0·031[121]
($T < 186$ K)		conduction[164]	
(TMPD)$_2$Ni(mnt)$_2$[177]	—	—	0·24[177]
...$(A^-A^-)(A^-A^-)$...			
Morpholinium–TCNQ[176]	1·2[57]a	0·32[35]	0·36[179, 118]
Rb–TCNQ(I)[128]	1·1[135]	0·44–0·53[163]	0·39[180]
($T \approx 110$ K structure)		($T < 370$ K)	
K–TCNQ[178]	0·96[59]	0·35[35]	~0·2[120]
		0·15–0·45[163]	
	1·1[57]a	($T < 400$ K)	

a Powder data.
b Magnetically inequivalent stacks.

F. Segregated Simple Alternating Stacks

Simple FR salts with dimerized stacks of D^+ or A^- ion-radicals are relatively common, with the entries in Table IV providing only some of the more extensively studied examples. The self-complexes now have two (or more) π overlaps along the stack, and the transfer Hamiltonian is

$$H_{alt} = \sum_{n\sigma} \{t(a_{2n\sigma}^+ a_{2n+1\sigma} + h.c.) + t'(a_{2n\sigma}^+ a_{2n-1\sigma} + h.c.)\} \qquad (63)$$

where h.c. denotes the hermitean conjugate. The limit $t' \to 0$ of non-overlapping dimers now corresponds to self-complex dimers A^-A^- or D^+D^+, rather than the DA dimers in Section III.D. Thus ion–radicals at sites $2n$, $2n + 1$ interact strongly, while those at sites $2n$, $2n - 1$ have smaller transfer integrals t'. In general, we expect $t' < t$ to imply a larger interplanar separation, although the relative orientation of the molecules is also important. Some approximate computations[175] probe the geometry of TCNQ$^-$ stacking, and favour having molecular planes either perpendicular to the stack axis or tilted about 30°, but the rich variety of observed degree of π-overlap catalogued by Herbstein[9] have not been correlated with the magnitude of the CT integral.

In simple dimerized salts like[137] TMPD–ClO$_4$ below 186 K, Rb–TCNQ(I)[128], or morpholinium–TCNQ,[176] there is again no question about having diamagnetic counterions and ion-radical stacks. If, in addition, the ion–radical sites are crystallographically equivalent, then a single site energy suffices and the argument in (61) for treating the site energy as an additive constant holds. It is a general feature of simple ($\frac{1}{2}$-filled) FR stacks that the site energy is exactly, or for non-equivalent sites very nearly, an additive constant. By contrast, the mixed simple stacking encountered in Section III.C for 1:1 CT crystals is characterized by very different energies for D and A sites.

The approximation (51) for using the self-consistent ground-state expectation value of the long-range Coulomb interaction then leads to

$$H_{SSA} = H_{alt} + U \sum_n a^+_{n\alpha} a^+_{n\beta} a_{n\beta} a_{n\alpha} \tag{64}$$

H_{SSA} is the alternating form of H_{SSR} and poses no additional theoretical difficulties, although there are fewer exact results. The magnetic properties are again expected to be given more accurately, since charge fluctuations are neglected in $\langle V_C \rangle^{(0)}$. Any more complicated FR stack is readily described by introducing additional transfer integrals in (63). However, EPR is sensitive only to the distinction between regular and alternating cases.[23, 66] The largest CT integral in an alternating stack defines the ion–radical dimer discussed in Section II.A. The triplet $^3|A^-A^-\rangle$ or $^3|D^+D^+\rangle$ states at energy ΔE_{ST} in (16) are thermally accessible. The smaller CT integrals lead to motion of the triplets along the stack, and these mobile triplets, or triplet spin excitons, have been reviewed elsewhere.[23] Regular stacks, by contrast, do not lead to triplet spin excitons, but show characteristic EPR spectra discussed in Section V.D.

Comparison of Table III and IV for SSR and SSA stacks shows that alternation is primarily reflected by an increased ΔE_p, which for SSA stacks is associated with ΔE_{ST} in (16). No important changes in ΔE_{CT} or ΔE_c are expected in a $\frac{1}{2}$-filled case of H_{CT}, since optical or charge-carrying excitations require a high-energy, doubly occupied sites in both cases. The absence of extensive electron reorganization on forming an alternating stack is consistent with our molecular–exciton hypothesis of weakly overlapping, largely unperturbed molecular sites. The similarity of ΔE_{CT} and ΔE_c data (except for NMP–TCNQ) in all three simple stacks discussed, including the MSR case in Table II, is quite remarkable in view of the different ion-radicals involved. The roughly similar ion–radical dimensions in Fig. 2 support similar values of the correlation energy U for creating a doubly occupied site.

G. Complex Regular FR Stacks: Fractional Charges

Complex TCNQ salts like Q(TCNQ)$_2$,[181] TMPD(TCNQ)$_2$,[33] and tetraphenylphosphonium $(TPP)(TCNQ)_2$[182] contain crystallographically equivalent, or nearly equivalent, TCNQ sites and help to establish[9] the dimensions of $TCNQ^{-0.5}$. A regular array[181] occurs in $Q(TCNQ)_2$, although the disordered Q^+ ions shown in Fig. 6 lead to slightly different site energies. The $TCNQ^{-0.5}$ stack in $(TMPD)(TCNQ)_2$ is regular,[33] while $(TPP)(TCNQ)_2$ probably corresponds[182] more closely to stacks of $(TCNQ)_2^-$ dimers. CT in complex salts poses several difficulties. First, the question of the crystallographic equivalence of the ion-radicals often requires very accurate structures. Equivalent sites lead to fractional charges, while markedly non-equivalent sites may be better represented as stacks with both A^- and A sites. The site energy ϵ_n in these self-complexes is entirely determined by the crystal field. Second, complex salts necessarily have several molecules per unit cell along the stack. There can then be a single t or several, perhaps quite different, CT integrals. Third, the enhanced mobility of the electrons makes the Madelung approximation (51) more tenuous. Complex salts are consequently less well understood than the three types of simple stacks already considered.

TABLE V. Physical properties of π-molecular FR crystals with segregated, complex regular stacks

Crystal structure^c	$\Delta E_{CT}/eV$	$\Delta E_c/eV$	$\Delta E_p/eV$
$\ldots A^{-\frac{1}{2}}A^{-\frac{1}{2}}A^{-\frac{1}{2}}A^{-\frac{1}{2}}\ldots$			
$Q(TCNQ)_2$[181] ^{d}	$<0.5^a$	0.023[140]	0.0[120]
	1.3[32] ^{a}	(80–150 K)	(not activated)
$(Et_2TCC)(TCNQ)_2$[26]	—	0.16[26]	<0.001[192]
$TEA(TCNQ)_2$[184]	<0.5[32] ^{a}	0.16[35]	0.041[120]
	1.3[32] ^{a}	0.13[163]	0.034[118]
$DTC(TCNQ)_2$[183]	—	0.08[193]	—
$\left\{ \begin{array}{l} \ldots A^{-\gamma}A^{-\gamma}A^{-\gamma}A^{-\gamma}\ldots \\ \ldots D^{+\gamma}D^{+\gamma}D^{+\gamma}D^{+\gamma}\ldots \end{array} \right\}$			
$TTF–TCNQ$[186] ^{b}	0.35[169]	0.0031[36]	~0.015[194]
		$T < 10$ K	~0.011[195]
			$T < 60$ K

^{a} Powder data.
^{b} Magnetically non-equivalent stacks.
^{c} Could be $\ldots A^{-1}AA^{-1}AA^{-1}AA\ldots$ for highly non-equivalent sites.
^{d} Disordered cations.

The crystals in Table V are definitely segregated and complex, but may not be exactly regular. None, however, has been reported to show resolved fine-structure splittings in the low-temperature EPR, which is a simple test for strongly alternating CT integrals and thus alternating exchange interactions. Ditoluenechromium $(DTC)(TCNQ)_2$ contains[183] inequivalent TCNQ sites, which probably cannot be assigned to be either fully neutral or ionic from the available data, but apparently has regular spacings along the stack. The detailed manner of TCNQ overlapping for these and related crystals is discussed elsewhere.[9] Triethylammonium (TEA) (TCNQ)$_2$ contains[184] tetrads, with almost equal spacing, and does not show fine structure. The crystallographic assignments of neutral and ionic TCNQ groups has been criticized[185] and the charge distribution in TEA(TCNQ)$_2$ is not accurately known. DTC(TCNQ)$_2$ and TEA(TCNQ)$_2$ thus illustrate small variations in site energies and CT integrals, respectively, and are naturally included among complex regular stacks.

Regular stacking implies a single t in H_{CT}, while equivalent sites lead to a single ϵ_n and to a single U_n in the FR stack. Thus the site energy is again an additive constant and H_{CT} reduces to

$$H_{SCR} = H_{SSR} + \frac{1}{2} \sum_{nn'}' |M(n, n')| \rho_n \rho_{n'} \tag{65}$$

for a segregated, complex, regular stack. H_{SCR} is $\frac{1}{4}$-filled for a $1:2$ salt, $\frac{1}{3}$-filled for a $2:3$ salt, etc. As shown in Section IV, the solution of Hubbard models for electron filling other than $\frac{1}{2}$ is relatively less well understood. An even greater difficulty centres on the treatment of the long-range Coulomb interactions in (65). A collection of A^- and A sites presents considerable orbital, as well as spin, degeneracy even in the limit of large U, with no doubly occupied A^{--} sites. The corresponding $\frac{1}{2}$-filled case is orbitally non-degenerate. Crystal equivalence may still be invoked to approximate the Coulomb interactions in (65) by their ground state expectation value (51). However, now charge and spin excitations are expected to be of comparable energies and the data in Table V generally support this view. It should be recalled that the idealized nature of Hubbard models in solid-state physics is associated with the potential inadequacy of applying the Madelung approximation to long-range Coulomb interactions. By contrast, the intrasite correlation U in H_{CT} arises from molecular considerations.

TTF–TCNQ also forms[186] segregated regular stacks, but now there are *two* ion–radical systems, one based on $TTF^{+\gamma}$, the other on $TCNQ^{-\gamma}$. The stacking is thus quite different from the mixed structure of TMPD–TCNQ, which also contains a strong donor that, like TTF,[187] forms an extensive series of ion–radical salts.[77, 9] The very high conductivity[36–38] and metallic behaviour of TTF–TCNQ down to about 60 K have resulted in intensive efforts at purification, crystal growth, and accurate physical studies.[124, 125] Both ESCA[188] and photoemission[189] studies suggest that there may be less than unit charge per site on either chain. Two complex FR stacks $...D^{+\gamma}D^{+\gamma}D^{+\gamma}D^{+\gamma}...$ and $...A^{-\gamma}A^{-\gamma}A^{-\gamma}A^{-\gamma}...$, with $0.5 < \gamma < 1.0$, probably occur in TTF—TCNQ. This is a rare CT complex in which chemical information about the stoichiometry does not suffice for the electron filling N_e of H_{CT}. By contrast, the mixed $1:1$ CT crystals, whether neutral or ionic, encountered in Section III.C are $\frac{1}{2}$-filled and the degree of ionicity γ follows from the solution of H_{MSR} in (52). TTF–TCNQ also differs from complex FR salts in that γ need not be a simple fraction. Both the TTF and the TCNQ systems thus potentially represent arbitrary electron filling and therefore require whole classes of solutions to H_{CT}, even if the previously noted difficulties with long-range Coulomb interactions in complex salts are overcome. The small interplanar separations in either the TTF or the TCNQ stack indicate considerable CT stabilization. The unusual electric properties of TTF–TCNQ, and of either methylated analogues[190] or of similar organic metals[191, 187] provide a principal focus for future studies, with CT interactions occurring merely as a pervasive crystal parameter t.

H. Complex Segregated Alternating FR Stacks

Complex ion-radical salts based on linear arrays of dimers, trimers, or tetramers provide all the complications of H_{SCR} and, in addition, have several CT integrals. The simplest electron-transfer term, with only two types of π overlap and two CT integrals, is H_{alt} in (63). Many of the examples of Table VI are based on tetramers and thus contain additional transfer integrals. Such systems as $(\phi_3AsCH_3)(TCNQ)_2$, $[\phi_3PCH_3](TCNQ)_2$ and

TABLE VI. Physical properties of π-molecular FR crystals with segregated, complex, alternating stacks (triplet exciton crystals)

Crystal structure	$\Delta E_{CT}/eV$	$\Delta E_c/eV$	$\Delta E_p/eV$
...$A^-A^-AA^-A^-A$... Cs_2TCNQ_3[31]b	$\begin{cases} 0\cdot5-0\cdot7^{(32)a,d} \\ 1\cdot4^{(32)a} \\ 1\cdot38^{(59)} \end{cases}$	$0\cdot30^{(35)}$ $0\cdot32-0\cdot35^{(163)}$	$0\cdot16^{(119)}$ $0\cdot14^{(120)}$
(Morpholinium)$_2$(TCNQ)$_3$[176]	$\begin{cases} 0\cdot5-0\cdot7^{(32)a} \\ 1\cdot4^{(32)a} \end{cases}$	—	$0\cdot31^{(179)}$
...$(A_2^-A_2^-)_n$... $(\phi_3XCH_3)(TCNQ)_2$[185] c	$\begin{cases} <0\cdot5^{(32)a} \\ 1\cdot3^{(32)a} \end{cases}$	$0\cdot30^{(35)}$ $0\cdot39^{(134)}$	$0\cdot065^{(118)}$
X = P, As			

[a] Powder data.
[b] Magnetically non-equivalent stacks.
[c] X = P undergoes a phase transition at 315 K; mixed crystal studies[120, 134, 198] are available.

$Cs_2(TCNQ)_3$ were nevertheless among the first triplet spin exciton systems studied,[23, 118–122] since the pairwise correlations of the spins lead to unmistakable and strongly temperature-dependent fine structure summarized in Section V.C. In effect, the ion–radical dimer discussed in Section II.A almost provides a sufficient description of the EPR. The only additional feature, that the triplet excitons are mobile, is required primarily to average out any hyperfine structure. The alternating Heisenberg antiferromagnet

$$H_{spin} = J \sum_n \{(1 + \delta)S_{2n} \cdot S_{2n+1} + (1 - \delta)S_{2n} \cdot S_{2n-1}\} \qquad (66)$$

for $\delta \approx 1$ (strong alternation) gives a surprisingly complete description of the magnetic properties, as summarized in Section V.C, with the spin vectors S_n assigned to the nth spin. The EPR requires strongly correlated spins, but not a precise assignment of electrons to one or two sites.

$Cs_2(TCNQ)_3$ provides the best illustration of charge localization arising from inequivalent TCNQ sites.[9, 31] The two noncentric sites in Fig. 7 have bond lengths corresponding to TCNQ⁻, while the centric site in the unit cell is close to TCNQ. The site energies are therefore of the form of (60), with $\epsilon < 0$ and $n = 0, 3, 6, \ldots 3m$ corresponding to the centric (largely neutral) site. Soos and Klein[196] have applied the modified Hubbard model to such a $\frac{1}{3}$-filled stack, with the further approximations of a single transfer integral and the ground-state Coulomb interactions. As expected, largely neutral and ionic sites emerge if the site energy difference, ϵ in (60), is comparable to or larger than the CT integral t. A spin Hamiltonian of the form of (66) then follows for the localized charges.

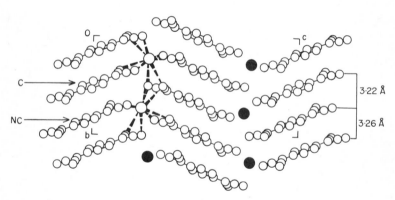

FIG. 7. The [100] projection of the $Cs_2(TCNQ)_3$ structure. The noncentric TCNQ sites are largely ionic, while the centric sites are largely neutral. The occurrence of magnetically non-equivalent TCNQ stacks is clearly shown. [Reproduced with permission from C. J. Fritchie, Jr. and P. Arthur, Jr., *Acta Cryst.*, **21**, 139 (1966)].

Complex FR salts, whether regular or alternating, provide too many parameters in H_{CT}, require additional work on long-range (three-dimensional) interactions, and even then lead to mathematically less tractable Hubbard models. The orbital degeneracy afforded by formally neutral sites in the lattice should clearly enhance conduction, since carriers no longer require forming doubly-occupied sites of energy $\sim U$. The greater conductivity of the complex stoichiometry of TCNQ systems that form both simple and complex salts was observed by Bierstedt, Siemons, and Kepler[35] and discussed further by LeBlanc.[34] Formally neutral sites should also permit additional low-energy CT bands, as indicated in (5), and Iida[32] has so assigned the reflectance spectra of many complex TCNQ salts. Thus the qualitatively different CT and electric properties of complex TCNQ salts are easily rationalized by H_{CT}. Indeed, the realization that the collective properties of many types of π-molecular CT crystals point to a strongly correlated model[22, 24] was a major incentive for developing a site representation.[25]

IV. ONE-DIMENSIONAL HUBBARD MODELS

A. General Comments: Scope of Hubbard Models

The applications of H_{CT} to the various possibilities in Fig. 4 for π-molecular CT crystals has, up to now, involved only symmetry considerations based on the *observed* crystal structure. It is emphasized that no effort has been made to *predict* the structure, since geometries are difficult even for the much simpler case of DA dimers. As indicated in Fig. 8, the general site representation leads, with certain approximations, to H_{CT}. The crystal structure in turn provides exact reductions which, when combined with additional restrictions, lead to modifications of the simple Hubbard model (6). The common features for the special cases of H_{CT} discussed in Section III are, as in (6), electron hopping in one dimension and intrasite correlations. The limit $t \ll U$ reduces, as shown in Fig. 8, to spin–exchange Hamiltonians that are basic to the paramagnetism of ion–radical crystals.

While a number of qualitative features follow at once from a strongly correlated model, accurate solutions are needed for detailed comparisons. Hubbard models have, in fact, been independently introduced in several different areas of chemistry and physics. We now examine some of theoretical results available for H_{Hu} and for related models.

Even the simple, one-dimensional Hubbard model in (6) encompasses the opposite limits of extreme delocalization and localization for different values of t and U. The limit $U \to 0$ leads to a simple metal, with a single nondegenerate band of orbital energies

$$\epsilon_k = 2t \cos kc \qquad (67)$$

where c is the lattice spacing and the N values of k are in the first Brillouin zone, with $-\pi < kc \leqslant \pi$. Any number of noninteracting electrons up to $2N$ can be accommodated by pairwise occupation of the delocalized (Bloch)

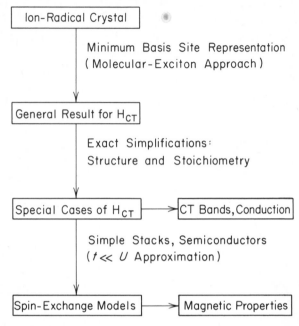

Schematic Analysis of π-Molecular CT
Crystals

Ion-Radical Crystal

Minimum Basis Site Representation
(Molecular-Exciton Approach)

General Result for H_{CT}

Exact Simplifications:
Structure and Stoichiometry

Special Cases of H_{CT} ——→ CT Bands, Conduction

Simple Stacks, Semiconductors
($t \ll U$ Approximation)

Spin-Exchange Models ——→ Magnetic Properties

FIG. 8. Theoretical approach to the molecular-exciton analysis of π-molecular CT crystals in the minimum-basis site representation.

orbitals of energy ϵ_k. The opposite limit of $t \to 0$ and U finite, on the other hand, provides a simple description of isolated, noninteracting sites. For intermediate values of t and U, the Hubbard model describes a continuous range from complete localization to complete delocalization in terms of a single parameter ratio t/U.

Hubbard[44] indeed investigated (6) as a means of obtaining quantitative interpolations between the metallic and atomic limits. In particular he sought to describe intermediate magnetic phases and the associated phase transitions, including the metal–nonmetal transition which had been previously described qualitatively by Mott.[46] Early investigators[44, 199, 200] thus sought conditions, for example on the ratio t/U and on the number of electrons, for either a metal–nonmetal transition or a ferromagnetic transition. However, the early self-consistent molecular-field approximations have been criticized, and general arguments[45, 48] were advanced against the occurrence of the ferromagnetic ground state, except in a very exceptional case.

This illuminating argument is quite analogous to the ion–radical self-complex treated in Section II.A. We begin with two electrons, one on each of two nearest-neighbour sites of a solid. If their spins are parallel (in a triplet state), the Pauli Principle precludes electron hops that place both electrons on one site. The antiparallel spins of a singlet state, on the other hand, permit electron hopping and the resulting configuration interaction lowers the original state $^1|\dot{R}\dot{R}\rangle$. As in (16) for the ion–radical self-complex, the preferential CT stabilization of the singlet state leads to an effective anti-ferromagnetic interaction between the spins, as indicated in (7) for the limit $t \ll U$. A ferromagnetic crystal ground state is therefore not expected, with a single exception. Near the limit $U \to \infty$, electron motion is still possible when the number of electrons is not equal to the number of sites. Now a ferromagnetic ground state occurs, as shown in Ref. 201 or in Chapter IV of Ref. 48, in a lattice in which the smallest cyclic paths for electron hopping involves an odd number of sites, as in a face-centred cubic lattice. Thus the one-dimensional Hubbard model (6) has a *singlet* ground state for an even number of electrons, or a doublet for an odd number of electrons.

Hubbard models remain the principal theoretical approach for describing metal–insulator transitions in transition-metal oxides.[47, 78] Classical Bloch–Wilson band theory predicts metallic behaviour whenever there is an odd number of electrons per metal ion. The insulating nature of such odd-electron compounds as NiO can be understood in terms of Hubbard models. Very little kinetic energy is available in a narrow (small t) band on delocalizing the electrons, which costs considerable electron–electron repulsion energy, and the $\frac{1}{2}$-filled solution (6) is a strongly-correlated, insulating state with roughly one electron localized on each site. The same reasoning applies to open-shell π-molecular CT crystals.

Other solid-state applications of Hubbard models, to systems with random interactions, include metal–ammonia solutions[202] and various spin glasses.[203] Herring's book[48] provides an extensive discussion of many additional applications (up to 1964) of related models, particularly for transition metals.

The Pariser–Parr–Pople[204] (PPP) model for π electrons in either aromatic molecules or linear polyenes provides a major chemical application. The PPP model is a modification of (6) in which nearest-neighbour intersite Coulomb interactions may also be included in addition to the U and t terms of a simple Hubbard model. Occasionally, the intersite Coulomb interactions are deleted[205-7] and a reinterpreted intrasite correlation U is introduced to obtain to a simple Hubbard model. The original formulation of the PPP model was not based on a second-quantized representation and did not employ exotic many-body methods. Second quantization and various generalized Hartree–Fock or Green's function techniques were introduced[208-210, 52, 53] in the last decade as the importance of exact results became evident. Many of the PPP arguments for forming π-electron molecular orbitals out of $2p_z$ atomic orbitals should be even better satisfied for the more weakly overlapping π-electron molecular orbitals out of which we construct crystal functions.[25] In other words, π-molecular CT crystals probably involve smaller ratios of t/U than infinite polyenes,[52] the one-dimensional molecules that are mathematically quite similar to an ion-radical stack. Hubbard models thus provide a natural approach to certain molecular problems as well as to either metallic or insulating open-shell solids.

A final application, closely related to our description of π-molecular CT crystals, involves crystals of partially oxidized tetracyanoplatinates.[77, 27] In $K_2Pt(CN)_4Br_{0.30} \cdot 3H_2O$ or $Rb_2Pt(CN)_4Br_{0.23} \cdot xH_2O$, the tetracyanoplatinate groups shown schematically in Fig. 9 occur in stacks.[211] The disordered Br^- ions between the stacks remove electrons from an otherwise filled band based primarily on the Pt d_{z^2} orbitals (with z along the stack axis).

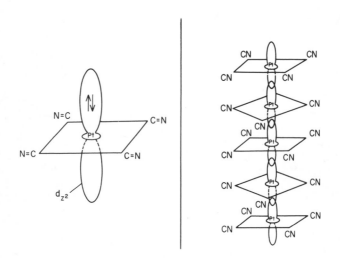

FIG. 9. Overlap of transition–metal d_{z^2} orbitals in Krogmann salts. [Reproduced with permission from M. J. Minot and J. H. Perlstein. *Phys. Rev. Lett.,* **26**, 371 (1971)].

Although Frenkel excitations (d–d transitions) occur at lower energy than the CT states, electron interchanges in the partly oxidized stack occur at very low energy in these good conductors.[27] The low-lying states of a complex Hubbard model, with disordered site energies, may be appropriate for Krogmann salts.

B. Exact Results for the Linear Hubbard Model

We now confine our attention to the simple one-dimensional Hubbard model, H_{Hu} in (6), and present mathematically rigorous results. H_{Hu} is among the very few nontrivial many-body problems whose ground-state solution[212] is known, and it thus serves as a test for approximate solutions in more complicated, usually three-dimensional cases. Such a restricted role in testing approximations has, in fact, provided much of the original interest in one-dimensional problems.[213] However, the intricacy of the mathematics has led to one-dimensional studies for their own sake. The growing evidence that many exchange-coupled transition metal ions, stacked square-planar complexes, and organic solids are realizations of one-dimensional systems has provided a quite different reason for interest in one-dimensional models.

The linear Hubbard model has several exact symmetry properties. Born–von Kármán cyclic boundary conditions lead to cyclic translational symmetry. There are reflections and rotations by π perpendicular to the stack axis. The model is spin-independent, a property retained in the general site representation H_{CT}, and both the total spin S and its z component S_z are good quantum numbers. H_{Hu} also commutes with simultaneous permutations of spin- and space-coordinate indices, and is restricted to a space antisymmetric under such permutations. The second-quantized representation (6) in terms of fermion operators $a_{n\sigma}^{+}, a_{n\sigma}$ guarantees this property, which also holds for H_{CT}. Time reversal and electron conservation are further symmetries. The eigenvalue spectrum turns out to be independent of the sign of t, so long as there are an even number of sites. This is readily seen on making the canonical transformation $a_{n\sigma}^{+} \rightarrow (-1)^n a_{n\sigma}^{+}$. Finally, if there are N sites and N_e electrons, then the eigenvalue spectra of (6) for $N_e = M$ and for $N_e = (2N - M)$ are merely shifted from one another by the constant $|N - M|U$. This electron–hole symmetry and the ensuing simple relation between the corresponding thermodynamic properties and eigenkets is demonstrated by the canonical transformation $a_{n\sigma}^{+} \rightarrow a_{n\sigma}$.

The ferromagnetic state, with all spins parallel, provides a simple, highly excited state of H_{Hu} for any t, $U > 0$, and N_e. Taking all spins parallel disallows doubly occupied sites, and there is no CT stabilization. The simple band solutions apply, with a single ("spin up") electron per occupied k state of energy ϵ_k. Filling the lowest energy orbitals leads to the minimum energy

$$E_0^{ferro} = -2N_e |t \sin 2\pi(N_e - N)/N| \qquad (68)$$

for N_e electrons on N sites with $N_e \leqslant N$. E_0^{ferro} does not depend on U for $U > 0$, and the ferromagnetic subspace thus gives the $U \rightarrow \infty$ limit in which CT is also inoperative. Other highly excited states with $S = \frac{1}{2}N_e - 1$ can be

c

generated for $N_e = N$ by a generalization of Bloch's treatment of spin waves in the Heisenberg ferromagnet.

In the limit $U \to 0$, (6) reduces to noninteracting electrons with energies ϵ_k in antisymmetrized products of Bloch orbitals. Each ϵ_k may now contain two electrons with opposite spin. The complete thermodynamic properties of a one-dimensional metal of noninteracting electrons can readily be obtained. Kommandeur and coworkers[214–16] have proposed that the $U = 0$ limit, or simple band theory, is adequate for even the semiconducting TCNQ salts. The transfer integrals for several sites per unit cell are fitted to yield the observed static susceptibility, while activated conduction is attributed to such extrinsic mechanisms as impurities, breaks in the one-dimensional stack, or disorder. The reliability of multi-parameter fits of χ_p and of the band model for TCNQ salts has been criticized elsewhere.[24] It is certainly true, however, that the complete mathematical analysis of (6) and of analogous Hubbard models is readily available in the limit $U \to 0$ of noninteracting fermions.

The limit $U \to \infty$ is also exactly soluble for open chains[217–19]. Now H_{Hu} cannot change the order of the spins in the chain, since the particles cannot pass each other without creating a (high-energy) doubly occupied site. The ferromagnetic eigenvalues in (68) merely becomes 2^{N_e}-fold spin degenerate. No preferential CT stabilization of states with low multiplicity is possible for $U \to \infty$, and all 2^{N_e} possible spin orientations are degenerate. The large-U limit, in conjunction with unequal site energies ϵ_n, has been proposed[196] for complex TCNQ salts. It leads to alternating exchange Hamiltonians of the type (66), to intrinsic semiconduction associated with electron hopping over the unequal ϵ_n, and to several CT bands. Since $U \to \infty$ leads to a *paramagnetic* solid with vanishing kinetic exchange, this idealized limit must be relaxed in actual cases.

The $t \to 0$ limit is also (trivially) soluble. It remains soluble at all temperatures, even with random site energies[220] or with nearest-neighbour Coulomb interactions.[221] The $U \to 0$, $U \to \infty$, and $t \to 0$ cases provide the limiting behaviour of H_{Hu} for any degree of electron filling and any temperature. For $U > 0$, the general argument that CT stabilizes low-spin configurations can be carried through for any N_e in (6). The minimum energy for a state with total spin S is[222] less than or equal to the minimum energy for total spin $S + 1$.

Lieb and Wu[212] showed that a generalization of the Bethe ansatz[223] for the linear Heisenberg antiferromagnet (7) leads to the exact ground-state energy of the $\frac{1}{2}$-filled ($N_e = N$) Hubbard model (6) for any t/U and $U > 0$. They further showed that the ground state is a singlet and is insulating, with an energy gap against conduction for finite $U > 0$. Ovchinnikov[224] has obtained the eigenvalues of the lowest single-particle (conducting) excitations which, in the atomic limit $t \to 0$, correspond to an electron–hole (CT) excitation with one doubly occupied and one empty site. The lowest triplet spin excitations, which do not involve conduction and do not have an energy gap, are also available.[224, 225] Finally, the lowest eigenvalue for any $S \leqslant N/2$ and the magnetization and susceptibility at absolute zero are known[225]. These results typically involve integrals over transcendental functions and require numerical integrations. It is worth emphasizing that these exact solutions hold only for a relatively few states and are restricted to absolute zero.

The $\frac{1}{2}$-filled Hubbard model reduces[45, 49] to the Heisenberg antiferromagnet (7) in the limit $t/U \to 0$ with $2t^2/U = J = $ constant. The ground-state energy,[226] the lowest triplet spin-waves,[227] the lowest energy-state of each total spin[228] and the susceptibility at absolute zero[228] were all obtained prior to the solution of H_{Hu}. The limit $t/U \to 0$ with t^2/U constant may be extended[229] to partly filled ($N_e \neq N$) Hubbard models and to a variety of modified Hubbard models as well. Here again, exact results are found for absolute zero only.

Shiba[230] has numerically computed exact results for general t/U for partially filled Hubbard models. The exact ground-state energy is shown in Fig. 10 for several values of t/U as a function of N_e/N, while Fig. 11 gives similar results for the susceptibility at absolute zero. The lowest state of each total spin S is also known.[230] Coll[231] shows plots of selected low-lying single-particle (CT) and spin-wave excitations for the partly filled linear Hubbard chain.

In addition to the symmetry properties and exact absolute zero results already mentioned, there are a variety of exact moments of the one-dimensional Hubbard model (6). Some of these results, together with numerical analyses, are deferred to Section IV.D. While the exact results are of limited value in actual comparisons with results on CT crystals or other systems, they provide a series of anchors for approximate or numerical methods that can be extended to finite temperature. The Lieb and Wu result rigorously shows, for example, that a segregated, simple, regular stack is a semiconductor at absolute zero.

C. Self-Consistent Approximations

It is evident from the extensive work summarized in Section IV.B that even the two-parameter Hubbard model (6) provides many different cases, especially if different electron densities N_e/N are included. It has been emphasized that exact results generally hold only for a few special states and for properties at absolute zero. Thus approximate solutions to (6) are also necessary. It turns out that more conventional many-body techniques, such

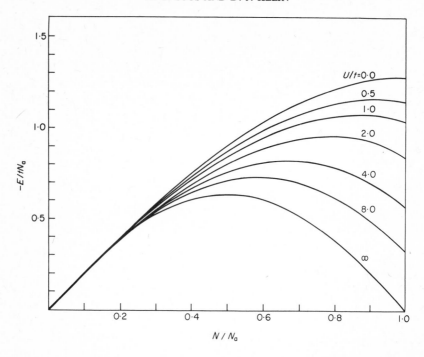

FIG. 10. The ground-state energy of the linear Hubbard model as a function of the number of electrons per site, N/N_a. The various curves are for different values of the ratio t/U. [Reproduced with permission from H. Shiba, *Phys. Rev.*, **B6**, 930 (1972)].

as restricted Hartree–Fock theory, do not work well for arbitrary values of t/U and of N_e/N. Self-consistent methods are particularly suspect for two neighbours in one-dimensional problems, since such methods become exact in the limit of weak $(1/N)$ interactions with N sites. It is possible, of course, that the poor reputation of self-consistent methods in one dimension in part reflects that exact results are available, while no such comparison is yet possible for the three-dimensional Hubbard model. Another difficulty with self-consistent methods for either the linear Heisenberg antiferromagnet[232] or the Hubbard model[22, 52] is that fermion representations lead to ground states which are not eigenfunctions of S^2, but only of S_z. The spherical spin symmetry of H_{CT}, of H_{Hu}, or of H_{spin} in (66) is thus lost, and is usually replaced by axial symmetry. The availability of exact results and the identification of various difficulties has naturally led to a large variety of unconventional self-consistent methods for improving one-dimensional analyses.

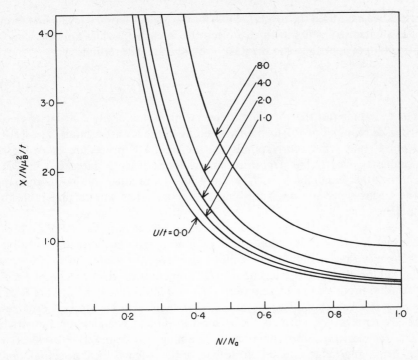

FIG. 11. The exact zero-temperature susceptibility of the linear Hubbard model as a function of the number of electrons per site, N/N_a. The different curves are for different values of the ratio t/U. [Reproduced with permission from H. Shiba. *Phys. Rev.*, **B6**, 930 (1972)].

The Bloch orbitals for the simple Hubbard model (6) are dictated by translational symmetry,

$$a_{k\sigma}^+ = N^{-\frac{1}{2}} \sum_n e^{iknc} a_{n\sigma}^+ \qquad (69)$$

with $-\pi < kc \leqslant \pi$ and $kc = 2\pi n/N$. The fermion operator $a_{k\sigma}^+$ creates an electron with spin σ in a crystal state with energy ϵ_k in (67). The Bloch orbitals are in fact the restricted Hartree–Fock (RHF) orbitals for the Hubbard model (6) and, with a different interpretation for the operators, for the Heisenberg antiferromagnet (7). The RHF ground state is

$$|\text{RHF}\rangle = \prod_{\substack{k\sigma \\ (\min)}} a_{k\sigma}^+ |0\rangle \qquad (70)$$

where $|0\rangle$ is the vacuum state and the product k, σ is over the lowest energies ϵ_k needed for N_e electrons. The expectation value $\langle \text{RHF} | H_{\text{Hu}} | \text{RHF} \rangle$ is

$$E_0^{\text{RHF}} = \sum_k^{\text{occ}} 2\epsilon_k + \frac{U}{N} \sum_{qkk'} \langle \text{RHF} | \cos qc \, a_{k+q\alpha}^+ a_{k'-q\beta}^+ a_{k'\beta} a_{k\alpha} | \text{RHF} \rangle \qquad (71)$$

where we have used the inverse of (69) for the quartic term H_{Hu}. The first term is evaluated as in (68), with two electrons in each ϵ_k, while only the $q = 0$ (direct) term contributes to the RHF ground state. The result is

$$E_0^{RHF}/N = -\frac{8|t|}{\pi} \sin \frac{N_e \pi}{N} + U \left(\frac{N_e}{2N}\right)^2 \tag{72}$$

The U term is just the correlation energy for N_e uniformly distributed electrons, as required in (70) for pairwise filling of the lowest orbitals. The RHF ground state, which corresponds to the usual HF procedure of pairwise occupancy, clearly has a *positive* energy for sufficiently large U/t in (72). On the other hand, for $N_e \leqslant N$, a simple orbital product with no more than one electron per site has vanishing energy and thus can be preferable to $|RHF\rangle$.

The failure of RHF theory occurs because the trial function $|RHF\rangle$, which is exact for $U = 0$, is independent of U and thus includes more double occupancy than it should. On a molecular level, the analogous difficulty is the failure of a single-determinant (HF) function to yield the proper dissociation products (which represent the $t \to 0$ limit). The difficulties of RHF theory can be rationalized by noting that such methods are best for one-electron potentials, whereas the U term in (6) is a two-electron potential. The one-electron approximation of retaining $q = 0$ in (71) introduces a uniform, long-range interaction instead of the original local potential, and this cannot be satisfactory. The difficulties with large U/t are, finally, not expected to be easily overcome by conventional perturbation theory starting with the $U = 0$ case in zeroth order, since the exact ground state is not analytic in U and there may also be lower-lying unrestricted HF solutions.

The energy difficulties with large U can be overcome by introducing an unrestricted Hartree–Fock (UHF) state in which selected Bloch orbitals are allowed to mix. The mixing of k and $(k + \pi)$ is especially convenient for the $\frac{1}{2}$-filled case $N_e = N$, but can be used for other electron densities as well. Several possibilities occur, depending on whether the new orbitals mix spin components. The choice

$$\zeta_{k\sigma}^+ = U_k a_{k\sigma}^+ + V_k a_{k+\pi\sigma}^+ \tag{73}$$

was adopted by Penn[233] and by Langer, Plischke, and Mattis.[234] The alternate choice of Strebel and Soos[22] is based on

$$\eta_k^+ = U_k a_{k\sigma}^+ + V_k a_{k+\pi\bar{\sigma}}^+ \tag{74}$$

and leads to identical energies, but with a trial function that is rotated in spin space.[235] Halperin and Rice[236] summarize the general possibilities for constructing UHF functions. In each case, the trial function is analogous to $|RHF\rangle$ in (70), but with creation operators corresponding to the new orbitals. The expectation value of $\langle UHF|H_{Hu}|UHF\rangle$ is then computed, and usually involves additional self-consistency conditions.[22, 234, 236]

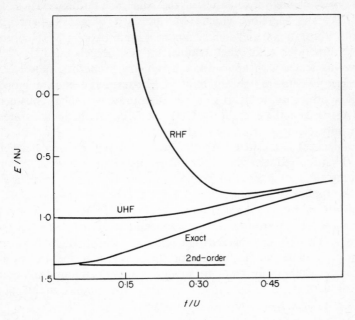

FIG. 12. The RHF, UHF and second-order ground-state energies of the half-filled linear Hubbard model compared with the exact result.

We compare in Fig. 12 the RHF and UHF results[234, 22] for either (73) or (74) with the exact ground-state energy for the $\frac{1}{2}$-filled ($N_e = N$) Hubbard model. The UHF approach clearly alleviates the difficulty for large U/t and, in fact, provides about 80 per cent of the exact CT stabilization of the ground state. Fig. 12 also contains a second-order perturbation result based on the U term as the zeroth-order part and t as the perturbation. This approach leads to the Heisenberg antiferromagnet (7), and is exact in the limit $t/U \to 0$ with $J = 2t^2/U = $ constant. Neither UHF solution preserves the S quantum number, both introduce long-range order, and neither has spin excitations, for small values of t/U, that resemble closely those of the Heisenberg antiferromagnet. The energy comparison in Fig. 12 is best for the $\frac{1}{2}$-filled case shown. For partial filling ($N_e < N$), $E_0^{UHF} \to 0$ as $U \to \infty$, while there is finite stabilization of order t available from electron delocalization. Thus the true ground-state is rigorously below the ferromagnetic energy in (68), with all spins aligned. The electron–hole symmetry of H_{Hu} leads to similar conclusions for $N < N_e \leqslant 2N$.

UHF methods are thus only moderately successful energetically and work best in the $\frac{1}{2}$-filled case. Unfortunately, UHF solutions invariably introduce various types of long-range order in the spin orientations, even at finite temperature, and therefore violate Mermin and Wagner's theorem.[237] Since both translational and spin symmetries are broken in UHF solutions, new wave functions with exact symmetries can be generated[210, 235, 238] by projecting states with appropriate symmetry properties. However, the new energies per site are typically only $N^{-\frac{1}{2}}$ better, and even orbital optimization after projection leads to similarly discouraging results. Thus neither the symmetries nor the energies of UHF methods are highly accurate, but are clear improvements over the RHF result.

The difficulties associated with the "atomic limit" $t/U \to 0$ have been extensively discussed elsewhere[239-41]. Hubbard models with attractive interactions ($U < 0$) have also received some attention[242] but are not likely to be applicable to π-molecular CT solids. The self-consistent methods discussed above provide approximate results for finite-temperature properties and will necessarily remain an important bridge between theory and experiment. We have emphasized that self-consistent methods have so far not been particularly successful in reproducing exact properties. Even so, self-consistent methods for Hubbard models represent an important body of theoretical work that can be applied, often with only minor modifications, to the question of CT in solid-state complexes.

D. Numerical Results: Static Properties

The linear Hubbard model (6) is a particularly simple special case of H_{CT}. No matter what interactions are retained in H_{CT}, the minimum-basis set introduced in Section II.E contains at most 4^N states for N sites, thus leading to $4^N \times 4^N$ secular determinants. Symmetry considerations may permit block diagonalization, thus reducing the work for a direct numerical attack. The most general spin Hamiltonian for $S_n = \frac{1}{2}$ sites results in $2^N \times 2^N$ secular determinants for N sites. Now direct numerical analysis is feasible for twice as many sites. Bonner and Fisher[243] demonstrated that numerical data up to $N = 10$ and 11 and extrapolation techniques are sufficient for computing the $N \to \infty$ thermodynamics of the linear Heisenberg chain (7). Thus direct numerical methods[244, 245] may be applied to obtain reasonably accurate thermodynamic functions for Hubbard models as well.

Table VII summarizes numerical results for some of the special Hubbard models encountered in Section III. The most commonly reported properties are the specific heat, C_v, and the magnetic susceptibility, χ_p. The average density of doubly occupied sites, $\bar{n}_D$, the entropy, and the internal energy are also reported sometimes, although not always listed in the table. It is evident

TABLE VII. Selected finite-temperature results for the Hubbard model

Type of stack	Results	Range of applicability	Method	Refs.
SSR	χ_p, C_v, n_D	$kT \gtrsim U$	Finite-chain extrapolation	245
	χ_p, C_v	$kT \gtrsim U/4, U \geqslant 4t$	Perturbative high-temp. expansion	260
	χ_p, C_v, n_D	$kT \gtrsim 4t^2/U, U \geqslant 5t$	Numerical-perturbative extrapolation	255
	χ_p, C_v	$kT \gtrsim U$	Equation of motion	263
	χ_p, C_v	larger t/U	Functional integral	266
SCR	χ_p, C_v	$kT \gtrsim U/3, U \geqslant 8t$	Perturbative high-temp. expansion	269
	χ_p, C_v	larger kT?	High-order Green's function decoupling	270
	χ_p, C_v	$kT \gtrsim t, U \gtrsim 5tN/N_e$	Numerical-perturbative extrapolation	229
	χ_p, C_v	$U \to \infty$	Exact	218, 219
SSA	χ_p ^{a}	$U \gtrsim 5t$	Self-consistent field	272, 274, 276
	χ_p, C_v	$kT \gtrsim 2J(1 + \delta)$	Finite-chain extrapolation	271
	χ_p		Triplet–exciton expansion $(t' \ll t)$	277, 278
	χ_p, C_v, n_D	$U = 0$	Exact	214

^{a} These SSA results are for the alternating Heisenberg chain, (66), which is the second-order perturbation expansion of the Hubbard model.

that most computations have dealt with the simpler Hubbard models. Only qualitative results are available for the more complicated SCA or MSR cases, which also contain the site energy ϵ_n as an additional parameter; self-consistent results for these cases have already been summarized. More reliable *static* thermodynamic properties, such as those in Table VII, for a wider variety of models should become available in the near future.

A number of treatments[246–54] have been omitted in Table VII. They sometimes do not contain explicit numerical results for thermodynamic properties; or they have been applied only to three-dimensional lattices; or they are inaccurate in the region near the atomic limit $(t/U \ll 1)$ of interest here. We now consider several of the entries in Table VII to evaluate the results and their regions of reliability.

1. The SSR Case

Figures 13 and 14 show the low-temperature χ_p and C_v results,[255] respectively, for the simple $\frac{1}{2}$-filled Hubbard model (6). Figure 13 also contains χ_p data[40] for NMP–TCNQ, which were originally thought to be an experi-

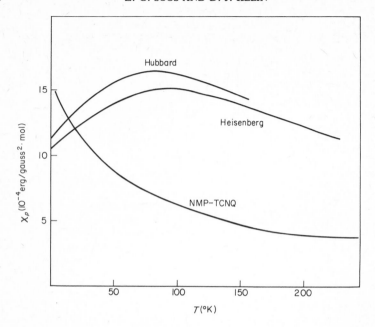

FIG. 13. The static susceptibility of the Hubbard model ($t = 0{\cdot}021$ eV, $U = 0{\cdot}14$ eV) and of the Heisenberg model ($J = 2t^2/U = 0{\cdot}063$ eV) as compared with the smoothed NMP–TCNQ data of Ref. 40.

mental realization of (6). The principal conclusion is that, at least at low temperature and small t/U, the simple Hubbard model is only a slight modification of the corresponding Heisenberg model (7), with $t/U \to 0$ and $J = 2t^2/U = $ constant, whose thermodynamic properties are known from the Bonner–Fisher extrapolation[243] of finite-chain computations. At higher temperature, for finite t/U, C_v develops a second maximum[244, 245] that is associated with single-particle excitations. Such charge-carrying excitations do not occur in the Heisenberg limit. The average density, n_D, of doubly-occupied sites is,[244, 255] as expected, small for lower t/U and lower temperature.

Since bulk properties (per site) usually approach a constant,[243, 256] with corrections of order N^{-1} or N^{-2}, the extrapolation of finite chain properties to $N \to \infty$ provides a direct numerical approach. The Hamiltonian matrices increase as $4^N \times 4^N$ for Hubbard models and have been diagonalized[257–8, 244–5, 162] only up to $N \leqslant 6$. Finite-chain computations preserve some of the lower moments of the Hamiltonian, and consequently yield extrapolations[245] that are most accurate at high temperature. Prospects for extending *direct* computations to larger N, and thus to lower temperature, or to Hubbard models with lower symmetry than (6) are not favourable.

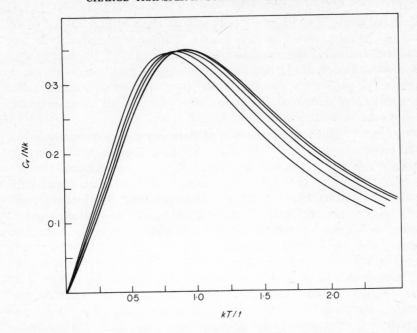

FIG. 14. Specific heats per site for the half-filled Hubbard model. At the higher temperature the $t/U = 0$ Heisenberg result lies highest with the ratios $t/U = 0.05$, 0.10, 0.15 and 0.20 lying progressively lower. Note that at low temperatures this order is reversed. [Reproduced with permission from W. A. Seitz and D. J. Klein. *Phys. Rev.*, **B9**, 2159 (1974)].

The high-temperature perturbative expansions of Bulaevskii and Khomski[259] and of Hone and Pincus[260] yield analytical results for the first few terms of the logarithmic expansion of the grand partition function, Z. The U term of (6) is taken in zeroth order, and corrections due to t are obtained by differentiating $\ln Z$. A difficulty with such low-order expansions is that, at low temperature, certain qualitative features of the zeroth-order system are incorrectly given. Extrapolation of additional higher-order terms should suffice, but would entail considerable effort if the present practice of evaluating the expansion term by term is retained. The radius of convergence for $N = 2$ is $|t|/U = \frac{1}{4}$, and restrictions of perturbation expansions to small $|t|/U$ can be expected in general.

Seitz and Klein[255] developed a numerical perturbative approach for low temperature. The idea is to introduce, *via* a linked degenerate perturbation expansion,[261–2] an extended effective spin Hamiltonian with, however, only 2^N states. The effects of single-particle (charge-carrying) excitations with energy of order U is included, up to $(2t/U)^7$, in the low-lying manifold of spin excitations. The transformed spin Hamiltonian is then solved for finite chains, of length $N \leqslant 12$, and results such as those in Figs 13 and 14 are obtained by extrapolating with respect to both the perturbation order and the chain length. The restriction to low temperature ($kT \ll U$) could be lifted by constructing effective Hamiltonians for higher manifolds, with one or more single-particle excitations. While lower-order results would suffice for such excited manifolds, the perturbative nature of the expansion again restricts the method to at least $2|t|/U < 1$.

Visscher[263] noted that derivatives of thermodynamic expectation values with respect to $\beta = (kT)^{-1}$ merely yield more complicated expectation values, which may in turn be differentiated. The resulting infinite set of linear, coupled, first-order differential equations is truncated and numerically integrated, starting from the exact high-temperature solution. Improved methods of numerically integrating the equations, such as iterated diagonalization or Magnus approximation,[264-5] would undoubtedly extend the range of utility to lower temperature.

The functional integral method of Kimball and Schrieffer[266] is to convert (6) into an N-fold integral over one-particle operators by using[267] an operator identity that expresses an (ordered) exponential of certain two-electron operators into (ordered) exponentials related to one-electron operators. Conventional many-body techniques may be used to solve the one-electron problem. Now approximations are introduced to carry out the infinite-fold integrals over the one-electron partition function. Bari[268] has shown that the particular approximations used are inadequate in the atomic limit $t \ll U$.

The many different theoretical approaches, either at low or high temperature, attest to the importance of the simple Hubbard model (6) as a many-body problem. The results in Table VII for the SSR stack may be applied to CT crystals in which H_{CT} reduces to a simple Hubbard chain.

2. The SCR Case

We now consider $N_e \neq N$ in (6). Figure 15 shows C_v results[229] for partial filling of the one-dimensional Hubbard model near the atomic limit. These curves are largely understood as arising from two contributions: a Heisenberg spin contribution with a low-temperature maximum and a motional contribution for electrons moving over singly occupied sites leading to the broad high-temperature maximum. The susceptibility for small t/U and low temperature is similar[229] to that of a Heisenberg chain with an effective exchange

$$J = \frac{2t^2 r_e}{U + (4t/\pi) \sin r_e \pi} \left(1 - \frac{\sin 2\pi r_e}{2\pi r_e}\right) \tag{75}$$

(75) is restricted to about $kT \leqslant 5J$. We note that as the electron charge r_e

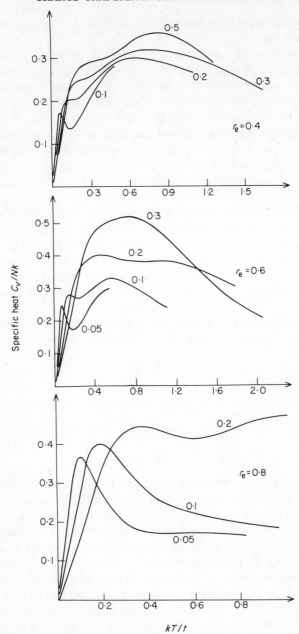

FIG. 15. Specific heats per site for partially filled linear Hubbard model with electron/site ratio r_e. The various curves are labelled by the appropriate t/U ratio. [Reproduced with permission from D. J. Klein and W. A. Seitz. *Phys. Rev.*, **B10**, 3217 (1974)].

per site approaches unity, J reduces to the familiar second-order perturbation result of the half-filled chain.

The high-temperature perturbative expansion of Beni, Pincus, and Hone[269] uses the $t = 0$ model in zeroth order to obtain properties to second order as in the analogous[259, 260] half-filled case. These expansions are, however, less accurate for the partially filled case because of the difficulty in describing electron motion among singly occupied sites. Bartel and Jarrett[270] employ a Green's function decoupling, which is effected at a rather high order, and solve the remaining equations numerically. While the assumptions invoked make it difficult to assess the region of validity, the preservation of lower moments suggests that the method is accurate at high temperature.

Klein and Seitz[229] introduced a perturbation expansion about the exactly soluble projected Hubbard model[219] in zeroth order. The transformed Hamiltonian is accurate to second order. The zeroth-order (projected) picture retains some t contributions exactly and is consequently more accurate than merely retaining the U term in (6). The resulting second-order Hamiltonian is evaluated for lower-lying manifolds of states and, less accurately, for high-lying manifolds. Thermodynamic properties are obtained for $N \to \infty$ exactly in terms of Heisenberg model properties. Extension to larger t/U is again limited by the perturbative nature of this approach.

3. The SSA Case

The half-filled Hubbard model with alternating CT integrals has been most often discussed in terms of the alternating Heisenberg chain (66) with exchange constants

$$J(1 + \delta) = 4t^2/U$$

$$J(1 - \delta) = 4(t')^2/U \tag{76}$$

The $\delta = 0$ limit reduces to the SSR case and, as before, (66) is a limiting case with both $t'/U \to 0$ and $t/U \to 0$ and both $(t')^2/U$ and t^2/U constant. Properties of the alternating Heisenberg chain are less reliably known and the probable occurrence of an energy gap for any $\delta \neq 0$ leads to qualitatively different χ_p behaviour than in the gapless regular chain shown in Fig. 14. The trivial limit of extreme alternation, $\delta \to 1$, leads to noninteracting ion-radical dimers with a singlet ground state and a triplet excited-state at $2J$. Then both C_v and χ_p vanish exponentially as $T \to 0$, and this general feature has been suggested[271] for any degree of alternation $\delta \neq 0$. Since the spin Hamiltonian (66) does not include single-particle excitations, it is restricted not only to Hubbard models near the atomic limit, but also to $kT \ll U$.

A variety of approaches to the alternating Heisenberg chain (66) has been proposed. Self-consistent-field methods[272–6] yield accurate ground-state energies, although questions[276] have been raised about the accuracy of χ_p computations. The difficulty is compounded by the occurrence of three distinct self-consistent ground states[232] with widely different properties. Duffy and Barr[271] have used extrapolations of exact finite-chain computations. Only even chains, with lower symmetry, are reasonable for the alternating case, and lead to less accurate extrapolations than when $\delta = 0$. A third approach[277, 278] is based on the $\delta \to 1$ limit of strong alternation, in which the triplet excitations on the dimers are preserved even when $t' \neq 0$. Such perturbative approaches are limited to small t'/t, or to strong alternation $\delta \sim 1$. Several of the extensions of the Heisenberg limit for SSR stacks to Hubbard models near the atomic limit can readily be applied to SSA stacks as well.

E. Single-Particle Excitations and Dynamic Properties

Few reliable results are currently available for such "dynamic" properties of (6) as spin or charge correlation functions, optical absorption coefficients, and electrical conductivity. There are serious difficulties, both mathematical and physical. First, dynamic properties require both eigenvalues and eigenkets in explicit finite-chain computations, while only eigenvalues are needed for static properties. Second, eigenkets are not only tedious to compute, but are less accurate. First-order errors in eigenkets lead to only second-order errors in eigenvalues, while still yielding only first-order expectation values for dynamic properties. The physical difficulties may be even more compelling. It was emphasized in Section III that H_{CT} reduces to modified Hubbard models only with additional approximations, e.g. $\langle V_C \rangle^{(0)}$ in (51), that may fail for CT excitations and for charge-carrying states. An electronic excitation like an electron–hole pair is likely to distort the molecular frameworks,[16] to polarize the adjacent molecules,[14] to couple to optical phonons[78] in an ion-radical crystal, and to interact with nearby Frenkel excitations,[279] just as Frenkel states in molecular solids are modified[17] by nearby CT levels. None of these contributions are included in H_{CT} and all may potentially be important. A polaron[14, 47, 78, 79] model for charge-carriers in ionic organic solids is quite attractive. Excitonic polarons[280–82] based on the high polarizability of aromatic molecules are one possibility; intramolecular vibrations[283] also lead to polaron formation, as does the standard[78] coupling to optical phonons. A major difficulty with such theoretical models is their present lack of specificity, since a plausible case can be made for several mechanisms and neither theory nor experiment is sufficiently refined to provide accurate parameters. Thus it is by no means certain that approximate numerical results for single-particle excitations of rigid-lattice Hubbard models should be quantitatively compared with experiment.

Spin correlation functions[67, 68, 256] are the dynamic properties of the Heisenberg linear chain that are most amenable to quantitative comparison with both magnetic resonance and neutron diffraction studies. The one-dimensional Hubbard model near the atomic limit $t \ll U$ is again expected to be similar, at least at low temperature where single-particle excitations are not thermally accessible. Hone and Pincus[260] have computed the first terms in the high-temperature perturbative expansion of the two-spin autocorrelation function needed to discuss NMR data. Accurate calculations of the spin correlation functions for any special case of H_{CT} encountered in Section III will be needed for a quantitative treatment of magnetic properties. It should be noted that the physical difficulties associated with excited states are minimized for low-energy spin excitations in ion–radical crystals.

The optical and electric properties require the site representation of the dipole moment and the current operators, respectively. The second-quantized expression for the dipole moment operator μ is

$$\mu = er = e \sum_{i\mu} \sum_{j\nu} \langle i\mu|r|j\nu\rangle c_{i\mu}^{+} c_{j\nu} \tag{77}$$

Here r is the (one-particle) position vector for all particles, including nuclei, and the indices i, j run over the molecular sites, while μ, ν sums over the molecular orbitals of a given site. Since μ is a one-particle operator, only one- and two-site terms occur when (77) is expanded using shift operators introduced in Section II.D. The general site representation for μ yields

$$\mu = e \sum_{irs} \langle ir|r_i|is\rangle X_i^{rs}$$

$$+ e \sum_{ij} {}^{'}\sum_{rstu} \langle ir\Lambda js|r_i + r_j|it\Lambda ju\rangle (-1)^{n_{it}(n_{js} - n_{ju})} X_i^{rt} X_j^{su} \tag{78}$$

where r_i is the position vector for an electron at molecular site i, the symbol Λ indicates an antisymmetrized product and n_{it} is the number of electrons for state t of site i. As in Section II, we now specialize (78) to the minimum basis (29) or (30) for D or A sites in a CT crystal.

All one-site matrix elements $\langle ir|r_i|is\rangle$ vanish in the minimum basis. The off-diagonal elements vanish in general, since r_i conserves both electrons and spin, while g, α, β, and $\alpha\beta$ all differ in either electrons or spin. The diagonal elements vanish for molecules with a centre of inversion, like PD, TMPD, TTF, chloranil, or TCNQ, since r_i is odd under inversion. This result also holds for crystals in which D and A sites retain a centre of inversion. Neglecting one-site contributions for symmetric molecules at a general crystal site should be a good first approximation, although polarization is neglected. Such an approximation is in the spirit of assuming weakly perturbed molecular states in π-molecular crystals.

The only non-vanishing two-site terms in (78) involve transferring a single electron between sites i and j. These matrix elements are therefore first-order in the intersite differential overlap and are significant only when sites i and j are neighbours in the stacks. Several different two-site terms are possible in principle, as shown in Section VI.B for the CT integral, depending on the number of electrons at i and j. We again retain a single contribution

$$M = \langle D^+ A^- | z | DA \rangle \tag{79}$$

The dipole–moment operator then reduces simply to

$$\mu = ez \sum_i M_{i,\,i+1} (a_{i\sigma}^+ a_{i+1\sigma} + a_{i+1\sigma}^+ a_{i\sigma}) \tag{80}$$

where z is a unit vector along the stack axis.

The corresponding expression for charge carriers along the stack involves the current operators, ev, where v is the velocity. The minimum-basis site representation again contains only two-site terms,

$$v = iz \sum_{j\sigma} V_{jj+1} (a_{j\sigma}^+ a_{j+1\sigma} - a_{j+1\sigma}^+ a_{j\sigma}) \tag{81}$$

where V is a representative velocity matrix element for two sites,

$$V = m_e^{-1} \langle D^+ A^- | P_z | DA \rangle \tag{82}$$

and m_e is the electron mass and P_z is the momentum along the chain. The expansion of various quantum-mechanical observables in terms of the minimum basis (29) and (30) is straightforward, and at least approximate matrix elements can be readily computed whenever the approximate crystal ground state is available.

Kubo[284] and Sadakata and Hanamura[285] have computed optical absorptivities for the half-filled linear and cubic Hubbard lattices, respectively. Green's function techniques were employed and special attention was directed toward the atomic limit. Generally for small t/U, one finds a CT absorption centred near U with a half-width of several t. As t/U increases the band appears to shift to slightly lower frequencies, to broaden and to increase slightly in total intensity. For the one-dimensional case[284] a sharp spike appears, as is typical[286] in one-dimensional models with a high density of states at band edges.

Harris and Lange,[239] Bari, Adler and Lange,[287] Brinkman and Rice,[288] Beni, Holstein and Pincus[218] and Sadakata and Hanamura[285] have carried out low-order perturbation expansions for a variety of charge correlation functions, including the electrical conductivity. The perturbation expansion is based on the atomic limit, and becomes exact as $t/U \to 0$. It is found that the zero-temperature conductivity of the half-filled model is zero, while it is finite for various partial fillings. The zero conductivity at half-filling is in agreement with the ideas of Mott[46] for narrow-band transition–metal oxides. The finite value at partial fillings is associated with the ability of the electrons to move about without creating additional doubly occupied sites. Suezaki[289] gives numerial plots for half-filled linear chains near the atomic limit as calculated via a Green's function technique. Some comparison to TEA(TCNQ)$_2$ is also presented. A contrasting analysis and summary of this system has been proposed by Brau and Farges.[290]

The site representation for H_{CT} obtained in Section II is based on strong intrasite (molecular) correlations. The qualitative features of the CT excitations (2)–(5) in ion–radical solids are immediately obvious, as expected, since the phenomenological theory was introduced to mimic the polarization of the CT band, the nature of the electron transfer, and the possible occurrence of several CT bands in complex salts. The general conductivity features of the paramagnetic semiconductors in Tables II–VI also follows simply. We expect $\Delta E_p < \Delta E_c$ in simple salts, since the former does not involve single-particle excitations; we also expect lower ΔE_c in complex salts, where only site-energy differences must be overcome, and such a correlation was an early observation.[35] On the other hand, no quantitative interpretation of CT bands, intensities, and lineshapes or of charge carriers, polarization and phonon interactions has yet been achieved, even for the idealized Hubbard model. Thus H_{Hu} and even H_{CT} may prove to contain serious oversimplifications for single-particle excitations in real complexes. As emphasized in Section II, however, a molecular exciton approach leads naturally to a site representation and H_{CT} is the *simplest* reasonable candidate for describing the qualitative features of the magnetic, optical, and electric properties of the different types of ion–radical stacks.

V. MAGNETIC PROPERTIES OF CT CRYSTALS

A. Antiferromagnetic Exchange and Collective Spin States

Although the importance of NMR investigations of DA dimers has been recognized,[3, 291] there has been little useful correlation of magnetic and optical properties in π-molecular solids. Previous work has primarily focused on optical properties, with excursions into the general problems of charge carriers, while magnetic studies in ion–radical crystals focused on the consequences of strong exchange, whether in regular chains as indicated in (7) or in alternating chains as in (66). Such many–spin problems are central to magnetic insulators and explicitly exclude optical or electric excitations. The identification of the CT contribution to the effective exchange interaction $J \approx t^2/\Delta E_{CT}$ provides a connection[23, 50] between magnetic and optical properties and was a primary motivation for constructing the site representation of H_{CT}. It must be emphasized, however, that it has always been difficult to enumerate[49, 85] various contributions to exchange interactions and that J is typically a phenomenological constant in the interpretation of magnetic data.

The smallness of the Mulliken integral t relative to CT excitations ΔE_{CT} led to the molecular–exciton approach outlined in Section II. The π-molecular crystals were treated as collections of weakly overlapping sites. CT interactions lift the 2^N spin degeneracy indicated in Table 1 for a nonoverlapping array of $S = \frac{1}{2}$ sites. Even a modest exchange $J \approx 10^{-3}$ eV, a negligible term in comparison to $\Delta E_{CT} \approx 1$ eV, is enormous in comparison with such purely magnetic terms as electron dipolar, hyperfine, and Zeeman interactions. A collective description is therefore required for the spins and the narrow, temperature-dependent EPR lines observed in solid-state complexes bear no resemblance to the rich hyperfine spectra of aromatic ion–radicals in solution.

It is convenient to treat the static susceptibility, χ_p, separately from the motional and relaxation data afforded by EPR and NMR data. Since χ_p is among the most easily computed exact properties discussed in Section IV, the accurate identification of antiferromagnetic exchange constants provides a method for evaluating transfer integrals t by using relations like (8). Thus CT in ion–radical systems is inherently easier to study experimentally than in closed-shell neutral complexes in which the lowest magnetic states occur at energies comparable to those for charge-carriers. As shown below, however, exact results for χ_p have often been more powerful in ruling out proposed models than in establishing an entirely consistent physical picture.[24] Ionic CT and FR crystals also fall into two classes,[66, 117] as indicated in Fig. 4, depending on whether the exchange is equally strong with both neighbours in the stack or is substantially stronger with one of the neighbours. The latter case corresponds to the triplet spin excitons first observed by Chesnut and Phillips[118] in TCNQ salts. Triplet spin excitons, in which two $S = \frac{1}{2}$ electrons of the triplet state are strongly correlated on adjacent sites, remain among the best understood and most versatile excitations in organic solids.[23, 118–122] Spin excitations in *regular* chains, in which two spins forming the triplet are uncorrelated,[66] have been identified in 1:1 CT crystals like PD–chloranil (PDC),[148] TMPD–TCNQ,[150] and TMPD–chloranil[292] through the temperature, pressure, and angular dependence of the EPR line. A phenomenological treatment of these uncorrelated triplets is presented in Section V.D. The analysis of exchanged–narrowed EPR spectra in the region $kT \lesssim J$ is far more difficult than in the high-temperature limit, $kT \gg J$. Thus magnetic resonance studies in π-molecular CT crystals provide an interesting series of problems which are related to CT through the antiferromagnetic exchange.

B. Static Susceptibility

Table VII shows that exact numerical results are available for the paramagnetic susceptibility, χ_p, and the magnetic specific heat, C_v, of either linear Heisenberg spin systems or linear Hubbard chains. Except at very low temperatures $\lesssim 5\,K$, the C_v contributions from lattice phonons are difficult to separate from magnetic contributions.[68, 126, 293] Then only χ_p data remain to be fitted, with the exchange parameters and/or t/U ratio treated as adjustable parameters. The experimental decomposition of the total static susceptibility into

$$\chi = \chi_p + \chi_d + \chi_i \tag{83}$$

is usually straightforward.[120, 294, 295] Here χ_d is a temperature-independent diamagnetic contribution that may be estimated from Pascal's constants[120, 295] or identified from the low-temperature diamagnetism of the sample; χ_i is a small impurity correction, usually taken to obey the Curie law, $\chi_i T = \text{constant}$, that may arise from isolated ion-radicals. Bulaevskii and coworkers[296] argue that the very low temperature susceptibility, $\chi_p \approx T^{-\alpha}$ with $\alpha = 0.6$, observed in TCNQ salts like NMP–TCNQ or Q(TCNQ)$_2$ is evidence for disordered exchange constants $\{J_n\}$ along the chain, including occasionally vanishing exchange. The point here is that, even if the "impurity" term in (83) may in special cases require further analysis, accurate experimental determination of χ_p is possible for most of the temperatures of interest. Less accurate direct resonance measurements of χ_p by the Schumaker–Slichter[297] technique, from integrated EPR intensity and even from NMR shifts[298] are also possible. Inorganic salts often require an additional, temperature-independent (Van Vleck) paramagnetic term in (83), but such contributions are negligible in organic solids.

Other reviews [67, 68] have examined in detail the low-temperature magnetic properties of such regular one-dimensional or nearly one-dimensional[299] exchange systems as $N(CH_3)_4MnCl_3(TMMC)$, $CuCl_2$–pyridine (CuCP), and $Cu(NH_3)_4SO_4 \cdot H_2O$ (CTS). Griffiths[300] showed that both χ_p and C_v in CTS were given by a *single* exchange constant, as required for a structure with a single type of water-bridged Cu–Cu interaction. The more extensive data on TMMC, an SSR chain based on $S = \frac{5}{2}Mn^{++}$ ions, include both $\chi_p^{\parallel}$ and $\chi_p^{\perp}$ data,[301] C_v data,[302] detailed neutron-diffraction studies,[303] and EPR data.[69] The CuCP systems is based on an SSR chain of spin $\frac{1}{2}Cu^{++}$ ions and, like TMMC, has provided[304] conclusive evidence for the Heisenberg linear chain (7). While the detailed description of other features involving spin correlations remain of considerable interest, such topics are beyond the scope of the present review. In favourable cases like $Cu(NH_3)_4$-$PtCl_4$, the angular and frequency dependences of the EPR line provide[305, 306] direct estimates of both inter- and intra-chain exchange, as well as Fourier components of specific spin correlation functions.

We have emphasized the formal connection between one-dimensional paramagnetic inorganic and organic crystals. The clear evidence for isotropic Heisenberg exchange in some inorganic SSR stacks,[66, 67, 300–306] the persuasive but less complete evidence in many other transition-metal chains,[73] and the success of triplet–exciton theory[23] in organic solids all point towards isotropic antiferromagnetic exchange between adjacent π-molecular free radicals in a one-dimensional chain, even though χ_p data for ionic CT and FR are often in partial disagreement with the results shown in Table VII. Many other mathematically simple models for χ_p have been proposed, ranging from simple (metallic) band models[214–5] with uncorrelated electrons, to insulating Ising chains and interchain Ising interactions.[307–8] *None* of these models explains the spin dynamics or specific heat or conductivity results, although they all provide χ_p fits for selected crystals, especially if an impurity contribution and several parameters (e.g. for an alternating stack) are included. The point is that, except for SSR chains, the ion–radical structures in Fig. 4 all permit at least two adjustable parameters. This is more than sufficient to fit the simple temperature dependence[120] of χ_p illustrated in Fig. 16 for TEA (TCNQ)$_2$.

Duffy[126, 127] has investigated χ_p for a number of neutral FR crystals which probably do *not* have a lowest CT excitation. Figure 17 shows the χ_p fit[126] for BDPA using $J = 4.6$ K in the Bonner–Fisher[243] numerical solution of (7). Equally good C_v fits are obtained for the same J. Unfortunately, the crystal structure of BDPA is not known and, since BDPA is probably not planar, may not even contain one-dimensional stacks. The familiar neutral FR solid DPPH forms at least six different structures,[159] some neat and others incorporating solvent molecules, and the available structure does not contain a one-dimensinal stack. While there is little doubt that these neutral FR solids show exchange, as was realized[309, 310] in early work from the absence of hyperfine structure, the lack of structural data precludes a definitive identification of SSR stacking in BDPA.

The rare SSR π-molecular CT crystals have been discussed in Section III.E and summarized in Table III. TMPD–ClO$_4$ dimerizes to an SSA stack below 186 K, while both Rb(TCNQ)(II) and the α form of K–chloranil undergo phase transitions that have not been fully characterized. This leaves NMP–TCNQ, with its disordered NMP cations and no apparent phase transitions. The conclusion[40] that NMP–TCNQ is, at low temperature, a quantitative realization of the one-dimensional Hubbard model is clearly incorrect, since the χ_p curve in Fig. 13 is not even qualitatively similar to either the Hubbard model, with the experimental values[41] of $t = 0.02$ eV and $U = 0.14$ eV obtained from other data, or to the Heisenberg model with

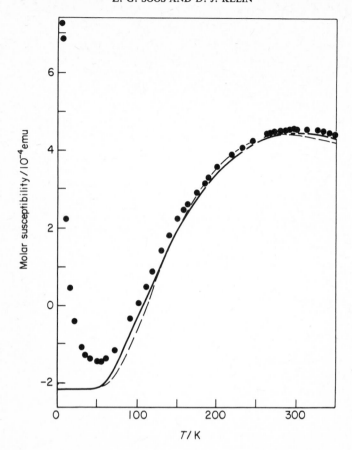

FIG. 16. Temperature dependence of χ for TEA(TCNQ)$_2$. The dashed line is for $\Delta E_p = 0.041$ eV in (85) and the solid line represents χ_p after correction for paramagnetic impurities. [Reproduced with permission from R. G. Kepler. *J. Chem. Phys.*, **39**, 3528 (1963)].

$J = 2t^2/U = 66$ K. The small Madelung energy[107] of NMP–TCNQ suggests that it may be very near the neutral–ionic transition and that it may show less than a complete transfer of an electron, since the sharp separation occurs only for t small compared to M and $(I_D - EA_A)$. In spite of extensive physical studies[26, 39–41] on NMP–TCNQ, such simple parameters as the ionization potential of NMP are not known, since individual NMP radicals have not been prepared,[312] and the detailed interpretation in terms of a simple Hubbard model must certainly be thoroughly reexamined.

Theoretical results for the alternating Heisenberg model, with $\delta \neq 0$ in (66), have been discussed in Section IV.D. There are now two exchange

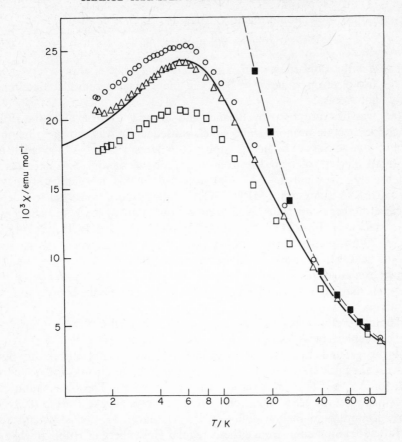

FIG. 17. Paramagnetic susceptibility for several differently prepared BDPA samples. The re-crystallized ($\triangle$) and unusually pure ($\bigcirc$) samples are close to the Bonner–Fisher solution to (7) with $|J| = 4 \cdot 55$ K. [Reproduced with permission from W. Duffy, Jr., J. F. Dubach, P. A. Pianetta, J. F. Deck, D. L. Strandburg and A. R. Miedema. *J. Chem. Phys.*, **56**, 2555 (1972)].

parameters $J(1 - \delta)$ and $J(1 + \delta)$. Thus χ_p fits in SSA stacks are inherently less convincing, unless independent methods are found for evaluating at least one of the exchange constants. The occurrence of an energy gap, ΔE_p, for paramagnetism is clearly seen in the limit $\delta \to 1$ of noninteracting dimers. Now the thermal equilibrium concentration ρ of triplet states at $2J = \Delta E_p$ is, for $\delta \to 1$,

$$\rho = \left[1 + \tfrac{1}{3} \exp(\Delta E_p/kT)\right]^{-1} \tag{84}$$

A chain of N ion–radicals, or $N/2$ dimers, thus has $N\rho/2$ triplet states, each with two unpaired spins, or just $N\rho$ unpaired spins. The paramagnetic susceptibility of the noninteracting triplets is

$$\chi_p = N\rho S(S + 1)g^2\beta^2/6kT \qquad (85)$$

where β is the Bohr magneton, $S = 1$, and the g value is always very close to the free-electron value $g_e = 2{\cdot}0023$ in these organic radicals containing only light atoms.

(85) was first proposed by Bleaney and Bowers[313] to explain the EPR of copper acetate monohydrate, which subsequently was shown to be one of a very large class of binuclear copper complexes.[71, 73, 74] The magnetism of small clusters of ligand-bridged transition–metal ions has since been extensively investigated.[71–76] (85) also fits[121] the low-temperature χ_p for such SSA stacks as TMPD–ClO$_4$, although deviations due to $\delta < 1$ occur at higher temperature, and accounts qualitatively for χ_p in such complex stacks as TEA(TCNQ)$_2$, $(\phi_3\text{AsCH}_3)(\text{TCNQ})_2$, $(\phi_3\text{PCH}_3)(\text{TCNQ})_2$, Cs$_2$(TCNQ)$_3$, etc. In these complex salts, the possible orbital degeneracy shown in Table IV provides additional complications. The weaker inter-dimer exchange $J(1 - \delta)$ is especially difficult to obtain, although EPR methods may be used in exceptionally favourable cases like copper acetate pyrazine.[314]

The interpretation of χ_p in alternating stacks requires fitting two parameters, while complex stacks require additional information about the site energies and the ground-state charge distribution. These complications do not obscure the basic feature of an *activated* χ_p, as indicated in (48) and given by (85) for $\Delta E_p \gg kT$, in *alternating* ion–radical stacks. The observation of activated χ_p in the MSR stacks of 1:1 ionic CT crystals like TMPD–TCNQ poses a more fundamental difficulty.[315] The large ΔE_p for MSR crystals in Table II cannot come from either a regular Heisenberg or Hubbard chain. There is no indication of dimerization at low temperature; such large ΔE_p would require substantial dimerization and would lead, contrary to experiment, to triplet spin exciton EPR spectra. We conjecture that the stack remains regular and that ΔE_p arises from the coupling of spin excitations to charge–density changes. As shown in Fig. 5, a finite CT integral t leads to slightly less than unit charge γ along the $\ldots A^{-\gamma}D^{+\gamma}A^{-\gamma}D^{+\gamma}\ldots$ stack in the $S = 0$ ground state. Since the (highly excited) ferromagnetic state with $S = N/2$ necessarily has $\gamma = 1$, it is evident that spin excitations in MSR stacks lead to charge–density changes,[22] in contrast to the behaviour of SSR stacks. The nature of the spin and charge coupling remains to be found and is an important problem in these strong-exchange systems.

C. Frenkel (Triplet) Spin Excitons

We turn now to the EPR of alternating FR crystals[23] and, for simplicity, discuss only the Heisenberg spin Hamiltonian (66) for half-filled SSA stacks. H_{CT} of course reduces to an alternating Heisenberg antiferromagnetic chain when the Hubbard model has two exchange constants t and t', both small compared to U. Effective spin Hamiltonians discussed in Section IV.D may easily be obtained for other special cases of H_{CT} encountered in Section III, although t now has to be small[238] in comparison with the lowest-energy CT excitation that increases the number of doubly occupied sites by one. We also use (66) to describe qualitatively the numerous *complex* TCNQ salts that show triplet spin excitons. When the site energies are sufficiently unlike, as is likely in $Cs_2 TCNQ_3$, H_{CT} can be shown[196] to reduce to (66).

The limit of complete alternation ($\delta \to 1$) leads to isolated dimers, each with probability ρ in (84) of being in the paramagnetic triplet excited state. For $\delta < 1$, the triplets are mobile and, for a rigid lattice, have a bandwidth of approximately $J(1 - \delta)$. Each triplet state is further split and shifted by the spin Hamiltonian

$$H_T = \beta S \cdot g \cdot B_0 + S \cdot D \cdot S \qquad (86)$$

where $S = 1$, B_0 is the applied field, and D is the electron dipolar interaction between the two electrons in the triplet state. The hyperfine interaction $S \cdot A \cdot I$ is important for ion-radicals in solution, but has been neglected in (86) because triplet motion along the ion–radical stack completely washes out hyperfine splittings. The narrow EPR lines[118] shown in Fig. 18 are typical of triplet spin excitons and immediately prove[23] that the exciton motion or delocalization is greater than the width of the solution hyperfine spectrum of about 10–200 gauss.

At very low temperature, when $kT < \Delta E_p$ and $\rho \ll 1$, there are few collisions between the triplets. Since motion along a crystallographically equivalent stack does not average out the fine structure D, the EPR spectrum corresponds to a density ρ of triplets described by (86). The ion–radical self-complex $A^- A^-$ or $D^+ D^+$ analysed in Section II.A thus provides the essential features. In terms of the spin densities ρ_μ and ρ_ν at the atoms of the adjacent radicals n and m, the components of the fine-structure tensor are[23, 316]

$$D = \tfrac{3}{4} g^2 \beta^2 \sum_{\mu \in n, \nu \in m} \rho_\mu \rho_\nu (r_{\mu\nu}^2 - 3z_{\mu\nu}^2) r_{\mu\nu}^{-5}$$

$$E = \tfrac{3}{4} g^2 \beta^2 \sum_{\mu \in n, \nu \in m} \rho_\mu \rho_\nu (y_{\mu\nu}^2 - x_{\mu\nu}^2) r_{\mu\nu}^{-5} \qquad (87)$$

in the principal-axis frame. For axial symmetry ($E = 0$) and high field ($g\beta B_0 \gg D$), the fine-structure splitting varies as $D(3 \cos^2\theta - 1)$ and is decisive proof of the triplet nature of spin excitons, just as the absence of hyperfine structure indicates either rapid motion or extensive delocalization.

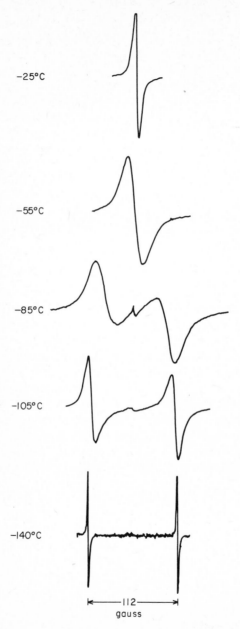

−25°C

−55°C

−85°C

−105°C

−140°C

|←———112———→|
gauss

FIG. 18. Temperature dependence of Frenkel (triplet) spin-exciton EPR in a single crystal of $(\phi_3PCH_3)(TCNQ)_2$ or $(\phi_3AsCH_3)(TCNQ)_2$ showing the collapse of the fine-structure splitting. The orientation is fixed, but the spectrometer gain settings are different at each T. [Reproduced with permission from D. B. Chesnut and W. D. Phillips, *J. Chem. Phys.*, **35**, 1002 (1961)].

The resolved fine structure typical of strongly alternating ion–radical stacks, whether simple or complex, provides direct information about the structure of the triplet exciton. The delocalized π-electron spin densities ρ_μ and ρ_v in (87) may be obtained, as discussed in II.G in connection with the Madelung energy, from either solution hyperfine[110, 111] or from approximate quantum-mechanical computations.[109] In simple salts, the crystal structure then provides the interionic separation $r_{\mu v}$ and the fine-structure constants follow from (87). Table VIII summarizes the rather good agreement between theory and experiment. It is again found that the detailed spin densities are not required, although it is important to delocalize the unpaired electrons.

The fine-structure constants in Table VIII show that the two unpaired electrons in the triplet state cannot be on the *same* site, as is obvious from the Pauli Principle. The two electrons in the triplet move together and thus resemble a Frenkel spin exciton.[66] This aspect of electron correlations is missed in a simple band model. Thus there is clear experimental evidence for strong electron correlations and against a band model for every ion–radical crystal with resolvable fine structure. And unless the ion–radical structures undergo a complete electronic reorganization on going from dimerized to regular stacking, a strongly correlated model like H_{CT} is needed for at least the paramagnetic insulators and semiconductors. The highly conductive TCNQ salts like TTF–TCNQ, NMP–TCNQ, Q(TCNQ)$_2$ perhaps show sufficient reorganizing to allow a band description, although there yet is no experimental evidence as decisive for band models as the fine-structure data in Table VIII for strong correlations. In further support of relatively similar and strong electron correlations in regular and alternating stacks, we note that Tables II–VI have similar energies for ΔE_{CT} in both regular and alternating stacks. EPR data in alternating chains thus provides important implications for optical and electric properties.

As the crystal is warmed, the thermal equilibrium density of excitons increases, as indicated by $\rho(T)$ in (84). Exciton–exciton collisions first broaden the fine-structure lines and eventually collapse the splitting, as shown in Fig. 18. Triplet exciton concentrations of a few per cent generally lead to a single, exchange-narrowed line at $g = 2$, which then narrows on further increase of T, and hence of ρ. The frequency with which a given triplet collides with another is, in the absence of specific interactions between triplets, just[23, 66, 117]

$$\omega_e = \rho\omega_e^0 \tag{88}$$

where ω_e^0 is a constant. Hence the residual exchange-narrowed Lorentzian line in Fig. 18 should have a width of about

$$\Gamma \sim \frac{D^2}{\omega_e} = \text{constant}/\rho \tag{89}$$

Z. G. SOOS AND D. J. KLEIN

TABLE VIII. Fine structure constants[a] for selected triplet exciton crystals[b]

Stoichiometry	Crystal	$\lvert D \rvert$ Experimental	$\lvert E \rvert$ Experimental	$\lvert D \rvert$ Theoretical	$\lvert E \rvert$ Theoretical	Ref. Exp.	Ref. Theo.
Simple							
1:1	$\ldots(D^+ D^+)_n \ldots$						
	TMPD–ClO$_4$	~0	212	13	236	121	331
	(TMPD)$_2$Ni(mnt)$_2$	602	79·2	670	70	177	177[c]
	$\ldots(A^- A^{2-})_n \ldots$						
	Morpholinium	448	54	465	44	331	331[d]
	Rb(TCNQ)(I)	400	49	617	70	180	180
	Li(TCNQ)	461	69			180	
Complex							
2:3	Cs$_2$(TCNQ)$_3$	281	45·3	262	34	119[c]	
	Morpholinium)$_2$(TCNQ)$_3$	282	45·3			179	180[e]
	Rb$_2$(TCNQ)$_3$	282	33·5			180	
1:2	$(\phi_3\mathrm{PCH}_3)(\mathrm{TCNQ})_2$ $(\phi_3\mathrm{AsCH}_3)(\mathrm{TCNQ})_2$	190	30	186	22·5	118	180[e]

[a] In MHz.
[b] Additional 1:1 TCNQ salts with unknown structures are given in Ref. 180.
[c] Less than 10 per cent variation for three different TMPD$^+$ spin densities at the observed separation.
[d] Hückel spin densities, assumed D_{2h} symmetry at 3·5 Å.
[e] For a scaled n-mer at 3·25 Å, as suggested in Refs. 119c and 23.

Jones and Chesnut[119] have shown that the exponential rate of line-narrowing is typically slightly larger than the activation energy for paramagnetism, ΔE_p. Thus the qualitative features of the temperature dependence is readily understood, but several quantitative problems remain. Lynden-Bell[317] has applied density-matrix methods to improve the description of exciton collisions, while Soos and McConnell[318] related the difference in activation energies to exciton–phonon interactions leading to a small spin-polaron. Recent EPR studies[198] on the mixed crystal $[\phi_3AsCH_3]_x[\phi_3PCH_3]_{1-x}$ $(TCNQ)_2$ for $0 \leqslant x \leqslant 1\cdot0$ show spectra typical of the neat crystals, an observation that is readily understood in terms of diffusing spin polarons, which would be insensitive to a disordered environment. The pressure dependence of the EPR linewidth of triplet excitons observed by Hughes[158] can be understood on the basis of (89). Since pressure increases the transfer integral t and thus J, the exciton density and EPR intensity are reduced and the line broadens.

Triplet spin excitons[177] based on dimerized $TMPD^+$ stacks in $(TMPD)_2$ $Ni(mnt)_2$, where mnt is the maleonitriledithiolato ligand, show that planar transition–metal anions may be used as counterions. Additional fine-structure lines[319] have been found in $(\phi_3AsCH_3)(TCNQ)_2$, while proton NMR studies[320-2] of TCNQ salts are providing further evidence for mobile triplet states. The morpholinium complexes[179] of TCNQ, which occur with $1:1$, $2:3$ and $3:4$ stoichiometries, all show triplet spin exciton behaviour. Such series of related complexes with different stoichiometries again emphasize the need for structural input in theoretical analysis, since it would be very difficult indeed to predict either the occurrence of several crystal phases or of several stoichiometries.

The resolved fine structure of Frenkel spin excitons and the rich angular, temperature, and pressure variations of the EPR spectrum provide a very detailed probe for CT interactions in ion–radical complexes with alternating stacks. Recent work supports the previous conclusion[23] that triplet spin excitons are relatively well understood. A variety of problems, especially if quantitative interpretations are desired, remain to be explored. The possibility of easily controlling the exciton density by merely changing temperature permits studying either isolated excitations or interactions among excitons in dense systems.

D. Uniform Magnetic Dilution: Regular Stacks

The structure of Frenkel (triplet) spin excitons indicates that the two spins forming the mobile triplet remain on adjacent, dimerized ion–radicals. This spatial correlation does not persist in regular chains, when interactions with both neighbours are equally strong. The weak paramagnetism of $1:1$

ionic CT crystals, with the inclusion of some impurities, follows the activated form indicated in (48) and leads to the ΔE_p values in Table II. The EPR of segregated stacks is also often activated, as illustrated by the salts in Table V, even when no fine structure is resolved at low temperature. The occurrence of an activated susceptibility and significant magnetic dilution, with only a few per cent of the available spins contributing to χ_p even at room temperature, are in conflict with theoretical results for the regular nearest-neighbour Heisenberg chain and raise questions about impurities. The latter seem improbable, except at the lowest temperatures, in view of the increasing crystal purity and the reproducibility of χ_p data. The theoretical problem remains, as discussed in Section V.B. Since no antiferromagnetic ordering has been observed in organic ion radicals, we conclude that the magnetic interactions are strongly one-dimensional and the strong observed narrowing is again attributed to isotropic Heisenberg exchange.

Soos[66] has developed a simple model for interpreting the EPR of regular chains in the region $kT \lesssim \Delta E_p$. The theoretical problem of accounting for ΔE_p is bypassed by defining an effective spin density

$$\rho(T) = \chi_p/\chi_c \tag{90}$$

Here χ_p is the *observed* paramagnetic susceptibility, and χ_c is the Curie-law result for an equivalent number of noninteracting spins,

$$\chi_c = N \frac{S(S + 1)g^2\beta^2}{3kT} \tag{91}$$

where $S = \frac{1}{2}$. Thus $\rho(T)$ is proportional to $\chi_p T$ and is a measure of the observed magnetic dilution whose origin remains to be identified, in contrast to the singlet–triplet case (84) in which the origin of ΔE_p is clearly understood. The absence of spatial correlations among the density ρ of unpaired spins then accounts for several features of the EPR spectrum. In effect, we apply the theory of magnetic dilution[323] to these spins and ignore the $N(1 - \rho)$ strongly-coupled spins that do not contribute to χ_p. Unlike the dilute systems of paramagnetic centres produced by substitutional doping or by X-ray damage, the spins are uniformly distributed and ρ is readily varied by changing T and thus χ_p.

In the absence of exchange, the electron dipolar linewidth of a FR solid is approximated by the square root of the Van Vleck second moment,[310]

$$M_2^{(0)} = \left(\tfrac{3}{4}\right)^2 \frac{g^4\beta^4}{\hbar^2} \sum_{mn}{}' (1 - 3\cos^2\theta_{mn})^2 r_{nm}^{-6} \tag{92}$$

Here θ_{mn} is the angle between the applied field B_0 and the vector r_{nm} connecting the local moments $S = \frac{1}{2}$ at sites n and m. The generalization[292] of (92) to delocalized π electrons in aromatic ion–radicals is straightforward and is again based on spin densities associated with each atom of the molecule. Uniform magnetic dilution reduces $M_2^{(0)}$ and leads to[66, 323]

$$M_2^{(0)\prime}(\rho) = \rho M_2^{(0)} \qquad (93)$$

for the approximate width of the Gaussian EPR line expected in the absence of exchange. When spin exchange ω_e is large compared to the dipolar field, an exchange-narrowed[310, 324] Lorentzian EPR absorption is predicted. In magnetically dilute systems, ω_e is expected to be proportional to ρ, as indicated in (88), and the linewidth is[66]

$$\Gamma \approx \frac{\rho M_2^{(0)} + M_N}{\rho \omega_e^0} \simeq \frac{M_2^{(0)}}{\omega_e^0} \qquad (94)$$

Here M_N is the second moment due to interactions with nuclear spins, which are not diluted and contribute at very small ρ only.

The absence of spatial correlations immediately accounts for the absence of fine structure (D, E vary as r^{-3} in (87)), while spin exchange along the stack accounts for the absence of hyperfine structure and the narrow Lorentzian lines. Comparison of (94) with the exchange-narrowed line (89) for Frenkel excitons reveals that the former is independent of ρ, while the latter goes as ρ^{-1}. In Fig. 19 we show the gentle temperature dependence of the EPR linewidth in several 1:1 crystals, which is in sharp contrast to the strong ρ^{-1} narrowing in triplet-exciton systems. The pressure dependence is similarly less strong, since in (94) it merely involves changes in $M_2^{(0)}$ and ω_e^0, without consideration of the excitation density. While triplet–exciton systems show increasing linewidth with increasing pressure, regular chains usually have decreasing linewidths.[158] Both effects are illustrated in Fig. 20 by the pressure dependence of the TMPD–ClO$_4$ phase transition. The low temperature dimerized phase contains Frenkel spin excitons, while the high-temperature phase has spin excitations for a regular lattice.

Isotropic exchange interactions result from CT stabilization or from other electrostatic forces in molecules with orbitally non-degenerate ground states and small spin-orbit constants. Both of these conditions are satisfied in π-molecular ion–radicals. The g values are usually very close to the free-electron value of 2·0023, with deviations in the third decimal place. Hence ω_e^0 in (94) is expected to be isotropic and the angular dependence of Γ results from the angular dependence of $M_2^{(0)}$. The spin densities, from either solution or theoretical data, together with the crystal structure permit a direct theoretical evaluation of $M_2^{(0)}$. The angular dependence[292] of Γ in the three

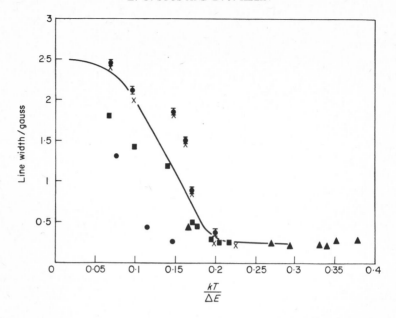

FIG. 19. Peak-to-peak EPR linewidth for three MSR CT complexes at Q band as a function of $kT/\Delta E_p$. The ■, × and ⦙ refer to three different PDC samples ($\Delta E_p = 0.13$ eV); ▲ refers to TMPD–TCNQ ($\Delta E_p = 0.07$ eV); ● refers to TMPD–chloranil ($\Delta E_p = 0.15$ eV). [Reproduced with permission from R. C. Hughes and Z. G. Soos, *J. Chem. Phys.* **48**, 1066 (1968).]

principal-axis planes of TMPD–chloranil is shown in Fig. 21 to be very well given by an isotropic ω_e^0 and a constant background term for other relaxation mechanisms. The TMPD–TCNQ angular variations[150] are in qualitative agreement with (94), but the inevitable twinning complicates a detailed analysis. The occurrence of three magnetically non-equivalent $\ldots D^+ A^- D^+ A^- \ldots$ stacks[148] in cosublimed PDC and the lack of a crystal structure preclude a direct test.

The EPR spectrum of regular stacks thus provides a single, narrow line near $g = 2$ and is far less rich than the triplet exciton spectra shown in Fig. 18. The temperature, pressure, and angular dependence of a signal, as well as its Lorentzian shape, may be understood in terms of uncorrelated spin excitations in a magnetically dilute crystal. Although the nature of ΔE_p in regular stacks remains a problem, the essential features of the resonance data for either regular or alternating stacks are consistent with spin Hamiltonians based on H_{CT}. The analysis of highly conducting TCNQ and TTF salts and more careful examination of various paramagnetic semiconductors will further test the importance of the antiferromagnetic exchange $J \approx t^2/\Delta E_{CT}$ arising from CT in open-shell systems.

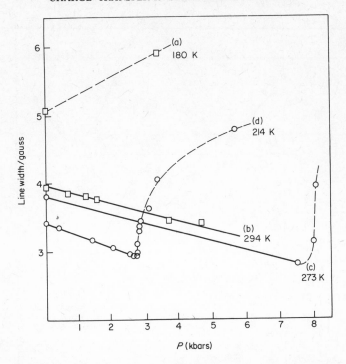

FIG. 20. The pressure dependence of the peak-to-peak EPR linewidth in TMPD–ClO$_4$. The solid lines indicate the high-temperature, regular stack, while the dashed lines correspond to the low-temperature, alternating phase. [Reproduced with permission from R. C. Hughes, PhD Thesis, Stanford University (1966)].

E. Magnetic One-Dimensionality and Interchain Interactions

Many π-molecular CT crystals crystallize in magnetically non-equivalent stacks, as shown in Fig. 7 for the SCA stacks in Cs$_2$(TCNQ)$_3$ and summarized in Tables II–VI. Magnetic non-equivalence provides a very simple direct test[148] for the assumed one-dimensionality of the exchange and of the π-overlaps leading to the CT integral. Resolving[119] two sets of fine-structure lines in Cs$_2$(TCNQ)$_3$ at low temperature is only possible if all interactions, including both exchange and dipolar interactions, between the non-equivalent chains is less than the splitting. Since splittings of $1G (\approx 10^{-4} cm^{-1})$ or less are readily resolved for the unusually sharp lines in these ion–radical solids, resolved splittings for magnetically non-equivalent chains provide stringent upper bounds on interchain exchange. The low-temperature phase of the SSA system TMPD–ClO$_4$ contains[121] two magnetically non-equivalent chains for each of two domains. The mosaic of twinned domains

D

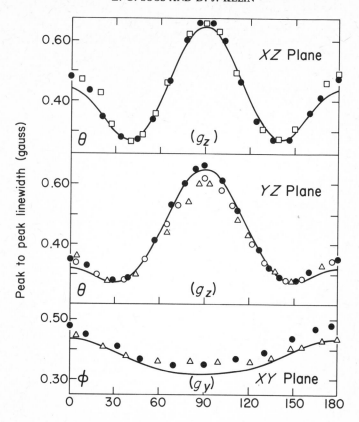

FIG. 21. Angular dependence of the 300 K EPR linewidth of TMPD–chloranil in the three principal-axis planes of the g tensor. The theoretical curve is based in (94). [Reproduced with permission from T. Z. Huang, R. P. Taylor and Z. G. Soos, *Phys. Rev. Lett.*, **28**, 1054 (1972).]

share a common axis and are resolved[121] for certain orientations of B_0. The non-equivalent chains within a domain were not resolved, thus yielding an estimate of 10^{-4}–10^{-1} cm^{-1} for interchain interactions, with the lower bound more likely. Three magnetically non-equivalent chains are found[148] in cosublimed PDC, with their g tensors related by rotations of $2\pi/3$ about the common chain axis. The angular variation of the three EPR lines in this MSR system are shown in Fig. 22. The complex regular salt (Et$_2$TCC)-(TCNQ)$_2$, where Et$_2$TCC = 3,3-diethylthiacarbocyaninium and is shown in Fig. 6, also has non-equivalent stacks[192, 26] with resolved uncorrelated spin excitations. These examples encompass the structural possibilities in Fig. 4 for ion–radicals, with the exception of the SSR case.

Since the electric, optical, and χ_p properties of these crystals are in no way different from those of related complexes with magnetically equivalent

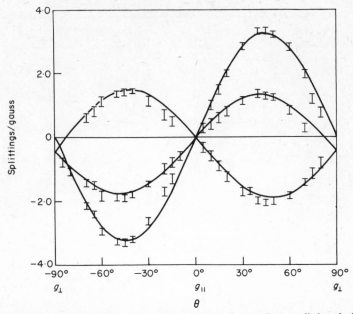

FIG. 22. Comparison of the angular dependence of the Q-band g-factor splittings in PDC at 263 K arising from magnetically inequivalent stacks with theory for three identical, noninteracting stacks. [Reproduced with permission from R. C. Hughes and Z. G. Soos, *J. Chem. Phys.*, **48**, 1066 (1968).]

chains, the magnetic one-dimensionality of π-molecular CT solids is, as expected from structural considerations, a general phenomenon. The interesting conductor TTF–TCNQ contains two magnetically non-equivalent chains[186] of both TTF and TCNQ and several EPR lines have been resolved[325] below the metal–insulator transition around 60 K. The somewhat broader lines due to the larger spin–orbit coupling of sulphur may be associated with separate chains, although it has not yet been possible to identify four lines based on two different *g* tensors and a crystal rotation. Thus magnetic non-equivalence provides a powerful method for ruling out interchain exchange without involving hypothetical arguments about the magnitude of small π-electron overlaps.

The delocalization of the uncorrelated spin excitations in PDC can be demonstrated by analysing the temperature dependence of the EPR splittings in Fig. 23. It is a general feature[324] of sudden hops between non-equivalent sites that, as the hopping frequency increases through the splitting, the lines first broaden, then coalesce, then sharpen. A particularly clear-cut example is given[326] by CO_2^- impurities in calcite, which are constrained to $2\pi/3$ rotations about the [111] axis and whose hopping rate can be varied over several orders of magnitude by increasing the temperature. By contrast,

Z. G. SOOS AND D. J. KLEIN

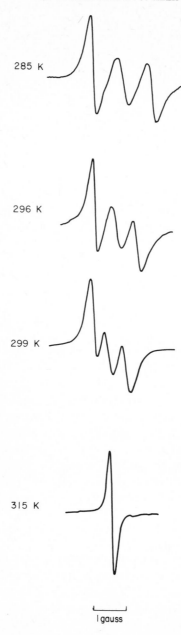

285 K

296 K

299 K

315 K

I gauss

FIG. 23. Temperature dependence of the EPR in PDC, showing the collapse of the g-factor splittings. The sweep, but not the gain, is the same for all four spectra. [Reproduced with permission from R. C. Hughes and Z. G. Soos, *J. Chem. Phys.*, **48**, 1066 (1968).]

the lines in Fig. 23 *narrow* slightly as they merge. The slight narrowing of each line is associated with suppression of the nuclear contribution M_N in (94) as ρ increases. The absence of broadening is interpreted[327] by considering the temperature dependence of the **g** tensor for *crystal* states on the inequivalent stacks. Weak interchain dipolar fields increase as $\rho^{\frac{1}{2}}$, as indicated in (93), and delocalize the excitations on the inequivalent chains, thus collapsing the splitting. The unusual line-merging in PDC is easily monitored through the temperature dependence of the dipolar fields arising from the temperature-dependent uniform magnetic dilution.

Hoffman and Hughes[150] exploited the spin-density dependence of the exchange, $\omega_e = \rho\omega_e^0$ in (88), in the MSR stack TMPD–TCNQ. The small ΔE_p provides larger spin densities and, in particular, permits a temperature variation of $\rho\omega_e^0$ through the Larmor frequency $\omega_0 = g\beta B_0\hbar^{-1}$. The "10/3" effect[324] is based on the observation that the truncated dipolar second-moment $M_2^{(0)}$ in (92) is appropriate for $\rho\omega_e^0 \ll \omega_0$, while the full dipolar second-moment is required for $\rho\omega_e^0 \gg \omega_0$. Instead of conventionally varying ω_0 for fixed exchange ω_e, the "10/3" effect in this magnetically dilute system was observed by raising T to achieve $\rho\omega_e^0 > \omega_0$. The increasing linewidth with temperature is shown in Fig. 24 and may be understood,[150] together with spin-lattice relaxation data, in terms of standard results for an exchange $\rho\omega_e^0$ in a magnetically dilute crystal.

The unusual EPR lineshape[328] for spin diffusion in one dimension, where correlations decay slowly as $t^{-\frac{1}{2}}$ in time, has recently been observed[69] in one-dimensional inorganic systems like TMMC and has since been extensively discussed.[67, 68, 299] Three-dimensional interactions, for example associated with interchain exchange, are very effective in spoiling the characteristic one-dimensional lineshape. Indeed, a Lorentzian EPR line with width approximately[328, 306]

$$\Gamma_{1d} \approx M_2^0(\rho)/(2\omega_e\omega_r)^{\frac{1}{2}} \tag{95}$$

is found when the intrachain exchange ω_e is much larger than the dipolar second-moment $M_2^{(0)}$ and the interchain exchange ω_r, which in turn are larger than the observed linewidth Γ. Other broadening contributions and randomizing rates are readily included in $M_2^{(0)}$ and in ω_r, respectively. Since $M_2^{(0)}(\rho) = \rho M_2^{(0)}$ in a magnetically dilute system and the previous assumption for non-interacting excitations leads to $\omega_e = \rho\omega_e^0$ and to $\omega_r = \rho\omega_r^0$, the ρ dependence in (95) is again lost, as in (94). Such weak temperature dependence is, as seen in Fig. 19, typical for MSR systems. Several rather different speculations about applying (95) to TMPD–TCNQ illustrate the difficulties of drawing detailed conclusions.

(*a*) The identification[315] of the increasing linewidth in Fig. 24 as due to a

$\rho^{+\frac{1}{2}}$ dependence in (95), rather than the "10/3" effect, is based on neglecting the probable ρ dependence of ω_r. Spin diffusion perpendicular to the chains, like ω_r, scales as ρ. The estimated interchain exchange ω_r is thus not convincing.

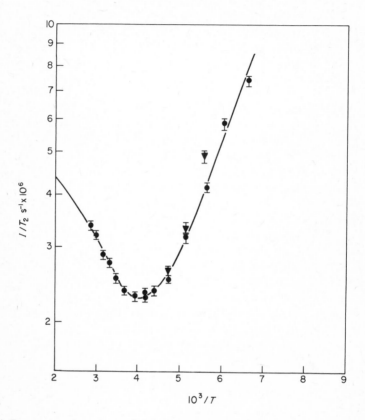

FIG. 24. Temperature dependence of the EPR linewidth (T_2^{-1}) in TMPD–TCNQ at X band showing the "10/3" effect as $\rho\omega_e^0$ exceeds the Larmor frequency at high temperautres. [Reproduced with permission from B. M. Hoffman and R. C. Hughes, *J. Chem. Phys.*, **52**, 4011 (1970)].

(b) The identification of $J(1 + \delta)$ with ΔE_p in regular chains is not yet possible. On factoring out the ρ dependence in (95) and computing $M_2^{(0)}$, the rate $(2\omega_e\omega_r)^{\frac{1}{2}}$ is found to be far smaller,[150, 292] by about 10^2, than $\Delta E_p \hbar^{-1}$ in TMPD–TCNQ, in TMPD–chloranil, and in PD–chloranil. This would set a bound $\omega_r^0 \approx 10^{-4}\omega_e^0$, if ω_e^0 were of the order of J, but it is precisely this relation that is lacking.

(c) The more rigorous analysis possible for a nearly one-dimensional spin system at *infinite* temperature, with $kT \gg J$, shows[299, 329] that the width of the non-Lorentzian line is

$$\Gamma \approx (M_2^{(0)})^{\frac{3}{4}}(\hbar/J)^{\frac{1}{4}} \qquad (96)$$

Introducing factors of ρ to describe uniform magnetic dilution now leads to a net $\rho^{\frac{3}{4}}$ dependence in (96), which is *not* observed in $1:1$ ionic CT crystals. Furthermore, while interchain exchange J' does produce a Lorentzian, it enters[299] as $J'(J'/J)^{\frac{1}{4}}$ rather than as $(J')^2/J$, and the net ρ dependence must be computed. There is no guarantee, of course, that the high-temperature theory can be extended to the region $kT < J$, and additional refinements may be required even at high temperature.[329]

We have dealt in some detail with the theoretical difficulties for estimating weak interchain exchange in these one-dimensional organic solids. When $kT \gg J$, as in $Cu(NH_3)_4PtCl_4$, the interchain constant may be deduced[305–6] from single-crystal EPR data. Similarly good resonance data should be possible in at least some aromatic ion–radicals, but must be interpreted with caution.

We emphasize that there currently is no fully satisfactory theory for exchange-narrowing or for spin correlations in the intermediate region $J \lesssim kT$. The preceeding qualitative discussion of magnetic resonance in either alternating or regular stacks has been largely based on judiciously invoking high-temperature results for a spin concentration $\rho(T)$ of either Frenkel or uncorrelated spin excitons. Rather different pictures emerge for spin excitations in strongly alternating and regular stacks. Frenkel spin excitons consist of strongly correlated pairs of spins whose motion[318] along the stack is probably best understood as the diffusion of a small spin polaron. The source of the exciton–phonon interaction is the formally antibonding nature of the triplet state of the ion–radical dimer A^-A^- or D^+D^+ in Section II.A, in contrast to the bonding singlet state. Thus excitation to $^3|D^+D^+\rangle$ is expected to increase the intermolecular separation.[318] The first term of the Taylor expansion of the intradimer exchange in (66) leads to

$$V_n = \left[\frac{\partial}{\partial R} J(1 + \delta)\right]_0 (d_{2n+1} - d_{2n}) f_n^+ f_n \qquad (97)$$

where d_n is the displacement of the nth molecular site and f_n^+, f_n are fermion operators which vanish unless the dimer $(2n, 2n + 1)$ is in the triplet state. Summing (97) over all dimers in an SSA stack leads to small spin-polarons when the exciton bandwidth $J(1 - \delta)$ is small, as previously shown.[318, 23] The uncorrelated spin excitations in regular stacks, by contrast, behave as independent $S = \frac{1}{2}$ excitations delocalized over the ion-radical chain and, as shown by PDC, delocalized even over the nearly degenerate stacks arising from the magnetic inequivalence. While coupling to phonons has been established[330] in $TEA(TCNQ)_2$, whose small ΔE_p allows large spin-densities at convenient temperatures, the nature of the exciton–phonon interaction has not been discussed in regular stacks.

The different limiting cases for strongly alternating and for regular stacks raise the possibility of intermediate regions. The strong correlation of Frenkel spin excitons is weakened and the two spins are not completely confined to neighbouring sites.[277] A preliminary analysis,[216] starting from a band model, suggests that some alkali TCNQ salts may contain such extended Frenkel spin excitons. The interpretation of magnetic resonance for ionic CT and FR complexes thus poses interesting problems quite independently of CT interactions and, in favourable cases, provides decisive information about such topics as the one-dimensionality, the structure of the thermal excitations, and electron correlations. Furthermore, a rich variety of effects peculiar to particular systems is found in addition to the broad separation into regular and alternating systems.

VI. COMPUTATION OF MODEL PARAMETERS

A. Crystal Perturbed States

The site representation for π-molecular CT crystals described in Section II provides a phenomenological approach to the physical properties of stacked open-shell ion–radical crystals, as well as to the related neutral crystals. The utility of the phenomenological theory has been demonstrated by the qualitative correlations indicated in Sections III and V. We found H_{CT} to provide a general framework for the electronic states of π-molecular CT solids. It is then natural to seek a formulation of H_{CT} in which the model parameters are expressed explicitly in terms of molecular wave functions. Although these expressions are rather complex and the molecular wave functions may currently not be known, such a formulation is still desirable. Explicit expressions verify some general claims already made for H_{CT}; they also provide further insight into the physical nature of the parameters, the approximations invoked, and possible extensions of the simpler forms already discussed. Finally the possibility of useful semiempirical calculation of some of the parameters arises.

It is not our intention to survey critically the many[332–4] current approaches to quantum calculations on individual molecules. The number of electrons, about 100 in the π-donors and acceptors shown in Fig. 2, rules out all but the simplest *ab initio* calculations. Semi-empirical methods such as extended Hückel theory have been important in some cases.[109] More recently results of CNDO/2 and Xα multiple scattering methods have become available for such molecules as TMPD$^+$ or TCNQ$^-$. There are occasionally even some *ab initio* Hartree–Fock results.[335, 336] Semi-empirical calculations for DA dimers have been critically summarized by Hanna and Lippert.[6] It is important to note that the various parameters arising from H_{CT} are quite small ($U \approx 1 \, \text{eV}$, $t \approx 0.2 \, \text{eV}$ and $J \approx t^2/U \approx 0.04 \, \text{eV}$), particularly in comparison with total electronic energies. Hence quite accurate calculations may be required in reliably evaluating the Mulliken CT integral t or the kinetic exchange J. Indeed, accurate J's for very weakly interacting atoms, even for two H atoms, can be a matter of great delicacy.[85] A number of problems characteristic of crystal computations are reviewed by Rice and Jortner[17] in their extensive treatment of the simplest Frenkel excitations.

When isolated-molecule molecular orbitals for different molecules are brought close to each other, their overlap will in general increase from zero. Although such nonorthogonal orbitals have been commonly used in previous dimer work,[7, 87, 93, 94] their solid-state use can give rise to many problems, and expansions[86, 337] in powers of the intermolecular overlap can be complicated. The general difficulties associated with the so-called "nonorthogonality catastrophe"[82] in explicitly exhibiting[85, 338, 339] the cancellation of nonphysical "unlinked" terms is easily avoided through the use of strongly orthogonal molecular states. Thus we orthogonalize the orbitals for different molecules to one another, as indicated in Section II.C for the dimer. The use of Löwdin's symmetric orthogonalization[95] procedure insures orthonormal orbitals bearing maximum similarity[340] to the original nonorthogonal orbitals. We presume the molecules to be weakly interacting and still speak of these orthonormal Wannier orbitals as isolated- or free-molecule orbitals, inasmuch as the usual descriptions of crystal field, induction and dispersion interactions are not significantly modified.

The next general problem is to relate the crystal-perturbed states (29) and (30) to isolated-molecule functions. This can be done[25] by developing a free-molecule site Hamiltonian h_n^{FM} similar to the crystal-perturbed site Hamiltonian referred to in (25). Such a choice has the added convenience that available gas-phase data on dimers might be incorporated, much as has been done[15–18] in the case of molecular-exciton theories of closed-shell molecular solids. With the μth Wannier spin-orbital for the ith molecule we associate fermion creation and annihilation operators $c_{i\mu}^+$ and $c_{i\mu}$. Then the Schrödinger Hamiltonian is

$$H = \frac{1}{2} \sum_{ij}' V_{ij} + \sum_{ij} \sum_{\mu\nu} \left\langle i\mu \left| -\frac{\hbar^2}{2m} \nabla^2 + \sum_k V_k(r) \right| j\nu \right\rangle c_{i\mu}^+ c_{i\nu}$$

$$+ \sum_{ijkl} \sum_{\mu\nu\pi\rho} \left\langle i\mu\Lambda j\nu \left| \frac{e^2}{r_{12}} \right| k\pi\Lambda l\rho \right\rangle c_{i\mu}^+ c_{j\nu}^+ c_{l\rho} c_{k\pi} \tag{98}$$

where the italic indices $ijkl$ label molecules, while the Greek indices $\mu\nu\pi\rho$ label orbitals for each molecule. $V_k(r)$ is the interaction of an electron with the nuclei of molecule k, V_{ij} $(i \neq j)$ is the intermolecular nuclear repulsion, which is constant for a rigid lattice, and the symbol Λ indicates an antisymmetrized product.

The free-molecule site Hamiltonian

$$h_i^{FM} \equiv \sum_{\mu\nu} \left\langle i\mu \left| -\frac{\hbar^2}{2m} \nabla^2 + V_i(r) \right| i\nu \right\rangle c_{i\mu}^+ c_{i\nu}$$

$$+ \frac{1}{2} \sum_{\mu\nu\pi\rho} \left\langle i\mu \otimes i\nu \left| \frac{e^2}{r_{12}} \right| i\pi \otimes i\rho \right\rangle c_{i\mu}^+ c_{i\nu}^+ c_{i\rho} c_{i\pi} \tag{99}$$

then consists simply of those terms of (98) that involve solely site i. The two-site interaction between free molecules is similarly obtained by selecting from (98) those terms involving exactly two sites. For the free-molecule interaction between sites i and j, we obtain

$$v_{ij}^{FM} \equiv V_{ij} + u_{i \leftarrow j}^{FM} + u_{i \to j}^{FM} + u_{i \leftrightarrow j}^{FM} \tag{100}$$

where $u_{i \leftarrow j}^{FM}$ involves only orbitals on site i (and nuclei on site j), $u_{i \to j}^{FM}$ involves only orbitals on site j, and $u_{i \leftrightarrow j}^{FM}$ involves orbitals on both sites i and j. We find

$$u_{i \leftarrow j}^{FM} \equiv \sum_{\mu\nu} \langle i\mu | V_j(r) | i\nu \rangle c_{i\mu}^+ c_{i\nu}$$

$$u_{i \leftrightarrow j}^{FM} = \sum_{\mu\pi} \left[\left\langle i\mu \left| -\frac{\hbar^2}{2m} \nabla^2 + V_i(r) + V_j(r) \right| j\pi \right\rangle c_{i\mu}^+ c_{j\pi} + \text{h.c.} \right]$$

$$+ \sum_{\mu\nu\pi\rho} \left\{ \left[\left\langle i\mu \otimes i\nu \left| \frac{e^2}{r_{12}} \right| i\pi\Lambda j\rho \right\rangle c_{i\mu}^+ c_{i\nu}^+ c_{j\rho} c_{i\pi} + \text{h.c.} \right] \right.$$

$$+ \left[\left\langle i\mu \otimes i\nu \left| \frac{e^2}{r_{12}} \right| j\pi \otimes j\rho \right\rangle c_{i\mu}^+ c_{i\nu}^+ c_{j\rho} c_{j\pi} + \text{h.c.} \right]$$

$$+ \left[\left\langle i\mu\Lambda j\nu \left| \frac{e^2}{r_{12}} \right| i\pi\Lambda j\rho \right\rangle c_{i\mu}^+ c_{j\nu}^+ c_{j\rho} c_{i\pi} + \text{h.c.} \right]$$

$$+ \left[\left\langle i\mu\Lambda j\nu \left| \frac{e^2}{r_{12}} \right| j\pi \otimes j\rho \right\rangle c_{i\mu}^+ c_{j\nu}^+ c_{j\rho} c_{j\pi} + \text{h.c.} \right] \right\} \tag{101}$$

$u_{i \leftarrow j}^{FM}$ involves interchanging i and j in $u_{i \leftarrow j}^{FM}$. (100) leads to the usual multipole expansion, at least in the limit of zero differential overlap (ZDO) which holds quite accurately for van der Waals, or longer, intersite separations. In the ZDO limit, the only contribution to $u_{i \leftrightarrow j}^{FM}$ in (101) is

$$\lim_{ZDO} u_{i \leftarrow j}^{FM} = \sum_{\mu\nu\pi\rho} \left[\left\langle i\mu \otimes iv \left| \frac{e^2}{r_{12}} \right| j\pi \otimes j\rho \right\rangle c_{i\mu}^+ c_{iv}^+ c_{j\rho} c_{j\pi} + \text{h.c.} \right] \quad (102)$$

and is used in (100). Matrix elements of v_{ij}^{FM} reduce at once to

$$\lim_{ZDO} \langle ir\Lambda js | v_{ij}^{FM} | it\Lambda ju \rangle = \langle ir \otimes js | \lim_{ZDO} v_{ij}^{FM} | it \otimes ju \rangle$$

$$= \left\langle ir \otimes js \left| \left\{ V_{ij} + \sum_{b \in j} V_i(r_b) + \sum_{a \in i} V_j(r_a) + \sum_{a \in i} \sum_{b \in j} \frac{e^2}{r_{ab}} \right\} \right| it \otimes ju \right\rangle \quad (103)$$

and involve only product functions $| ir \otimes js \rangle$ for the two sites. The ensuing multipole expansion[25] then includes both crystal-field and permanent induced moments. Dispersion contributions are included in (103) only in an average way, just as the neglect of three- and four-site terms in H is partially accounted[97] for by retaining suitable averages in the one- and two-site terms. It should be emphasized that (103) leads to only a modest number of matrix elements in the minimum basis g, α, β, and $\alpha\beta$ defined in (29) and (30), and that it already represents a potentially far more accurate treatment of intersite Coulomb forces than the choice $\langle V_C \rangle^{(0)}$ in (51) for the ground-state value of the crystal Coulomb interactions.

Next we construct the crystal perturbed site Hamiltonian, h_i, and the associated intersite interactions. The multipole perturbation of site i induced by all other sites is included directly in h_i through

$$h_i \equiv h_i^{FM} + \sum_{\substack{j \\ (j \neq i)}} \left\{ u_{i \leftarrow j}^{FM} + \sum_{\mu\nu} \text{Tr} \left[D_j c_{i\mu} u_{i \leftrightarrow j}^{FM} c_{iv}^+ \right] c_{i\mu}^+ c_{iv} \right\} \quad (104)$$

The trace is over the states of the neighbouring molecules j and D_j is a density matrix for the perturbing site j. In the minimum basis for a neutral DA crystal, D_j could be chosen as $|jD\rangle\langle jD|$ or $|jA\rangle\langle jA|$ for a donor or an acceptor at site j, respectively. A more accurate choice for D_j might include a small admixture of excited states or even of ionic states, possibly in a self-consistent manner dependent on the solution of H_{CT}, and even a temperature-dependent choice could be made. However, fairly simple choices for D_j are indicated in practice. It seems especially desirable to separate the development of H_{CT} from its final solution, since each step seems to tax current methods. The corresponding two-site interactions are

$$v_{ij} = V_{ij} + u_{i \leftrightarrow j}^{FM} - \sum_{\mu\nu} Tr \left[\mathbf{D}_j c_{i\mu} u_{i \leftrightarrow j}^{FM} c_{iv}^+ \right] c_{i\mu}^+ c_{iv}$$

$$- \sum_{\pi\rho} Tr \left[\mathbf{D}_i c_{j\pi} u_{i \leftrightarrow j}^{FM} c_{j\rho}^+ \right] c_{j\pi}^+ c_{j\rho} \qquad (105)$$

Summing h_i and v_{ij}, as indicated in (24) for the crystal Hamiltonian, demonstrates the important result that various choices for $\mathbf{D}_i$ merely change the partitioning of H into one- and two-site terms, without introducing any approximations.

It is straightforward to exhibit the matrix elements of (105) in the ZDO limit. After some manipulations we find

$$\lim_{ZDO} \langle ir\Lambda js | v_{ij} | it\Lambda ju \rangle = \delta_{rt} \delta_{su} \left\{ V_{ij} - Tr \left(\mathbf{D}_i \otimes \mathbf{D}_j \sum_{a \in i} \sum_{b \in j} \frac{e^2}{r_{ab}} \right) \right\}$$

$$+ Tr \left\{ (|it\rangle\langle ir| - \mathbf{D}_i) \otimes (|js\rangle\langle ju| - \mathbf{D}_j) \sum_{a \in i} \sum_{b \in j} \frac{e^2}{r_{ab}} \right\} \qquad (106)$$

The intersite interaction is now composed of two parts. The first contains internuclear repulsions and average electron–electron repulsions. Its primary monopole–monopole contribution need not be small if the sites have a net charge, but it is scalar and merely shifts all energy levels. The second non-scalar part of (106) takes into account instantaneous charge fluctuations, as occur in dispersion interactions. We note that this dispersion term vanishes when the expectation value over v_{ij} is taken with sites i and j in states $\mathbf{D}_i$ and $\mathbf{D}_j$. This demonstrates the inclusion of average intersite interactions in h_i, with a correspondingly weaker v_{ij}.

The free-molecule choice h_i^{FM} is natural for possible comparison with gas-phase data. The crystal-perturbed choice (104) for h_i and (105) for v_{ij}, in which the maximum average crystal perturbation is included in h_i, is advantageous for the phenomenological theory developed in Section II.

B. Improved Matrix Elements

The matrix elements of H_{CT} were introduced in Section II as adjustable parameters, with additional approximations to reduce their number. The resulting model was judged to be the simplest phenomenological model preserving the essential features of the CT problem. Here we investigate these approximations more thoroughly. We also seek more accurate expressions for the matrix elements in view of possible explicit computations of crystal parameters.

The Mulliken CT integral t defined in (43) connects sites with a pair of electrons above the dimer vacuum state $|D^{++}A\rangle$. In general, different numbers of electrons may be involved and the CT integral depends on the occupancies of each of the sites. Thus the integrals

$$
\begin{aligned}
t' &\equiv \langle D^+A \,|\, H \,|\, D^{++}A^-\rangle \\
t'' &\equiv \langle DA \,|\, H \,|\, D^+A^-\rangle \\
t''' &\equiv \langle D^+A^- \,|\, H \,|\, D^{++}A^{--}\rangle \\
t'''' &\equiv \langle DA^- \,|\, H \,|\, D^+A^{--}\rangle
\end{aligned}
\tag{107}
$$

are in general all distinct. If we postulate a common "frozen" core for D, D^+ and D^{++} and for A, A^- and A^{--}, as well as a single fixed orbital above these cores, then these integrals do in fact become identical. In a more accurate picture, we believe that the core and "valence" orbitals will both contract as the number of electrons on the molecule is decreased. This leads us to expect

$$
t' \leqslant t'' \leqslant t''' \leqslant t''''
\tag{108}
$$

The relative order in (108) should persist in even more accurate treatments not based on a core and valence electron picture at all. We note that correlations between the positions of pairs of electrons above a core can invalidate the simple valence-orbital picture, especially for extended molecules where these electrons can be partially localized in different portions of the molecule. Some authors[40] have suggested strong correlation of electrons in $TCNQ^{--}$, with each electron largely localized on the pair of cyanide groups at opposite ends of the molecule. This useful picture is probably oversimplified, as is indicated by solution hyperfine data[110] and by calculations,[335] which both show some of the relevant molecular orbitals to be spread over the quinoidal double bonds also.

In some cases, a single average t can be used in place of the distinct values indicated in (107). For instance, in a paramagnetic $\frac{1}{2}$-filled case with t/U small, we expect the ground state usually to have one electron on any given site, and the most important excited configurations to result from a CT yielding nearest-neighbour empty and doubly-occupied sites. Thus the CT integral connecting the more important configurations in the $\frac{1}{2}$-filled ground state will be t'' $(= t'''$ in self-complex FR stacks) or just (43). We then expect a single CT integral $t \approx t''$ to give nearly the same result as if different CT integrals were used, at least for the low-lying magnetic states. More quantitatively, we could specify

$$
t \equiv \langle (a^+_{i\alpha} a_{i+1\alpha} + a^+_{i\beta} a_{i+1\beta})(t' P_{0,i} P_{1,i+1} + t'' P_{1,i} P_{1,i+1} \\
+ t''' P_{0,i} P_{2,i+1} + t'''' P_{1,i} P_{2,i+1}) \rangle
\tag{109}
$$

where $P_{p,j}$ is the projector onto configurations in which site j is occupied by p ($= 0, 1, 2$) valence electrons.

In some cases the different CT integrals of (107) may play an important role. For less than half-filled FR stacks, t' will govern the single occupancy motion while $t'' = t'''$ still governs the "magnetic" splittings; these two effects can be[229] approximately separated in theoretical treatments so that these different CT integrals, if known, could readily be taken into account. Even for the half-filled case, t' and t'''' are expected to play an important role in conduction processes; indeed, the ratio t'/t'''' should be a direct indication whether hole or electron conduction is dominant.

The intersite Coulomb interaction and the attendant approximations deserve some further comment also. The Coulomb interaction between sites n and m was given in (42) for *ionic* molecules only, since the charge operators (40) vanish at a neutral site. Neutral sites nevertheless may contain partial charges, with vanishing sums, and an approximation like (42) leads to more rapid convergence[87] than a simple multipole expansion. Such partial charges in diamagnetic molecules must be computed rather than obtained from solution hyperfine data. The improvement to (42) is, still in the minimum basis,

$$V_C'(m, n) = \sum_{p, q = 0}^{2} M_{pq}(m, n)\, P_{p, m} P_{q, n} \tag{110}$$

where $P_{p, m}$ are again the projectors onto a configuration with $p = 0, 1$, or 2 valence electrons and $M_{pq}(m, n)$ is the corresponding sum over the delocalized charge distributions for sites with p and q valence electrons. The previous choice for V_C in (42) is regained if only the $M_{11}(m, n)$ term is retained and, in addition, the average value of $P_{1, n}$ is simply taken to be the charge operator ρ_n in (40). The intermediate approximation of scaling the energy difference between ionic and neutral sites leads to

$$M'(m, n) = \sum_{\mu \in m, \nu \in n} (\rho_\mu \rho_\nu - \rho_\mu' \rho_\nu')r_{\mu\nu}^{-1} \tag{111}$$

and is a straightforward improvement on (41) in which ρ_μ' and ρ_ν' refer to neutral sites. The assumption of identical geometries for both forms is inevitable, since only the actual structure will be available experimentally.

The Coulomb interactions were further simplified by introducing the Madelung approximation (51) in which we took

$$\langle \rho_i \rho_j \rangle = \langle \rho_i \rangle \langle \rho_j \rangle \tag{112}$$

for the ground-state expectation value of any two charges. (112) neglects Coulomb exchange stabilization[108] for adjacent sites in the stack. A quite different objection to $\langle V_C \rangle^{(0)}$ or to (112) is that the interaction is effectively smeared over the whole crystal, as occurs in self-consistent treatments. Thus the Madelung approximation replaces the various intersite Coulomb interactions by an average value. Consequently, the stronger short-range Coulomb interactions are under-represented, while the weaker long-range Coulomb interactions are over-represented.

An ionic MSR stack illustrates some of the consequences of the Madelung approximation. The energy required to create a doubly occupied D site adjacent to an A site in a $\ldots D^+ A^- D^+ A^- \ldots$ crystal is just

$$U' = U_D + (\epsilon_D - \epsilon_A) + 2M - M(1, 2) \tag{113}$$

Here U_D is the intrasite correlation (37), $(\epsilon_D - \epsilon_A)$ is the change in site energy, and $[2M - M(1, 2)]$ is the change in the lattice electrostatic energy. Inspection of (51) shows that the decrease of ρ_A from unity by $2/N$ leads to a change of $2M$, as if the resulting neutral sites were infinitely far apart.[22] The electron–hole interaction $-M(1, 2)$ is similarly missed in CT excitations of a neutral MSR stack.[22] The inclusion of $M(1, 2)$ is essential for quantitative computations of ΔE_{CT} and such a large term is missed in $\langle V_C \rangle^{(0)}$, which is basically a ground-state approximation. A simple, but partial remedy is to use the Madelung approximation to compute the ground-state charge density and then estimate additional corrections for excited states. The polaron stabilization of the individual electron and hole excitations, which is neglected in H_{CT}, must in fact be comparable to $M(1, 2)$, which is neglected in the Coulomb approximation, since we often find $\Delta E_c \approx \frac{1}{2} \Delta E_{CT}$ in Tables II–IV for simple stacks.

Thus the phenomenological parameters appearing in H_{CT} may depend in part on the approximation in which the model is to be solved. Nevertheless, we see important advantages in separating the problem into two separate parts: first, the evaluation of the parameters; second, the solution of the resulting model. The *ab initio* evaluation of the relevant matrix elements could be used in the refined definitions of the parameters for simpler models, for which solutions are already available. Should the reliability of the *ab initio* matrix elements be high enough and the initial comparisons using the simplified models be favourable, more extensive solutions, *via* many of the methods of section IV, would be feasible. We conclude that the theoretical evaluation of at least some of the parameters occurring in H_{CT} is possible. Calculations of Coulomb interactions, intrasite correlations and site energies are likely to be the most reliable, while CT and exchange integrals are more problematical.

VII. CONCLUSION

We noted at the beginning that the infinite molecular stacks encountered in solid-state CT crystals have hindered the generalization of the Mulliken model of configuration interaction in a single DA complex. Since π-molecular CT and free-radical crystals contain nearly nonoverlapping molecular species, a molecular–exciton approach was adopted. The general site representation developed in Section II and applied to representative CT and FR structures in Section III generalizes molecular–exciton theory to open-shell systems. The minimum-basis result for the CT Hamiltonian, H_{CT} in (45), is the natural extension of configuration interaction *via* CT states to an infinite crystal. Thus H_{CT} provides the general crystal analogue for the Mulliken treatment of a DA dimer.

Just as DA dimers have eluded a complete quantum-mechanical analysis, the accurate computation of the crystal-perturbed states occurring in H_{CT} has yet to be achieved. The phenomenological nature of H_{CT} is characteristic of molecular–exciton theories. The collection in Section III of representative CT excitation energies and activation energies for conduction and for paramagnetism indicates that the qualitative features of quite different crystal structures are contained in H_{CT} if strong intramolecular correlations and narrow bands are assumed. The experimental determination of the crystal parameters in H_{CT}, for example by comparing related structures, will be greatly facilitated by additional physical studies of single crystals.

The reduction of H_{CT} to various one-dimensional Hubbard models led us to summarize solid-state results for Hubbard models. As shown in Section IV, there are many results, ranging from nearly exact static properties to more approximate dynamic properties. These theoretical results can often be applied directly to H_{CT} and offer a variety of tests. The additional features of long-range Coulomb interactions, various electronic polarization effects, and electron–phonon interactions are neglected in Hubbard models, but can in principle be partially included in H_{CT} by permitting a deformable lattice and by including additional electronic features discussed in Section VI.

The reduction of H_{CT} to various spin systems in the limit of weak overlap and strong correlations ($t \ll U$) plays a central role in the magnetic properties of the semiconducting and insulating CT and FR crystals. Now CT interactions are associated with strong one-dimensional antiferromagnetic exchange $J \approx t^2/\Delta E_{CT}$ along the ion-radical stack. The magnetic resonance of both alternating and regular stacks is shown in Section V to reflect strong exchange, but a quantitative analysis is not yet available in the intermediate temperature region $kT \lesssim J$. The relation of π-molecular ion-radical stacks and exchange-coupled inorganic chains has been emphasized.

As has been emphasized throughout, H_{CT} is the simplest general Hamiltonian for the electronic states of molecular crystals in which a CT excitation occurs below any Frenkel excitation. The site representation of H_{CT} provides a typical phenomenlogical model: it is clearly related to both accurate molecular quantum mechanics and to observed optical, electric and magnetic properties. Detailed comparisons must, however, await developments in the quantum theory of intermolecular forces, as well as in the experimental techniques for studying π-molecular CT crystals. Refinements and extensions may safely be anticipated in view of current interest in organic photoconductors, in highly conducting organic solids, and in one-dimensional magnetic systems.

Acknowledgements:

We thank Drs A. J. Silverstein, R. C. Hughes, W. A. Seitz and Professor F. A. Matsen for many valuable discussions, criticisms, and assistance in the preparation of this manuscript. The financial support of the National Science Foundation and of the Alfred P. Sloan Foundation is gratefully acknowledged.

REFERENCES

1. H. A. Benesi and J. H. Hildebrand. *J. Amer. Chem. Soc.* **71**, 2703 (1949).
2. R. S. Mulliken. *J. Amer. Chem. Soc.* **74**, 811 (1952); note added in proof, *ibid.*, **72**, 600 (1950); *J. Phys. Chem.*, **56**, 801 (1952); *J. Chim. Phys.*, **61**, 20 (1964).
3. R. Foster, "Organic Charge-Transfer Complexes", Academic Press, London and New York (1969).
4. R. S. Mulliken and W. B. Person, "Molecular Complexes: A Lecture and Reprint Volume", Wiley, New York (1969).
5. J. Rose. "Molecular Complexes", Pergamon, Oxford (1967).
6. M. W. Hanna and J. L. Lippert. In "Molecular Complexes", Vol. I (ed. R. Foster); Paul Elek, London; Crane, Russak and Co., New York (1973), pp. 1–48.
7. M. J. S. Dewar and C. C. Thompson, Jr., *Tetrahedron Suppl.* **7**, 97 (1966).
8. G. Briegleb. "Electronen-Donator-Acceptor-Komplexe," Springer-Verlag, Berlin, (1961).
9. F. H. Herbstein. In "Perspectives in Structural Chemistry", Vol. IV (eds J. D. Dunitz and J. A. Ibers), Wiley, New York (1971), pp. 166–395.
10. C. K. Prout and B. Kamenar. In "Molecular Complexes", Vol. I (ed. R. Foster), Paul Elek, London; Crane, Russak, and Co. New York (1973), pp. 151–208.
11. C. K. Prout and J. D. Wright. *Angew. Chem. Int. Ed., Engl.*, **7**, 659 (1968).
12. K. Nakamoto. *J. Amer. Chem. Soc.*, **74**, 390, 392, 1739 (1952).
13. Y. Ohashi, H. Iwasaki and Y. Saito. *Bull. Chem. Soc. Japan*, **40**, 1789 (1967).
14. F. Gutmann and L. E. Lyons. "Organic Semiconductors", Wiley, New York (1967). Includes a comprehensive analysis and compilation of electric and optical properties of organic solids.

15. D. P. Craig and S. H. Walmsley. "Excitons in Molecular Crystals", Benjamin, New York (1968); R. S. Knox. "Theory of Excitons", Academic Press, New York (1963).
16. G. W. Robinson. *Ann. Rev. Phys. Chem.*, **21**, 429 (1970); R. M. Hochstrasser. *ibid.*, **17**, 457 (1966); O. Schnepp. *ibid.*, **14**, 35 (1963).
17. S. A. Rice and J. Jortner. *In* "Physics and Chemistry of the Organic Solid State", Vol. III (eds D. Fox, M. M. Labes, and A. Weissberger), Interscience, New York (1967), pp. 199–497.
18. A. S. Davydov. "Theory of Molecular Excitons", Plenum Press, New York (1971).
19. R. Foster and T. J. Thomson, *Trans. Faraday Soc.*, **59**, 296 (1963).
20. Y. Matsunaga. *J. Chem. Phys.*, **41**, 1609 (1964); *ibid.*, **42**, 1982 (1965).
21. H. M. McConnell, B. M. Hoffman and R. M. Metzger. *Proc. Natl. Acad. Sci. U.S.*, **53**, 46 (1965).
22. P. J. Strebel and Z. G. Soos. *J. Chem. Phys.*, **53**, 4077 (1970).
23. P. L. Nordio, Z. G. Soos and H. M. McConnell. *Ann. Rev. Phys. Chem.*, **17**, 237 (1966).
24. Z. G. Soos, *Ann. Rev. Phys. Chem.*, **25**, 121 (1974).
25. D. J. Klein and Z. G. Soos. *Mol. Phys.*, **20**, 1013 (1971).
26. I. F. Shchegolev. *Phys. Status Solidi*, (a) **12**, 4 (1972).
27. H. R. Zeller. *Adv. Sol. State Phys.*, **13**, 31 (1973).
28. L. R. Melby, R. J. Harder, W. R. Hertler, W. Mahler, R. E. Benson and W. E. Mochel. *J. Amer. Chem. Soc.*, **84**, 3374 (1962).
29. J. J. André, J. Clementz, R. Jesser and G. Weill. *Comptes Rendu*, **266**, 1057 (1968); J. J. André, and G. Weill. *ibid.*, **269**, 499 (1969).
30. K. Hausser and J. N. Murrell. *J. Chem. Phys.*, **27**, 500 (1957).
31. C. J. Fritchie, Jr. and P. Arthur, Jr. *Acta Cryst.*, **21**, 139 (1966).
32. Y. Iida. *Bull. Chem. Soc. Japan*, **42**, 637 (1969).
33. A. W. Hanson. *Acta Cryst.*, **B24**, 768 (1968).
34. O. H. LeBlanc, Jr. *In* "Physics and Chemistry of the Organic Solid State", Vol. III (eds D. Fox, M. M. Labes and A. Weissberger), Interscience, New York (1967), pp. 133–198.
35. W. J. Siemons, P. E. Bierstedt and R. G. Kepler. *J. Chem. Phys.*, **39**, 3523 (1963).
36. J. P. Ferraris, D. O. Cowan, V. Walatka, Jr. and J. H. Perlstein. *J. Amer. Chem. Soc.*, **95**, 948 (1973).
37. L. B. Coleman, M. J. Cohen, D. J. Sandman, F. G. Yamagishi, A. F. Garito and A. J. Heeger. *Sol. State Comm.*, **12**, 1125 (1973).
38. D. E. Schafer, F. Wudl, G. A. Thomas, J. P. Ferraris and D. O. Cowan. *Sol. State Comm.*, **14**, 347 (1974).
39. L. R. Melby. *Can. J. Chem.*, **43**, 1448 (1965).
40. A. J. Epstein, S. Etemad, A. F. Garito and A. J. Heeger. *Phys. Rev.*, **B5**, 952 (1972).
41. L. B. Coleman, J. A. Cohen, A. F. Garito and A. J. Heeger. *Phys. Rev.*, **B7**, 2122 (1973).
42. A. N. Bloch. *In* "Charge and Energy Transfer in Organic Semiconductors", (eds. M. Masuda and M. Silver), Plenum, New York (1974), p. 159; A. N. Bloch, D. O. Cowan and T. O. Poehler, *ibid.*, p. 167 (1974).
43. A. F. Garito and A. J. Heeger, in "Collective Properties of Physical Systems", (eds. B. Lundqvist and S. Lundqvist), Nobel Foundation and Academic Press, (1973), p. 129.
44. J. Hubbard, *Proc. Roy. Soc.*, *Ser. A*, **276**, 238 (1963); **277**, 237 (1964); **281**, 401 (1964).

45. J. H. Van Vleck. *In* "Quantum Theory of Atoms, Molecules, and the Solid State" (ed. P. O. Löwdin), Academic Press, New York (1966), pp. 475–84.
46. N. F. Mott. *Proc. Roy. Soc. Ser. A*, **62**, 416 (1949); *Advan. Phys.* **13**, 325 (1964); *Rev. Mod. Phys.*, **40**, 677 (1968).
47. D. Adler. *In* "Solid State Physics", Vol. 21 (eds. F. Seitz, D. Turnbull and H. Ehrenreich), Academic Press, New York (1968), pp. 1–113.
48. C. Herring. "Magnetism, IV: Exchange Interactions Among Itinerant Electrons" (eds G. T. Rado and H. Suhl), Academic Press, New York (1966).
49. P. W. Anderson. *In* "Solid State Physics", Vol. 14 (eds F. Seitz and D. Turnbull,) Academic Press, New York, (1963), pp. 99–214.
50. Z. G. Soos and P. J. Strebel. *J. Amer. Chem. Soc.*, **93**, 3325 (1971).
51. O. H. LeBlanc, Jr. *J. Chem. Phys.*, **42**, 4307 (1965).
52. A. A. Ovchinnikov, I. I. Ukranski and G. V. Kventsel. *Usp. Fiz. Nauk* **108**, 81 (1972); [Engl. Trans. *Sov. Phys. Uspekhi*, **15**, 575 (1973)].
53. J. Linderberg and Y. Öhrn "Propagators in Quantum Chemistry", Academic Press, New York (1973).
54. G. F. Kokoszka and R. W. Duerst. *Coord. Chem. Rev.*, **5**, 209 (1970).
55. L. J. DeJongh and A. R. Miedema. *Adv. in Phys.*, **23**, 1 (1974).
56. Y. Iida and Y. Matsunaga. *Bull. Chem. Soc. Japan*, **41**, 2615 (1968).
57. Y. Iida. *Bull. Chem. Soc. Japan*, **42**, 71 (1969).
58. Y. Iida. *Bull. Chem. Soc. Japan*, **43**, 2772 (1970).
59. S. Hiroma, H. Kuroda and H. Akamatu. *Bull. Chem. Soc. Japan*, **44**, 9 (1971).
60. Y. Sato, M. Kinoshita, M. Sano and H. Akamatu. *Bull. Chem. Soc. Japan*, **43**, 2370 (1970).
61. H. Kuroda, S. Hiroma and H. Akamatu. *Bull. Chem. Soc. Japan*, **41**, 2855 (1968).
62. T. Amano, H. Kuroda and H. Akamatu. *Bull. Chem. Soc. Japan*, **41**, 83 (1968).
63. T. Sakata and S. Nagakura. *Bull. Chem. Soc. Japan*, **42**, 1497 (1969).
64. H. W. Offen. *In* "Molecular Complexes", Vol. I (ed. R. Foster), Paul Elek, London; Crane, Russak, and Co., New York (1973), pp. 117–149.
65. J. Kommandeur. *In* "Physics and Chemistry of the Organic Solid State", Vol. II (eds D. Fox, M. M. Labes, A. Weissberger), Interscience, New York (1965), pp. 1–67.
66. Z. G. Soos. *J. Chem. Phys.*, **46**, 4284 (1967).
67. D. W. Hone. *AIP Conference Proceedings*, **5**, 413 (1972).
68. D. W. Hone and P. M. Richards. *Ann. Rev. Mat. Sci.*, **4**, 337 (1974).
69. R. E. Dietz, F. R. Merritt, R. Dingle, D. Hone, B. G. Silbernagel and P. M. Richards. *Phys. Rev. Letters*, **26**, 1186 (1971).
70. R. Colton and J. H. Canterford. "Halides of the First Row Transition Metals", Wiley-Interscience, London (1969).
71. W. E. Hatfield and R. Whyman. *Transition Metal Chem.*, **5**, 47 (1969).
72. N. Kato, H. B. Jonassen and J. C. Fanning. *Chem. Rev.*, **64**, 99 (1964).
73. G. F. Kokoszka and G. Gordon. *Transition Metal Chem.*, **5**, 181 (1969).
74. R. L. Martin. *In* "New Pathways of Inorganic Chemistry", (eds E. A. V. Ebsworth, A. G. Maddock and A. G. Sharpe), Cambridge Univesrity Press, Cambridge (1968) Chapter 9.
75. E. Sinn. *Coord. Chem. Rev.*, **5**, 313 (1970).
76. A. P. Ginsberg. *Inorg. Chim. Acta Rev.*, **5**, 45 (1971).
77. K. Krogmann. *Angew. Chem. Intl. Edt., Eng.*, **8**, 35 (1969).
78. I. G. Austin and N. F. Mott. *Advan. Phys.*, **18**, 41 (1969).

102   Z. G. SOOS AND D. J. KLEIN

79. D. Emin. *In* "Electric and Structural Properties of Amorphous Semiconductors" (eds P. G. LeCombre and J. Mort), Academic Press, New York (1973), pp. 261–329.
80. A. N. Bloch, R. B. Weisman and C. M. Varma. *Phys. Rev. Letters*, **28**, 753 (1972), A. N. Bloch and C. N. Varma, *J. Phys. C: Solid State Physics.*, **6**, 1849 (1973).
81. J. H. Perlstein, M. J. Minot and V. Walatka, Jr. *Mat. Res. Bull.*, **7**, 309 (1972).
82. D. C. Mattis. "The Theory of Magnetism", Harper and Row, New York (1965), pp. 46–55.
83. R. H. Boyd and W. D. Phillips. *J. Chem. Phys.*, **43**, 2927 (1965).
84. S. Nakayama and K. Suzuki. *Bull. Chem. Soc. Japan*, **46**, 3694 (1973).
85. C. Herring. *In* "Magnetism", Vol. II.B (eds G. T. Rado and H. Suhl), Academic Press, New York (1963), pp. 1–185.
86. J. N. Murrell, M. Randic and D. R. Williams. *Proc. Roy. Soc., Ser.* A, **284**, 566 (1965).
87. J. L. Lippert, M. W. Hanna and P. J. Trotter. *J. Amer. Chem. Soc.*, **91**, 4035 (1968); M. W. Hanna and D. E. Williams, *ibid.*, **90**, 5358 (1968); M. W. Hanna, *ibid.*, **90**, 285 (1967).
88. P. R. Certain and L. W. Bruch. *In* "International Review of Science, Physical Chemistry, Series One, Volume 1, Theoretical Chemistry" (eds A. D. Buckingham and W. Byers Brown), Butterworth and Co., London (1972), pp. 113–165.
89. D. M. Chipman, J. D. Bowman and J. O. Hirschfelder. *J. Chem. Phys.*, **59**, 2830 (1973).
90. J. O. Hirschfelder, C. F. Curtis and R. B. Bird. "Molecular Theory of Gases and Liquids", Wiley, New York (1954).
91. H. Margenau and N. R. Kestner. "Theory of Intermolecular Forces", Pergamon, New York, (1969).
92. A. D. Buckingham and B. D. Utting. *Ann. Rev. Phys. Chem.*, **21**, 287 (1970).
93. R. J. W. Le Fèvre, D. V. Radford and P. J. Stiles. *J. Chem. Soc. B*, 1297 (1968).
94. E. G. Cooke and J. C. Schug. *J. Chem. Phys.*, **53**, 723 (1970).
95. P. O. Löwdin, *J. Chem. Phys.*, **18**, 365 (1950).
96. I. Fischer-Hjalmers. *J. Chem. Phys.*, **42**, 1962 (1965).
97. J. Hubbard. *Proc. Roy. Soc., Ser. A*, **285**, 542 (1965).
98. V. M. Agranovich. *Soviet Phys–JETP*, **10**, 307 (1960); V. M. Agranovich and Y. M. Konobeev. *Soviet Phys. Solid State*, **3**, 260 (1961)].
99. R. Hoffman. *Radiat. Res.*, **20**, 140 (1963).
100. J. I. Krugler, C. G. Montgomery and H. M. McConnell. *J. Chem. Phys.*, **41**, 2421 (1964).
101. D. M. Hanson. *CRC Crit. Rev. in Solid State Sci.*, **3**, 243 (1973).
102. M. Moshinsky. "Group Theory and the Many-Body Problem", Gordon and Breach, New York (1968).
103. V. G. Wybourne. "Symmetry Principles and Atomic Spectroscopy," Wiley, New York (1970).
104. J. Kommandeur and G. T. Pott. *Mol. Phys.*, **11**, 93 (1966).
105. M. Tosi. *In* "Solid State Physics", Vol. 16 (eds F. Seitz and D. Turnbull), Academic Press, New York (1964), pp. 1–120.
106. R. M. Metzger. *J. Chem. Phys.*, **57**, 1870, 1876 (1972).
107. R. M. Metzger. *J. Chem. Phys.*, **57**, 2218 (1972).
108. Z. G. Soos and A. J. Silverstein. *Mol. Phys.*, **23**, 775 (1972).
109. A. D. McLachlan. *Mol. Phys.*, **1**, 233 (1958). For a compilation of theoretical and experimental results on spin densities (to about 1969), see R. M. Metzger, Ph.D. Thesis, Stanford, 1970 (unpublished) Tables I–VIII in Chapter II.

110. P. H. Rieger and G. K. Fraenkel. *J. Chem. Phys.*, **37**, 2795 (1962).
111. J. R. Bolton, A. Carrington and J. Dos Santos-Veiga. *Mol. Phys.* **5**, 615 (1962).
112. C. Kittel. "Quantum Theory of Solids", Wiley, New York (1963).
113. P. W. Anderson. "Concepts in Solids", Benjamin, New York (1963).
114. T. Sakata and S. Nagakura. *Bull. Chem. Soc. Japan*, **43**, 1346 (1970).
115. J. Kommandeur and F. R. Hall. *J. Chem. Phys.*, **34**, 129 (1961).
116. L. S. Singer and J. Kommandeur. *J. Chem. Phys.*, **34**, 133 (1961).
117. H. M. McConnell and R. Lynden-Bell. *J. Chem. Phys.*, **36**, 2393 (1962).
118. D. B. Chesnut and W. D. Phillips. *J. Chem. Phys.*, **35**, 1002 (1961).
119. M. T. Jones and D. B. Chesnut. *J. Chem. Phys.*, **38**, 1311 (1963); M. T. Jones, *ibid.*, **40**, 1837 (1964); D. B. Chesnut and P. Arthur, Jr., *ibid.*, **36**, 2969 (1962).
120. R. G. Kepler. *J. Chem. Phys.*, **39**, 3528 (1963).
121. D. D. Thomas, H. Keller and H. M. McConnell. *J. Chem. Phys.*, **39**, 2321 (1963).
122. D. D. Thomas, A. W. Merkl, A. F. Hildebrandt and H. M. McConnell. *J. Chem. Phys.*, **40**, 2588 (1964).
123. M. J. Hove, B. M. Hoffman and R. J. Loyd. *J. Phys. Chem.*, **76**. 1849 (1972).
124. *Bull. Amer. Phys. Soc. Series II*, **19**, No. 3 (1974). Program of March 1974 APS Meeting in Philadelphia, in which several sessions were devoted to TTF–TCNQ.
125. 167th A.C.S. National Meeting, Los Angeles (1974). Abstracts of papers include many TTF–TCNQ studies.
126. W. Duffy, Jr., J. F. Dubach, P. A. Pianetta, J. F. Deck, D. L. Strandburg and A. R. Miedema. *J. Chem. Phys.*, **56**, 2555 (1972).
127. W. Duffy, Jr. and D. L. Strandburg. *J. Chem. Phys.*, **46**, 456 (1967); for another linear neutral FR crystal, see M. Mukai, N. Azuma and K. Ishizu. *Bull. Chem. Soc. Japan*, **43**, 3618 (1970).
128. A. Hoekstra, T. Spoelder and A. Vos. *Acta Cryst.*, **B28**, 14 (1970).
129. I. Shirotani and H. Kobayashi. *Bull. Chem. Soc. Japan*, **46**, 2595 (1973).
130. C. J. Fritchie, Jr. *Acta Cryst.*, **20**, 892 (1965).
131. L. B. Coleman, S. K. Khanna, A. F. Garito, A. J. Heeger and B. Morosin. *Phys. Letters*, **42A**, 15 (1972).
132. M. Konno, H. Kobayashi, F. Marumo and Y. Saito. *Bull. Chem. Soc. Japan*, **46**, 1987 (1973).
133. J. G. Vegter, T. Hibma and J. Kommandeur. *Chem. Phys. Letters*, **3**, 427 (1969).
134. Y. Iida. *J. Chem. Phys.*, **59**, 1607 (1972); *Bull. Chem. Soc. Japan*, **46**, 320 (1972); **43**, 3685 (1970); **44**, 3344 (1971).
135. N. Sakai, I. Shirotani and S. Minomura. *Bull. Chem. Soc. Japan*, **45**, 3321 (1972).
136. J. L. DeBoer and A. Vos. *Acta Cryst.*, **B28**, 835 (1972).
137. J. L. DeBoer and A. Vos. *Acta Cryst.*, **B28**, 839 (1972).
138. B. G. Anex and E. B. Hill, Jr. *J. Amer. Chem. Soc.*, **88**, 3648 (1966).
139. S. Hiroma and H. Kuroda. *Bull. Chem. Soc. Japan*, **46**, 3645 (1973).
140. V. Walatka, Jr. and J. H. Perlstein. *Mo. Cryst. and Liq. Cryst.*, **15**, 269 (1971).
141. A. R. Blythe, M. R. Boon and P. G. Wright. *Discuss. Faraday Soc.*, **51**, 110 (1971).
142. R. E. Rundle. *J. Phys. Chem.*, **61**, 45 (1957); J. R. Miller. *Adv. Inorg. Chem. Radiochem.*, **4**, 133 (1962); *J. Chem. Soc.*, 713 (1965).
143. P. S. Gomm, T. W. Thomas and A. E. Underhill. *J. Chem. Soc.(A)*, 2154 (1971); P. Day, A. F. Orchard, A. J. Thomson and R. J. P. Williams. *J. Chem. Phys.*, **43**, 3763 (1965).
144. L. V. Interrante. *Chem. Commun.*, 302 (1972).

145. L. V. Interrante and R. P. Messmer. *Inorg. Chem.*, **10**, 1174 (1971); *L. V. Interrante and F. B. Bundy. *ibid.*, 1169 (1971).
146. A. W. Hanson. *Acta Cryst.*, **19**, 610 (1965).
147. J. L. DeBoer and A. Vos. *Acta Cryst.*, **B24**, 720 (1967).
148. R. C. Hughes and Z. G. Soos. *J. Chem. Phys.*, **48**, 1066 (1968).
149. M. Kinoshita and H. Akamatu. *Nature*, **207**, 291 (1965).
150. B. M. Hoffman and R. C. Hughes. *J. Chem. Phys.*, **52**, 4011 (1970); *Solid State Comm.*, **7**, 895 (1969).
151. M. Ohmasa, M. Kinoshita, M. Sano and H. Akamatu. *Bull. Chem. Soc. Japan*, **41**, 1998 (1968).
152. G. T. Pott and J. Kommandeur. *Mol. Phys.*, **13**, 373 (1967). The molionic interpretation is incorrect[9, 63] and TMPD–chloranil shows uncorrelated spin excitations.
153. A. Ottenberg, C. Hoffman and J. Osiecki. *J. Chem. Phys.*, **38**, 1898 (1963).
154. K. Hatano, Y. Fujita and T. Kwan. *Bull. Chem. Soc. Japan*, **43**, 3022 (1970).
155. H. Masai, K. Sonogashira and N. Hagihara. *J. Organometal. Chem.*, **34**, 397 (1972).
156. V. Y. Krivnov and A. A. Ovchinnikov. *Fiz. Tver. Tela*, **15**, 172 (1973) [English Trans: *Soviet Phys.—Solid State*, **15**, 118 (1973)].
157. W. H. Bentley and H. G. Drickamer. *J. Chem. Phys.*, **42**, 1573 (1965).
158. R. C. Hughes, Ph.D. Thesis, Standford University, 1966 (unpublished); A. W. Merkl, R. C. Hughes, L. J. Berliner and H. M. McConnell. *J. Chem. Phys.*, **43**, 953 (1965); R. C. Hughes, A. W. Merkl and H. M. McConnell. *J. Chem. Phys.*, **44**, 1720 (1966).
159. D. E. Williams. *J. Amer. Chem. Soc.*, **89**, 4280 (1967).
160. H. Inokuchi, Y. Harada and Y. Maruyama. *Bull. Chem. Soc. Japan*, **35**, 1559 (1962).
161. D. J. Klein, W. A. Seitz, M. A. Butler and Z. G. Soos. unpublished.
162. D. Cabib and T. A. Kaplan. *Phys. Rev.*, **7B**, 2199 (1973).
163. N. Sakai, I. Shirotani and S. Minomura. *Bull. Chem. Soc. Japan*, **45**, 3314 (1972).
164. G. T. Pott and J. Kommandeur. *J. Chem. Phys.*, **47**, 395 (1967). See Refs. 9, 63 and 137 for a refutation of the proposed disproportionation.
165. J. J. André and G. Weill. *Chem. Phys. Letters*, **9**, 27 (1971).
166. J. L. DeBoer, A. Vos and K. Huml. *Acta Cryst.*, **B24**, 542 (1968).
167. J. Tanaka, M. Inoue, M. Mizuno and K. Horai. *Bull. Chem. Soc.*, *Japan*, **43**, 1998 (1970).
168. D. B. Chesnut. *J. Chem. Phys.*, **40**, 405 (1964). This model fails for the phase transitions in TCNQ salts.[23]
169. J. B. Torrance, personal communication. J. B. Torrance, B. A. Scott, D. C. Green, and P. Chaudhari. *Symposium on Superconductivity and Lattice Instability, Gatlinburg, Tenn.* (1973). A quite different (metallic rather than CT) interpretation is offered for TTF–TCNQ by P. M. Grant, R. L. Green, G. C. Wrighton and G. Castro, *Phys. Rev. Lett.*, **31**, 1311 (1974); A. A. Bright, A. F. Garito and A. J. Heeger. *Phys. Rev.*, **B10**, 1328 (1974).
170. H. R. Zeller. *Phys. Rev. Lett.*, **28**, 1452 (1972).
171. R. E. Peierls. "Quantum Theory of Solids", Oxford University Press, Oxford (1955), p. 108; M. Menyhard and J. Solyom, *J. Low Temp. Phys.*, **12**, 529 (1973); J. Solyom, *ibid.*, **12**, 547 (1973).
172. D. B. Chesnut. *J. Chem. Phys.*, **45**, 4677 (1966).
173. P. Pincus. *Solid State Comm.*, **9**, 1971 (1971); G. Beni and P. Pincus. *J. Chem. Phys.*, **57**, 3531 (1973).

174. H. Chihara, M. Nakamura, and S. Seki. *Bull. Chem. Soc. Japan,* **38**, 1776 (1965).
175. D. B. Chesnut and R. W. Mosely. *Theoret. Chim Acta,* **13**, 230 (1969).
176. T. Sundaresan and S. C. Wallwork. *Acta Cryst.,* **B28**, 3507 (1972).
177. M. J. Hove, B. M. Hoffman and J. A. Ibers. *J. Chem. Phys.,* **56**, 3490 (1972).
178. G. R. Anderson and C. J. Fritchie, Jr. *Paper 111, Second National Meeting, Society for Applied Spectroscopy, San Diego* (1963).
179. J. C. Bailey and D. B. Chesnut. *J. Chem. Phys.,* **51**, 5118 (1969).
180. T. Hibma, P. Dupuis and J. Kommandeur. *Chem. Phys. Lett.,* **15**, 17 (1972).
181. H. Kobayashi, F. Marumo and Y. Saito. *Acta Cryst.,* **B27**, 373 (1970).
182. P. Goldstein, K. Seff and K. N. Trueblood. *Acta Cryst.,* **B24**, 778 (1968).
183. R. P. Shibaeva, L. O. Atovmyan and M. N. Orfanov. *J. Chem. Commun.,* 1494 (1969).
184. H. Kobayashi, Y. Ohashi, F. Marumo and Y. Saito. *Acta Cryst.,* **B26**, 459 (1970).
185. A. T. McPhail, G. M. Semeniuk and D. B. Chesnut, *J. Chem. Soc.* (*A*), 2174 (1971).
186. T. Kistenmacher, T. E. Phillips and D. O. Cowan. *Acta Cryst.,* **B30**, 763 (1974).
187. F. Wudl and E. W. Southwick. *Chem. Commun.* 254 (1974); F. Wudl (personal communication).
188. M. A. Butler, J. P. Ferraris, A. N. Bloch and D. O. Cowan. *Chem. Phys. Lett.* **24**, 600 (1974).
189. W. D. Grobman, R. A. Pollak, D. E. Eastman, E. T. Maas, Jr. and B. A. Scott. *Phys. Rev. Lett.* **32**, 534 (1974).
190. J. P. Ferraris, T. O. Poehler, A. N. Bloch and D. O. Cowan. *Tetrahedron Lett.,* 2553 (1973); A. N. Bloch, J. P. Ferraris, D. O. Cowan and T. O. Poehler. *Sol. State Comm.,* **13**, 753 (1973).
191. E. M. Engler and V. V. Patel. *J. Amer. Chem. Soc.,* **96**, 7376 (1974).
192. D. N. Fedutin, I. F. Shchegolev, V. B. Stryukov, E. B. Yagubskii, A. V. Zvarykina, L. O. Atovmyan, V. F. Kaminskii and R. P. Shibaeva. *Phys. Stat. Sol.* (*b*), **48**, 87 (1971).
193. E. B. Yagubskii, M. L. Khidekel, I. F. Shchegolev, L. I. Buravov, B. G. Gribov and M. K. Makova. *Izv. Akad. Nauk SSSR Ser. Khim.,* 2124 (1968).
194. J. C. Scott, A. F. Garito, and A. J. Heeger. *Phys. Rev.,* **B10**, 3131 (1974).
195. Y. Tomkiewicz (personal communication). Y. Tomkiewicz, B. A. Scott, L. J. Tao and R. S. Title, *Phys. Rev. Lett.* **32**, 1363 (1974).
196. Z. G. Soos and D. J. Klein. *J. Chem. Phys.,* **55**, 3284 (1971).
197. T. Sundaresan and S. C. Wallwork. *Acta Cryst.,* **B28**, 491 (1972).
198. Y. Suzuki and Y. Iida. *Bull. Chem. Soc. Japan,* **46**, 2056 (1973).
199. M. C. Gutzwiller. *Phys. Rev.,* **134**, A923 (1964); *ibid.,* **137**, A1726 (1965).
200. J. Kanamori. *Progr. Theor. Phys.,* **30**, 275 (1963).
201. Y. Nagoaka. *Phys. Rev.,* **147**, 392 (1966).
202. M. Cohen and J. C. Thompson. *Adv. Phys.,* **17**, 857 (1968).
203. H. O. Hooper and A. M. DeGraaf. "Amorphous Magnetism", Plenum, New York, (1973).
204. J. A. Pople. *Trans. Faraday Soc.,* **42**, 1375 (1953); R. Pariser and R. G. Parr. *J. Chem. Phys.,* **21**, 446, 767 (1953).
205. J. N. Murrell and L. Salem. *J. Chem. Phys.,* **34**, 1914 (1961).
206. H. C. Longuet-Higgins and L. Salem. *Proc. Roy Soc. Ser. A.,* **257**, 445 (1960).
207. J. Linderberg and Y. Öhrn. *J. Chem. Phys.,* **49**, 716 (1968).
208. J. Cizek. *J. Chem. Phys.,* **45**, 4256 (1966).
209. R. A. Harris and L. M. Falicov. *J. Chem. Phys.,* **51**, 5034 (1969).

210. Y. A. Kruglyak and I. I. Ukranksy. *Intl. J. Quantum Chem.*, **4**, 57 (1960).
211. M. J. Minot and J. H. Perlstein. *Phys. Rev. Lett.*, **26**, 371 (1971).
212. E. H. Lieb and F. Y. Wu. *Phys. Rev. Lett.*, **20**, 1445 (1968).
213. E. H. Lieb and D. C. Mattis. "Mathematical Physics in One Dimension", Academic Press, New York (1966).
214. P. A. Fedders and J. Kommandeur. *J. Chem. Phys.*, **52**, 2014 (1970).
215. J. G. Vegter, J. Kommandeur and P. A. Fedders. *Phys. Rev.* **B7**, 2929 (1973).
216. T. Hibma, G. A. Sawatzky and J. Kommandeur. *Chem Phys. Lett.*, **23**, 21 (1973). The band limit is modified here to include some correlations.
217. J. B. Sokoloff. *Phys. Rev.*, **B2**, 779 (1970).
218. G. Beni, T. Holstein and P. Pincus. *Phys. Rev.*, **B8**, 312 (1973).
219. D. J. Klein. *Phys. Rev.*, **B8**, 3452 (1973).
220. T. A. Kaplan, S. D. Mahanti and W. M. Hartman. *Phys. Rev. Lett.*, **27**, 1796 (1971).
221. G. Beni and P. Pincus. *Phys. Rev.*, **B9**, 2963 (1974).
222. E. H. Lieb and D. C. Mattis. *Phys. Rev.*, **125**, 164 (1962).
223. H. A. Bethe. *Z. Phys.*, **71**, 205 (1931); C. N. Yang and C. P. Yang. *Phys. Rev.*, **147**, 303 (1966); **150**, 321 (1966); **151**, 258 (1967).
224. A. A. Ovchinnikov. *Zh. Eksp. Teor. Fiz.*, **57**, 2137 (1969). [English translation: *Sov. Phys. JETP*, **30**, 1100 (1970)].
225. M. Takahashi, *Progr. Theor. Phys.*, **42**, 1098); **43**, 1619 (1970).
226. L. Hulthen. *Arkiv. Mat. Astron. Fysik.*, **26A**, No. 11 (1938).
227. J. des Cloizeaux and J. J. Pearson. *Phys. Rev.*, **128**, 2131 (1962).
228. R. B. Griffiths. *Phys. Rev.*, **133**; A 768 (1964).
229. D. J. Klein and W. A. Seitz. *Phys. Rev.*, **B10**, 3217 (1974).
230. H. Shiba. *Phys. Rev.*, **B6**, 930 (1972).
231. F. Coll. *Phys. Rev.*, **B9**, 2150 (1974).
232. A. J. Silverstein and Z. G. Soos. *J. Chem. Phys.*, **53**, 326 (1970).
233. D. R. Penn. *Phys. Rev.*, **142**, 350 (1966).
234. W. Langer, M. Plischke and D. C. Mattis. *Phys. Rev. Lett.*, **23**, 1448 (1969).
235. D. J. Klein and Z. G. Soos, unpublished work.
236. B. I. Halperin and T. M. Rice. *In* "Solid State Physics", Vol. 21 (eds F. Seitz, D. Turnbull and H. Ehrenreich), Academic Press, New York (1968), pp. 115–192.
237. N. Mermin and H. Wagner. *Phys. Rev. Lett.*, **17**, 1133 (1966).
238. P. J. Strebel, Ph.D. Thesis, Princeton University, (1971), (unpublished).
239. A. B. Harris and R. V. Lange. *Phys. Rev.*, **157**, 295 (1967).
240. D. M. Esterling. *Phys. Rev.*, **B2**, 4686 (1970).
241. D. R. Penn. *Phys. Lett.*, **A26**, 509 (1968).
242. M. Takahashi. *Progr. Theor. Phys.*, **43**, 917 (1970).
243. J. C. Bonner and M. E. Fisher. *Phys. Rev.*, **135**, A640 (1964).
244. H. Shiba and P. Pincus. *Phys. Rev.*, **B5**, 1966 (1972).
245. H. Shiba. *Progr. Theor. Phys.*, **46**, 77 (1971).
246. T. Arai. *Phys. Rev.*, **B4**, 216 (1971).
247. R. A. Bari. *Phys. Rev.* **B7**, 2128 (1973).
248. A. A. Berlin, G. A. Vinogradov and A. A. Ovchinnikov. *Intl. J. Quantum Chem.*, **6**, 263 (1972).
249. D. M. Esterling and H. C. Dubin. *Phys. Rev.*, **B6**, 4276 (1972).
250. H. Hasegawa and J. Kanamori. *J. Phys. Soc. Japan*, **31**, 382 (1971).
251. J. Hubbard and K. P. Jain. *J. Phys. Soc.*, **C1**, 1650 (1968).
252. P. Richmond and G. L. Sewell. *J. Math. Phys.*, **9**, 349 (1968).

253. J. B. Sokoloff. *Phys. Rev.*, **B1**, 1144 (1970); **B2**, 3707 (1970); **B3**, 3826 (1971).
254. R. A. Tahir-Kheli and H. S. Jarrett. *Phys. Rev.*, **180**, 544 (1968).
255. W. A. Seitz and D. J. Klein. *Phys. Rev.*, **B9**, 2159 (1974).
256. F. Carboni and P. M. Richards. *Phys. Rev.*, **177**, 889 (1969).
257. O. Heilman and E. H. Lieb. *Trans. N.Y. Acad. Sci.*, **33**, 116 (1971).
258. K. H. Heinig and J. Monecke. *Phys. Status Solidi*, **B49**, K139, K141 (1972); **B50**, K117 (1972).
259. L. N. Bulaevskii and D. I. Khomski. *Phys. Lett.*, **41A**, 257 (1972); *Fiz. Tuer. Tela*, **14**, 3594 (1972); [English trans: *Soviet Phys. Solid State*, **14**, 3015 (1973)].
260. D. W. Hone and P. A. Pincus. *Phys. Rev.*, **7B**, 4889 (1973).
261. L. N. Bulaevskii. *Zh. Eksp. Teor. Fiz.*, **51**, 230 (1966); [English trans: *Soviet Phys. JETP*, **24**, 154 (1967)].
262. D. J. Klein and W. A. Seitz. *Phys. Rev.*, **B8**, 2236 (1973).
263. P. B. Visscher. *Phys. Rev.*, **B10**, 932 (1974).
264. W. Magnus. *Commun. Pure Appl. Math.*, **7**, 649 (1954).
265. P. Pechukas and J. C. Light. *J. Chem. Phys.*, **44**, 3897 (1966).
266. J. C. Kimball and J. R. Schrieffer. *Proceedings of the Seventh Conference on Magnetism and Magnetic Materials, Chicago*, (1971). *Amer. Inst. Phys.*, *New York* (1973).
267. M. Cryot. *J. de Phys.* (*Paris*), **33**, 125 (1972).
268. R. A. Bari. *Phys. Rev.*, **5B**, 2736 (1972).
269. G. Beni, P. Pincus and D. Hone. *Phys. Rev.*, **B8**, 3389 (1973).
270. L. C. Bartel and H. S. Jarrett. *Phys. Rev.* **B10**, 942 (1974).
271. W. Duffy, Jr. and K. P. Barr. *Phys. Rev.* **165**, 647 (1968).
272. Z. G. Soos. *J. Chem. Phys.* **43**, 1121 (1965).
273. Z. G. Soos. *Phys. Rev.* **149**, 330 (1966).
274. L. N. Bulaevskii. *Zh. Eksp. Teor. Fiz.*, **44**, 1008 (1963); [English Trans: *Soviet Phys.*, *JETP*, **17**, 684 (1963)].
275. D. B. Abraham and A. D. McLachlan. *Mol. Phys.*, **12**, 301, 319 (1967).
276. D. B. Abraham. *J. Chem. Phys.*, **51**, 3795 (1969).
277. R. Lynden-Bell and H. M. McConnell. *J. Chem. Phys.*, **37**, 794 (1962).
278. A. B. Harris. *Phys. Rev.*, **B8**, 2166 (1973).
279. S. Donaich, B. Roulet and M. E. Fisher. *Phys. Rev. Lett.*, **27**, 262 (1971).
280. P. M. Chaikin, A. F. Garito and A. J. Heeger. *J. Chem. Phys.*, **58**, 2336 (1973).
281. P. M. Chaikin, A. F. Garito and A. J. Heeger. *Phys. Rev.*, **B5**, 4966 (1972).
282. R. A. Bari, *Phys. Rev. Lett.*, **30**, 790 (1973).
283. P. Pincus. *Solid State Comm.*, **11**, 51 (1972); G. Beni, P. Pincus and J. Kanamoti. *Phys. Rev.* **B10**, 1896 (1974).
284. K. Kubo. *J. Phys. Soc. Japan*, **31**, 30 (1971).
285. I. Sadakata and E. Hanamura. *J. Phys. Soc. Japan*, **34**, 882 (1973).
286. W. L. Pollens and S. Choi. *J. Chem. Phys.*, **52**, 3691 (1970).
287. R. A. Bari, D. Adler and R. V. Lange. *Phys. Rev.*, **B2**, 2898 (1970).
288. W. Brinkman and T. M. Rice. *Phys. Rev.*, **B2**, 1324 (1970).
289. Y. Suezaki. *J. Phys. Soc. Japan*, **34**, 89 (1973).
290. A. Brau and J. P. Farges. *Phys. Status Solidi*, **B61**, 257 (1974). See also J. P. Farges, Thesis, University of Nice (1974) (unpublished) for an application of the modified Hubbard model to this system.
291. C. A. Fyfe. in "Molecular Complexes", Vol. I (ed. R. Foster), Paul Elek, London; Crane, Russak and Co., New York (1973), pp. 209–299.
292. T. Z. Huang, R. P. Taylor and Z. G. Soos. *Phys. Rev. Lett.*, **28**, 1054 (1972).

108   Z. G. SOOS AND D. J. KLEIN

293. S. Etemad, A. F. Garito and A. J. Heeger. *Phys. Lett.*, **40A**, 45 (1972).
294. Z. G. Soos and R. C. Hughes. *J. Chem. Phys.*, **46**, 253 (1967).
295. K. Siratori and T. Kondow. *J. Phys. Soc. Japan*, **27**, 301 (1969).
296. L. N. Bulaevskii, A. V. Zvarykina, Y. S. Karimov, R. B. Lyoborskii and I. F. Shchegolev. *Soviet Phys. JETP*, **35**, 384 (1972).
297. R. T. Schumacher and C. P. Slichter. *Phys. Rev.*, **101**, 58 (1956).
298. A. Kawamori and K. Suzuki. *Mol. Phys.*, **8**, 95 (1964).
299. M. J. Henessey, C. D. McElwee and P. M. Richards. *Phys. Rev.*, **B7**, 930 (1973).
300. R. B. Griffiths. *Phys. Rev.*, **135**, A659 (1964).
301. L. R. Walker, R. E. Dietz, K. Andres and S. Darack. *Solid State Comm.*, **11**, 593 (1972); R. Dingle, M. E. Lines and S. L. Holt. *Phys. Rev.*, **187**, 643 (1969).
302. R. E. Dietz, F. S. L. Hsu and W. H. Haermmerle. *Bull. Amer. Phys. Soc.*, **17**, 268 (1972).
303. M. T. Hutchings, G. Shirane, R. J. Birgeneau and S. L. Holt. *Phys. Rev.*, **B5**, 1999 (1972).
304. K. Takeda, S. Matsukawa and T. Haseda. *J. Phys. Soc.*, *Japan*, **30**, 1330 (1971).
305. Z. G. Soos, T. Z. Huang, J. S. Valentine and R. C. Hughes. *Phys. Rev.*, **B8**, 993 (1973).
306. T. Z. Huang and Z. G. Soos. *Phys. Rev.*, **B9**, 4981 (1974).
307. M. Inoue, S. Emori and M. Kubo. *Inorg. Chem.*, **7**, 1427 (1967); S. Emori, M. Inoue and M. Kubo. *Bull. Chem. Soc. Japan*, **45**, 2259 (1972).
308. M. Inoue and M. Kubo. *J. Mag. Res.*, **4**, 175 (1971); S. Emori, M. Inoue, and M. Kubo. *Bull. Chem. Soc. Japan*, **44**, 3299 (1971).
309. C. H. Townes and J. Turkevich. *Phys. Rev.*, **77**, 148 (1950); A. N. Holden, C. Kittel, F. R. Merrit and W. A. Yager. *Phys. Rev.*, **75**, 1614 (1949); **77**, 147 (1950).
310. G. E. Pake. "Paramagnetic Resonance", W. A. Benjamin, New York (1962), Chapter 4.
311. A. J. Heeger and A. F. Garito. *A.I.P. Conf. Proc. 18th Intl. Conf. on Magnetism and Magnetic Materials* (eds C. D. Graham, Jr. and J. J. Rhyne), A.I.P., New York (1972), pp. 1476–92.
312. H. M. McIlwain. *J. Chem. Soc.*, 1704 (1937); NMP decomposed under the conditions used and NEP was prepared instead.
313. B. Bleaney and K. D. Bowers. *Proc. Roy. Soc. Ser. A*, **214**, 451 (1952).
314. J. S. Valentine, A. J. Silverstein and Z. G. Soos. *J. Amer. Chem. Soc.*, **96**, 97 (1974).
315. S. Etemad and E. Ehrenfreund. *A.I.P. Conf. Proc. No. 18, Magnetism and Magnetic Materials.* (eds C. D. Graham, Jr., and J. J. Rhyne), A. I. P., New York (1972), pp. 1499–503.
316. A. Carrington and A. D. McLachlan. "Introduction to Magnetic Resonance", Harper and Row, New York (1967), Chapter 8.
317. R. Lynden-Bell. *Mol. Phys.*, **8**, 71 (1964).
318. Z. G. Soos and H. M. McConnell. *J. Chem. Phys.*, **43**, 3780 (1965).
319. J. M. Brown and M. T. Jones. *J. Chem. Phys.*, **51**, 4687 (1969).
320. G. Nyberg, D. B. Chesnut and B. Crist. *J. Chem. Phys.*, **50**, 341 (1969).
321. A. Kawamori. *J. Chem. Phys.*, **47**, 3091 (1967).
322. F. Devreux and M. Nechtschein. *Phys. Lett.*, **46A**, 49 (1973).
323. C. Kittel and E. Abrahams. *Phys. Rev.*, **90**, 238 (1953).
324. A. Abragam. "The Principles of Nuclear Magnetism", Clarendon, Oxford (1961), Chapter IV and X; P. W. Anderson, *J. Phys. Soc. Japan*, **9**, 316 (1954).
325. W. M. Walsh, Jr., L. W. Rupp, Jr., F. Wudl, D. E. Schafer and G. A. Thomas. *Bull. Amer. Phys. Soc.*, **19**, 296 (1974); Y. Tomkiewicz (personal communication).

326. R. C. Hughes and Z. G. Soos. *J. Chem. Phys.*, **52**, 6302 (1970).
327. Z. G. Soos. *J. Chem. Phys.*, **49**, 2493 (1968); P. J. Strebel and Z. G. Soos. *J. Chem. Phys.*, **50**, 2911 (1969).
328. Z. G. Soos. *J. Chem. Phys.*, **44**, 1729 (1966).
329. G. Reiter. *Phys. Rev.*, **B8**, 5311 (1973).
330. Y. Maréchal and H. M. McConnell. *J. Chem. Phys.*, **43**, 4126 (1965).
331. M.-A. Maréchal and H. M. McConnell. *J. Chem. Phys.*, **43**, 497 (1965).
332. H. F. Schaeffer. "The Electronic Structure of Atoms and Molecules", Addison-Wesley, Reading, Mass. (1972).
333. J. A. Pople and D. L. Beveridge. "Approximate Molecular Orbital Theory", McGraw-Hill, New York (1972).
334. O. Sinanoglu and K. A. Brueckner. "Three Approaches to Electron Correction in Atoms", Yale University Press, New Haven, Conn. (1970).
335. H. T. Jonkman and J. Kommandeur. *Chem. Phys. Lett.*, **15**, 496 (1972); H. T. Jonkman, G. A. Van der Velde and W. C. Nieuwpoort. *Chem. Phys. Lett.*, **25**, 62 (1974).
336. P. S. Bagus, I. P. Batra and E. Clementi. *Chem. Phys. Lett.*, **23**, 305 (1973); E. Clementi, *Proc. Natl Acad. Sci. U.S.*, **69**, 2942 (1972); M. Ratner, T. R. Sabin and E. E. Ball. *Mol. Phys.*, **26**, 1177 (1973); F. Herman, A. R. Williams and K. H. Johnson. *J. Chem. Phys.*, **61**, 3508 (1974).
337. F. A. Matsen and D. J. Klein. *J. Phys. Chem.*, **75**, 1860 (1971).
338. T. Arai. *Phys. Rev.*, **134**, A824 (1964).
339. W. J. Mullin. *Phys. Rev.*, **136**, 1126 (1964).
340. B. C. Carlson and J. M. Keller. *Phys. Rev.*, **105**, 102 (1957).

2 Dielectric Properties of Molecular Complexes in Solution

NORMAN KULEVSKY

Chemistry Department, University of North Dakota, Grand Forks, North Dakota, 58201, U.S.A.

I. INTRODUCTION

The dipole moment (μ) of a solute may be obtained from measurements on dilute solutions of the refractive indices (n), densities (d) and dielectric constants* (ϵ') in a low frequency electric field, while the relaxation time (τ) for the polarization process as well as μ can be calculated from the dielectric constant and loss (ϵ'') at microwave frequencies.[1, 2] These properties are useful in studying molecular complexes and electron–donor-acceptor (EDA) interactions for two reasons.[3] Firstly, the existence of a complex between a solute and solvent caused by weak EDA interaction may be inferred from comparisons of ϵ' and τ of the solute in a donor or acceptor solvent with ϵ' and τ found in a non-interacting solvent or expected on the

---

* Dielectric constant is symbolised by ϵ' in this and other chapters to distinguish it from ϵ which is used to represent the optical molar absorption coefficient (extinction coefficient).

basis of the molecular structure of the solute. Secondly, the measured value of μ for a complex (μ_c) can give information as to the electron distribution in the complex and is often used to discuss the theories of complex formation.

Until a few years ago the dipole moments of molecular complexes were almost exclusively interpreted on the basis of Mulliken's charge-transfer theory.[4] According to this theory the ground-state wavefunction (ψ_N) of the complex may be written as

$$\psi_N = a\psi_0(D, A) + b\psi_1(D^+ - A^-) \tag{1}$$

where $\psi_0(D, A)$ and $\psi_1(D^+ - A^-)$ are the wavefunctions for the no-bond structure and the dative or charge-transfer structure in which an electron has been donated from D to A. The coefficients a and b measure the relative contributions of the two structures and satisfy the normalization condition $(a^2 + b^2 + 2abS) = 1$, in which S is the overlap integral of ψ_0 and ψ_1. The dipole moment of the complex in the ground state (μ_N) is then

$$\mu_N = a^2\mu_0 + b^2\mu_1 + 2ab\mu_{01} = a^2\mu_0 + b^2\mu_1 + abS(\mu_0 + \mu_1) \tag{2}$$

where μ_0 and μ_1 are the dipoles of the no-bond and dative structures and $\mu_{01} = \langle\psi_1|\hat{\mu}|\psi_2\rangle$. Briegleb has taken $\mu_c = \mu_N + \mu_{ind}$, where μ_{ind} is the dipole induced in one molecular by the polar groups of the other molecule.[3] On the other hand, Le Fèvre and his coworkers[5] believe that μ_{ind} should be included in μ_0 and that μ_c and μ_N are identical. Which ever way μ_N is calculated, it is clear that equation (2) can be solved for a, b and $(b^2 + abS)$, (the fractional contribution of the dative state) from μ_N, μ_1, μ_0, and S. These relations have been applied to the three major types of complexes σ^*-acceptor–n-donor, σ^*-acceptor–π-donor and π^*-acceptor–π-donor. However, in the last few years a good deal of doubt has been expressed regarding the contribution of the charge-transfer state to the ground state of the latter two types of complexes.[5–13] It has been suggested that important contributions to both the energy of formation and the dipole moments arise from electrostatic interactions which would normally be included in the no-bond function. This in effect means that μ_0 has been underestimated and the degree of charge transfer in the ground state overestimated. These ideas will be discussed in more detail after the dipole moment data have been presented.

In the period since Briegleb's review[3] the amount of data published on this topic has been considerable. Some of the more recent work concerning the dipole moments of iodine complexes has been briefly reviewed by Hanna and Trotter[14] in connection with the theories of complex formation. In this chapter the data obtained from measurements of both static dielectric constants and dielectric dispersions and their theoretical implications will be discussed for a variety of types of interactions. For the most part, studies of hydrogen-bonded systems or stable complexes of metal halides with organic donors will not be considered.

Since the calculation of μ_c and τ for complexes from dielectric data involves some modification of the procedures used for stable molecules, it will be advantageous to start with a discussion of the procedures used to investigate complexes.

II. METHODS FOR OBTAINING μ_c AND τ FROM DIELECTRIC DATA

A. General Procedures

1. Calculations From Static Dielectric Constants

Standard methods for calculating μ and τ from the dielectric constants of solutions have been given in several books.[1, 2] Since in some studies of molecular interactions these procedures are used with little or no modification, a brief résumé of them is given here. For a solute the Debye equation (3)

$$\mu = \frac{3}{2} \sqrt{\frac{(P_{2\infty} - P^a - P^e)kT}{\pi N_0}} \tag{3}$$

gives the relationship between μ and the total polarization at infinite dilution $(P_{2\infty})$, the atomic, and electronic polarizations (P^a and P^e). Two procedures are commonly used to obtain values for $P_{2\infty}$ and $P^a + P^e$ (the distortion polarization). In the first, the polarization of the solution ($P_{12} = P_{sol}$ for a binary solution) is calculated from the dielectric constant (ϵ'_{sol}) and density (d_{sol}) of the solution by means of equation (4)

$$P_{sol} = \frac{(\epsilon'_{sol} - 1)}{(\epsilon'_{sol} + 2)} \frac{\Sigma(X_i M_i)}{\alpha_{sol}} \tag{4}$$

where X_i and M_i are the mole fraction and mole weight of the ith component respectivity. For a binary solution, P_2 may then be found from equation (5) (P_1 is the polarization of the pure solvent) which is based upon the principle that for solutions, the polarizations of the components are additive.

$$P_2 = (P_{12} - P_1)/X_2 + P_1 \tag{5}$$

A similar pair of equations (6) and (7) can also be used to relate the refractive index of the solution (n_{sol}), molar refraction of the solution (R_{sol}), and molar refractions of the components (R_1 and R_2).

$$R_{sol} = \frac{(n^2_{sol} - 1)}{(n^2_{sol} + 2)} \frac{\Sigma(X_i M_i)}{\alpha_{sol}} \tag{6}$$

$$R_2 = (R_{12} - R_1)/X_2 + R_1 \tag{7}$$

Values of R_2 and P_2 over a range of concentrations are then extrapolated to $X_2 = 0$ in order to obtain $R_{2\infty}$ and $P_{2\infty}$. To calculate μ (in debye) from equation (3) it is common to assume that $R_{2\infty}$ is equal to the distortion polarization, or that P^a is approximately 5 to 10 per cent of $R_{2\infty}$.

More reliable values of $P_{2\infty}$ and $R_{2\infty}$ may be obtained through a second procedure in which advantage is taken of the linear relationships (in dilute solutions) found between the weight fraction of solute (w_2) and ϵ_{12}, n_{12}^2 and the specific volume (v_{12}) or d_{12}:

$$\epsilon_{12}' = \epsilon_1' + \alpha w_2 \tag{8}$$

$$v_{12} = v_1 + \beta w_2 \quad \text{or} \quad d_{12} = d_1 + \beta' w_2 \tag{9}$$

$$n_{12}^2 = n_1^2 + \gamma w_2 \tag{10}$$

The slopes α, β, and γ found in these relations are then used in equations (11) and (12) to find $P_{2\infty}$ and $R_{2\infty}$.

$$P_{2\infty} = \frac{3\alpha v_1 M_2}{(\epsilon_1' + 2)^2} + M_2(v_1 + \beta)\frac{(\epsilon_1' - 1)}{(\epsilon_1' + 2)} \tag{11}$$

$$R_{2\infty} = \frac{3\alpha v_1 M_2}{(n_1^2 + 2)^2} + M_2(v_1 + \beta)\frac{(n_1^2 - 1)}{(n_1^2 + 2)} \tag{12}$$

A variation of this method which avoids the need to measure the density of each solution has been developed by Guggenheim. Following this procedure, $(P_{2\infty} - P^e - P^a)$ may be calculated from equation (13)

$$P_{2\infty} - P^e - P^a = \frac{3M_2}{v_1}\frac{\alpha - \beta'}{(\epsilon_1' + 2)^2} \tag{13}$$

2. Calculations From Dielectric Loss
Evaluation of μ and τ from dielectric loss (ϵ'') values as a function of frequency can be made from the Kirkwood and Fuoss equation (14)

$$\epsilon_m'' = \epsilon'' \cosh \beta \ln f_m/f \tag{14}$$

where f_m is the frequency of maximum loss ϵ_m'' and β is a distribution factor. The value of ϵ_m'' is chosen to give the best straight-line plot of $\cosh^{-1}(\epsilon_m''/\epsilon'')$ against $\ln f$. The gradient of the line is β and the intercept is $\ln f_m$. Equations (15) and (16) can then be used to calculate μ and τ.

$$\tau = \frac{1}{2\pi f_m} \tag{15}$$

$$\mu = \left[\frac{13\,500\,kT\,\epsilon''_m}{\beta(\epsilon'_0 + 2)^2 \pi C N_0} \right]^{\frac{1}{2}} \tag{16}$$

where k, T, ϵ'_0, C and N_0 are the Boltzmann constant, absolute temperature, static dielectric constant, molar concentration of polar species and Avagodro's number respectively.

B. Complexes in Binary Solutions

1. Stable Complexes

In dielectric studies of molecular complexes, binary solutions may be used in two different ways. For complexes stable enough to be prepared as pure samples and that do not dissociate in solvents suitable for dielectric measurements, the equations given in Section IIA may be used without any modifications. Unfortunately, not many molecular complexes have the requisite stability to be studied in this way.

One group of complexes which appear to meet these criteria are the interhalogen complexes of some substituted pyridines.[14] Spectroscopic observations indicate that in carbon tetrachloride the dissociation constants for these species range from 10^{-4} to 10^{-7} mol l^{-1}. Therefore it is possible to calculate $P_{2\infty}$ from ϵ'_{12} using the formal concentrations of the solutions and equations (8), (9) and (11). Assuming that formation of the complex does not have any large effect on the atomic and electronic polarizations, the molar refractions can be calculated from tables of atomic refractions. Unfortunately these methods have also been applied to complexes that form pure solids but which are probably also extensively dissociated in solution. An example of such a case is encountered in the picric acid–naphthalene complex. The dissociation constant of this complex in carbon tetrachloride is reported[15] to be 0·28 mol l^{-1}, however μ_c values for it have been calculated by procedures that ignore the dissociation.[16] Results of studies in ternary solutions show that for some complexes in which one component is polar the value of μ_c for the complex may be smaller than μ of the polar component.[17] If this is the case for the picric acid–naphthalene complex, dissociation will produce the more polar picric acid and neglecting the dissociation would make the value of μ_c calculated from the experimental data higher than its true value.

2. Complexes Between Solute and Solvent

A second and more common use of binary solutions is the study of complexes formed by a solute interacting with a solvent. In these types of solutions, questions arise concerning the concentration of complex present at equilibrium. In many cases the assumption is made that all of the solute is complexed so that the equations in Section II.A can be used without modification.

E

However, for some of these complexes, values of the association constants have been obtained and the equilibrium concentrations of the complexes used to find μ. Examination of the results obtained for the iodine complexes of benzene and p-dioxan using both methods of calculating the concentration of complex reveals that there are no large or consistent deviations in μ_c. For benzene, $\mu_c = 0.70D$ is obtained from dielectric loss measurements when equilibrium concentrations are used,[18] while from the static dielectric constants and formal concentrations $\mu_c = 0.60D$ is found.[19, 20] For dioxan the deviations of the results are slightly larger, i.e. $\mu_c = 1.0D$[18] using equilibrium concentrations and $\mu_c = 0.95D$[21] and $1.3D$[19] using the formal concentrations.* Another question that may arise, concerns ter-molecular or higher complexes that might be formed when one component is present in excess. This problem will be discussed in Section II.C.2.

For some of the complexes studied by these methods, results reported by different groups may have larger differences than would be expected for normal dielectric measurements. Aside from questions arising because of the method used to calculate the concentrations of the complex, there are two other possible reasons for any discrepancies between the results. Firstly, those values of μ_c obtained from dielectric loss measurements employing equation (16) may differ from those obtained from static dielectric measurements because of the atomic polarization term (P^a). When μ is calculated from static dielectric constants alone, a value of P^a obtained by some type of approximation is used, while it is not needed at all with calculations from the dielectric loss. Secondly, some of the reported μ values are based upon measurements of ϵ' at only one concentration which means that any effects of concentration on μ are ignored. From the above it is clear that if trends and differences in μ for different systems are to be used in drawing conclusions regarding the relationship of μ to the molecular structure, donor ability or other parameters of the components, then only those values obtained by similar procedures should be compared.

Many organic compounds with π-electron systems that act as electron acceptors are considered to be nonpolar because of their molecular symmetry. Compounds of this type that commonly appear in studies of molecular complexes are p-benzoquinone (p-BQ), p-chloranil (p-CA), tetracyanoethylene (TCNE), p-dinitrobenzene (p-DNB), 1,3,5-trinitrobenzene (TNB), and hexafluorobenzene. Ideally, the appearance of a dipole moment for any of these substances dissolved in a nonpolar donor solvent could be interpreted as arising from a solute-solvent interaction. Unfortunately, for some of these

---

* The values of dipole moment are quoted, as in the original literature, in debye units (D). These units are related to the *SI* unit, the coulomb metre, (C m) by:

$$D = 3,335\,640 \times 10^{-30}\,C\,m$$

compounds there is a good deal of uncertainty regarding the correct values of P^a to use in the calculations, both in the vapour state and in solution. Thus values of μ_c for these π^*-acceptors in binary mixtures with donor solvents are not as easily obtained or interpreted as was the case for iodine solutions. For example, the values of $(P_{2\infty} - R_{2\infty})$ for p-BQ are 9·6 cc and 8·7 cc in carbon tetrachloride and n-hexane, respectively,[22] while similar large values are found for p-CA,[23] TNB,[24] and p-DNB[24] in benzene and dioxan. In order for the dipole moments to be zero for these molecules, values of P^a which are an extraordinarily large fraction of $R_{2\infty}$ (30 per cent of $R_{2\infty}$) must be assumed, and it has been argued that these molecules do indeed have nonzero dipoles.[25] However, Charney and Becker[22] have shown from measurements of the infrared intensities of p-BQ in benzene that $P^a \simeq (P_{2\infty} - R_{2\infty})$. This problem can be avoided by obtaining μ from measurements of dielectric loss or by examining the interactions in ternary solutions.

Complexes between solute and solvent have also been investigated by methods based upon the deviations from additivity of the refractive indices.[26, 27] For mixtures of two liquids, A and D the additive refractive index of the solutions (n_a) calculated as if there were no intermolecular interactions would be $n_a = [V_A n_A + (1 - V_A) n_D]$, where n_A and n_D are the refractive indices of A and D and V_A is the volume fraction of the solution. For many of the binary systems in which spectroscopic or other methods indicate that molecular interactions occur, it is found that $\Delta n = (n_{sol} - n_a) > 0·004$. Therefore these authors conclude that $\Delta n > 0·004$ is indicative of complex formation. It is also observed that Δn is a function of concentration, having its maximum value at a mole ratio of A:D corresponding to the stoichiometric ratio in the complex. This technique can be used to indicate the formation of a complex and give some idea as to the relative strength of the interaction but it cannot give the dipole moment of the complex. It is also a useful method for studying the interaction between polar components.

Baur and his coworkers have developed a procedure by which deviations from additivity of the total polarization and refraction of a binary solution can be related to the dipole moment and association constant of a complex (AD) formed from two nonpolar liquids (A and D).[28, 29] This method is applicable only if the polarization and refraction of the pure components can be measured. If ΔP and ΔR are the differences between the P and R values obtained for a solution and those values calculated for a solution of the same formal concentration on the basis of the additivity of polarization and refraction, the following equations can be derived.

$$\Delta P = \frac{4\pi N_0 C_{AD}}{3(C_D^0 + C_A^0)} \left[\alpha_{AD} - \alpha_D - \alpha_A + \frac{\mu_{AD}^2}{3kT} \right] \tag{17}$$

$$\Delta R = \frac{4\pi N_0 C_{AD}}{3(C_D^0 + C_A^0)} \left[\alpha_{AD}^e - \alpha_D^e - \alpha_A^e \right] \tag{18}$$

$$\Delta = \Delta P - \Delta R = \frac{4\pi N_0 C_{AD}}{3(C_N^0 + C_D^0)} \left[\alpha_{AD}^a - \alpha_D^a - \alpha_A^a + \frac{\mu_{AD}^2}{3kT} \right] \tag{19}$$

In these equations, C_A^0 and C_D^0 are the formal molar concentrations of A and D, C_{AD} the molar concentration of complex, α_i^e and α_i^a the electronic and atomic polarizabilities of the ith species, and $\alpha_i = \alpha_i^e + \alpha_i^a$. Since formation of a complex can be reflected in either ΔP or ΔR, additivity of either P or R does not prove that a complex is absent. In particular a complex with $\mu_{AD} > 0$ can be present even if $\Delta P = 0$ as long as $\Delta > 0$. Secondly, only if $\alpha_{AD}^a = (\alpha_A^a + \alpha_D^a)$ will ΔP be directly proportional to μ_{AD}^2. However, measurement of ΔP by itself can give a reliable estimate of the dipole moment of the complex if its magnitude is 1D or greater. Introduction of the mole fractions (X_i) and mole-fraction association constant (K_x) into equation (19) leads to equation (20). This equation is valid only at small concentrations of A and ignores the atomic polarizability terms.

$$\Delta = \frac{4\pi N_0 \mu_{AD}^2 K_x}{9kT(K_x + 1)} \left[X_A^0 - (K_x + 1)^{-1} X_A^{0\,2} \right] \tag{20}$$

In principle, equation (20) allows both K_x and μ_{AD} to be simultaneously determined from linear plots of Δ/X_A^0 vs X_A^0. However if K_x and μ_{AD} are small (e.g. hexafluorobenzene with aromatic hydrocarbons) the experimental uncertainty of the data may be too great for this method to be accurate. More importantly however, this equation implies that any of the linear extrapolations of dielectric constants vs mole fraction or weight fraction to infinite dilution (methods given in Section IIA) will be in error because of the presence of an equilibrium process. Only at very low X_A^0 will such curves be truly linear, and measurements that do not take values of K into account must be made at greater dilutions than usual. Baur and his coworkers estimate that there is a 10 per cent error in the linear fit if $K_x = 1$ and $X_A^0 = 0$ to 0·1, which decreases to 1 per cent if the range of X_A^0 is 0 to 0·01. The high values of μ_c that were initially reported for hexafluorobenzene–benzene by other workers[30, 31] are presumably due to this factor.

C. Complexes in Ternary Solutions

1. Binary Complexes
A solution prepared from two interacting substances A and D and an inert solvent (S) must be treated as a four component system containing S, A, D, and AD. If the equilibrium concentrations of each species (X_i) can be com-

puted from a known value of the association constant, equation (4) can be used to evaluate P_{sol} while P_{AD} may be obtained from P_{sol} using equation (21). Equation (21) is an extension of equation (5) to a four component system.[32, 33] The values of P_A and P_D

$$P_{AD} = \frac{1}{X_{AD}} [P_{sol} - X_S P_S - X_D P_D - X_A P_A] \tag{21}$$

used in these calculations are evaluated for binary solutions of A and D in the solvent, at the same concentration they have in ternary solutions. As some of the $\pi-\pi^*$ complexes investigated have very little systematic variation of P_{AD} with concentration, the average values of P_{AD} rather than P_{AD} have been used in equation (3).[34] On the other hand, iodine–amine complexes have larger values of P_{AD} and linear extrapolations of P_{AD} can be used to obtain $P_{AD\infty}$.[33] Since for the most part, refractive indices of these solutions have not been measured, in order to calculate μ_c from equation (3), it is necessary to approximate the distortion polarization for the complexes as the sum of the polarizations for the free nonpolar components or as the sum of the atomic refractions. Toyoda and Person have tested the validity of this approximation for the iodine complexes with benzene and pyridine.[33] Calculations based upon spectroscopic data indicate that in both cases complex formation causes only negligible changes in the distortion polarization when compared to the orientation polarization of the complex.

If the concentrations of A and D are chosen so as to ensure that one of the components is completely complexed, then it is no longer necessary to have good values of the association constants in order to calculate X_i. Since the concentrations are $X_{AD} = X_A^0, X_D = X_D^0 - X_A$, and $X_A = 0$, when the ratio of X_D^0 to X_A^0 is very large the calculation is reduced to the problem of finding P_{AD} in a mixed solvent (S + D).[35] In these cases linear relations similar to equations (8) and (9) are found if ϵ' and d are plotted vs w_{AD} for solutions with a fixed concentration of D. The values of ϵ' and d for the mixed solvent (ϵ'_{SD} and d_{SD}) replace ϵ'_S and d_S in these equations and the slopes are α and β. P_{AD} can then be obtained from equation (22)

$$P_{AD} = M_{AD} p_{SD} \left[1 + \frac{3\alpha}{(\epsilon'_{SD} + 2)(\epsilon'_{SD} - 1)} - \frac{\beta'}{d_{SD}} \right]$$

$$- M_{AD}(p_{SD} - p_S) + M_{AD}(p_D - p_S) \left[w_D + \frac{M_D}{M_{AD}}(1 - w_D) \right] \tag{22}$$

where p_{SD}, p_S, and p_0 are the specific polarization of the mixed solvent, pure solvent, and pure donor, respectively, and M_{AD} and M_D are the molecular weights of AD and D. The first term in the equation corresponds to the ordinary Halverstadt–Kumler equation (11), while the other terms are corrections for the decrease in the free donor concentration caused by complex formation. In many cases these latter terms can be ignored. P_{AD} is then used in conjunction with the previously mentioned approximations regarding the distortion polarization to calculate μ from equation (3). However, the necessity for approximations regarding the distortion polarization can be avoided if the refractive indices of these solutions are also measured. Chan and Liao (36) have made such measurements, and found that n^2_{sol} is a linear function of w_{AD} for solutions with high ratios of X^0_D to X^0_A. In this case it is possible to calculate μ by the Guggenheim procedure involving equation (13).

The accuracy of μ_{DA} calculated by these last two methods depends on the validity of the assumption that A is entirely complexed. Kobinata and Nagakura were able to use the electronic absorption spectra and association constants to prove its validity for the strong iodine–amine complexes.[35] On the other hand, for the weaker tetracyanoethylene (TCNE)–aromatic hydrocarbon complexes in carbon tetrachloride, the observed linearity of plots of ϵ'_{sol} and d^2_{sol} vs w_{AD} has been taken as proof of this assumption by Chan and Liao.[36] However, if the value of K given by Briegleb and his coworkers[34] for this complex is used to calculate the equilibrium concentrations in the solutions used by Chan and Liao, it is found that only about 90 per cent of the TCNE is complexed. The data reported for the iodine–triethylamine interaction by Funatsu and Toyoda also indicate that ϵ'_{sol} and d_{sol} vs w_{AD} can be linear even if there are other species present besides AD and excess D.[37] They have shown that this interaction leads to the formation of other products besides molecular complexes. However, it is also observed that when the behaviour of ϵ and d with time is followed, plots of ϵ'_{sol} and d_{sol} vs w_{AD} are still linear, even though the secondary reactions are only partially completed. This suggests that the linearity of these plots is not a very sensitive criterion for determining the species actually present. Comparisons of the results obtained using the approximation that all A is complexed and using an experimental association constant are possible in only a few cases and do not give a clear cut answer as to the validity of the assumption. Values of μ for TCNE–durene and TCNE–hexamethylbenzene (HMB) complexes are 1·26D and 1·35D if K values are used,[34] while they are 1·32D and 1·65D if the assumption is made that all TCNE in solution is complexed.[36] The agreement is good for the durene complex, but poor for the HMB complex.

A procedure similar to those discussed above, has been developed by a Russian group.[38, 39] The advantage of this method is that by extending the measurements to a concentration region where plots of ϵ'_{sol} and d_{sol} vs concentration are no longer linear, the value of K may also be determined. In this titration technique, values of ϵ'_{sol} and d_{sol} are measured after small increments of one component (say D) are added to a solution containing the other component (A). At first all of the added D is complexed and the plots of ϵ'_{sol} and d_{sol} vs concentration are linear. However as the concentration of D increases, a significant fraction of D remains uncomplexed and the linear plot begins to curve. The curves become linear again after all of the A present has been complexed. P_{AD} and μ_{AD} can be obtained from the initial linear portions of the curves using equations (8), (9), and (11), while the intersection of the two linear portions of the curves indicates the stoichiometry of the complex. The value of P_{sol} can be calculated from the experimental values of ϵ' and d at the concentration equivalent to the stoichiometric point. It is then possible to calculate the weight fractions of A and D in the complex (Δw_A and Δw_D) from equations (23) and (23a).

$$P_{sol} = P_A(w_A - \Delta w_A) + P_D(w_D - \Delta w_D) + P_{AD}(\Delta w_A + \Delta w_D) + P_S w_S \qquad (23)$$

$$\frac{\Delta w_A}{\Delta w_D} = \frac{M_A}{M_D} \qquad (23a)$$

Once values of Δw_A and Δw_D have been found it is then a simple matter to obtain K_x.

This last procedure is similar to but easier to apply than the method developed by Smith and Few in which values of μ_{AD} and K are also simultaneously found.[40] The calculation is based upon equation (24)

$$\Delta P/(P_A^* - P_A) = 1 + M_A/(K_C w_D d) \qquad (24)$$

where P_A^* is the apparent polarization of A in mixture of S and D calculated on the basis that no complex is formed ($w_A P_A^* = P_{sol} - w_D P_D - w_S P_S$) and $\Delta P = (P_{AD} - P_A - P_D)$. A plot of $(P_A^* - P_A)^{-1}$ vs $(w_D d)^{-1}$ is a straight line whose slope is $M_A/K\Delta P$ and intercept ΔP^{-1}. Because the limited solubility of many acceptors precludes the easy determination of $P_{A\infty}$, which should be used in equation (24), this method is not often used in studying molecular associations that do not involve hydrogen bonding. For hydrogen-bonding studies, several variations on this technique have been reported.[41, 42]

2. Ternary Complexes

All of the methods described in the previous section are based upon the assumption that only $1:1$ complexes exist in solution. However, many recent studies[43] of the interactions between π^*-acceptors and π-donors using ultraviolet spectra and nmr techniques have led to the proposition that termolecular complexes such as AD_2 are also present in solutions with high ratios of X_D^0 to X_A^0. For solutions containing AD_2, the total polarization may be obtained from ϵ'_{sol} and equation (4) if the equilibrium concentrations of all species (A, D, AD, AD_2) can be calculated from the association constants.[44] Expressing the polarization of the solution as the sum of the

122   N. KULEVSKY

polarizations of its components, equation (25) may then be used to calculate P_{AD} and P_{AD_2}.

$$P_{sol} - (P_A + P_D X_D + P_S X_S) = P_{AD} X_{AD} + P_{AD_2} X_{AD_2} \qquad (25)$$

This equation which contains two unknowns can only be solved by using sets of simultaneous equations. Each of the solutions for which polarization data are available can be used to obtain these simultaneous equations, any pair of solutions giving a set of simultaneous equations. Unfortunately there is usually only a limited range of experimentally attainable concentrations and ϵ'_{sol} values available for solutions of molecular complexes. Therefore the equations for some of the pairs are not completely independent of each other and the small differences between large numbers which occur in the calculations cause some of the results to be negative. Eliminating all sets of equations that give negative polarizations, a large set of values can still be obtained if there are as many as ten measured ϵ'_{sol} values. The average values of P_{AD} and P_{AD_2} obtained in this way can then be used in equation (3) with the usual approximations regarding the distortion polarization to obtain μ_{AD} and μ_{AD_2}. This method has been used with success on only one complex, TCNE–HMB, however it does have implications as to the reported values of other π^*–π complexes. Complexes which are less associated than TCNE–HMB may still have an appreciable amount of AD_2 present in solution. Thus, μ_{DA} values calculated without taking the presence of AD_2 into consideration would be lower than they should be, since μ_{AD_2} has been found to be lower than μ_{AD}. In studies where π^*-acceptors are dissolved in π-donors such as benzene or mesitylene the formation of higher complexes is also likely to occur even if the equilibrium constants are small.

III. HALOGEN COMPLEXES

A. Experimental Results

Of the halogens, iodine is the one most frequently used in studies of the dielectric properties of complexes. Because it is a homonuclear diatomic, $\mu = 0$ in an inert solvent and values of $\mu_c > 0$ in a second solvent should indicate a solute–solvent molecular interaction. Since the publication of Briegleb's tabulation of the dipole moments of I_2 in donor solvents[3] several studies have appeared in which the measurements in some of these same solvents are repeated as well as in several other solvents. These data are presented for both nonpolar and polar solvents in Table 1.

That polar complexes are indeed formed between iodine and the nonpolar solutes: benzene,[18–20] mesitylene[18, 20, 45] and p-dioxan,[18–21] is indicated by the dipole moments listed in Table 1. Since the results obtained from both

TABLE I. Dipole moments of the complex (μ_c) and of the Donor (μ_D) and relaxation times (τ) of iodine in donor solvents.

	μ_c/D	μ_D/D	τ/ps	Ref.
Diethyl ether[a]	1·93	1·37		20
p-Dioxan	0·95	0		21
p-Dioxan[b]	1·3	0		19
p-Dioxan[c]	1·0	0	5·8 ± 1	18
p-Dioxan		0	3·2	45
Benzene[a]	0·38	0		20
Benzene[b]	0·6	0		19
Benzene[c]	0·70	0	3·0	18
Benzene			3·4	45
Mesitylene[c]	0·98	0	6·6 ± 1	18
Mesitylene	1·1			45
Mesitylene[a]	1·38			20
p-Xylene[a]	0·79	0		20
p-Xylene[b]	0·9	0		19
Toluene[a]	0·78	0·35		71
o-Xylene[a]	0·93	0·54		71
m-Xylene[a]	1·04	0·35		71
Tetralin[a]	1·14	0·63		71

[a] Calculation based upon assumption that all iodine is complexed; using data from only one solution.
[b] Calculation based upon assumption that all iodine is complexed.
[c] Calculation based upon the equilibrium concentration of complex.

the dielectric loss and static dielectric measurements are in approximate agreement, it may be concluded that large changes in P^a cannot account for the apparent dipole moments found by the latter method. In turn, this conclusion may generate added confidence for the belief that values of $\mu_c > 0$ obtained only from measurements of the static dielectric constant are not caused by changes in P^a.

Relaxation times for these complexes, where available, are also given in Table 1. These times, which are shorter than would be expected for rigid polar molecules of the same size as the complex, are interpreted by consideration of a relaxation mechanism in which the dissociation of the complex to its nonpolar constituents is coupled to an orientational relaxation of the polar complex.[17, 45, 46] The kinetics of such a mechanism in an electric field lead to an observed relaxation time, $\tau = 1/(k_1 + 2k_2)$, where k_1 is the rate constant for dissociation of the complex and k_2 is the rate of rotation of the complex.[46] This equation implies that k_1 is of the same order of magnitude as k_2, since only then will the observed relaxation time differ from that expected for a stable polar molecule, i.e. $\tau = \frac{1}{2} k_2^{-1}$

Dielectric studies of the interaction of lone-pair donors with halogens have most often been carried out in ternary solutions. Among the donors for

E*

TABLE II. Dipole moments of iodine complexes in ternary solutions.

Donor	Solvent	$T/°C$	μ_c/D	μ_D/D	Ref.
Trimethyamine	p-Dioxan	25	10·05	0·63	33
Trimethyamine	Toluene	−50	6·5		50
Triethylamine	p-Dioxan	25	12·44	0·80	33
Triethylamine	p-Dioxan	20	11·3		47
Triethylamine	n-Heptane	19·5	5·6		49
Triethylamine	Cyclohexane	25	5·6		49
Triethylamine	Cyclohexane	25	5·5		37
Triethylamine	Benzene	25	6·7		37
Triethylamine	Benzene	25	10·5		48
Triethylamine	Toluene	−40	6·9		50
Triethylamine	Toluene	23	8·9		50
Isopropylamine	Benzene	20	6·2	1·3	35
Isopropylamine	p-Dioxan	20	7·2		35
Ethylamine	Benzene	20	6·2–6·5	1·4	35
Diethylamine	Benzene	20	6·2	1·1	35
Diethylamine	p-Dioxan	20	7·0		35
Ammonia	p-Dioxan	20	6·1–6·7	1·47	35
Tripentylamine	Benzene	25	8·50	1·3	48
Trioctylamine	Benzene	25	8·20	2·2	48
Butylamine	Benzene	25	6·07	1·3	48
Tribenzylamine	Benzene	25	2·80	0·9	48
Pyridine	n-Heptane	25	4·9	2·22	33
Pyridine	Cyclohexane		4·5		32
Pyridine	Cyclohexane	31·5	4·44		51
Pyridine	Benzene	25	5·70		48
Pyridine	Carbon tetrachloride	25	6·24		52
Quinoline	Benzene		4·45	2·3	21
Quinoline	Benzene	25	5·70		48
2-Picoline	Cyclohexane	31·5	2·83	1·92	51
2-Picoline	Carbon tetrachloride	25	6·26		52
4-Picoline	Carbon tetrachloride	25	6·96	2·6	52
2,6-Lutidine	Carbon tetrachloride	25	5·95	1·9	52
2,4,6-Collidine	Carbon tetrachloride	25	5·98	2·0	52
2,2′-Dipyridyl	Cyclohexane	31·5	2·15	0·9	51
Triphenylamine	Carbon tetrachloride	30	3·13	0·5	58
Triphenylarsine	Carbon tetrachloride	30	8·58	1·1	58
p-Dithian	Carbon tetrachloride	30	2·15	0·0	57
p-Oxathian	Carbon tetrachloride	30	3·20	0·42	58
Dimethyl sulphide	Carbon tetrachloride	30	3·62	1·4	57
Diethyl sulphide	n-Octane	25	4·62	1·63	55
Dipropyl sulphide	n-Octane	25	4·90	1·72	55
Dibutyl sulphide	n-Octane	25	5·00	1·64	55
Dipentyl sulphide	n-Octane	25	5·37	1·67	55
Dioctyl sulphide	n-Octane	25	5·07	—	55
Dinonyl sulphide	n-Octane	25	5·23	—	55
Didecyl sulphide	n-Octane	25	4·78	—	55

TABLE II.—*continued*

Donor	Solvent	$T/°C$	μ_c/D	μ_D/D	Ref.
Dihexadecyl sulphide	n-Octane	25	4·67	—	55
Diisobutyl sulphide	n-Octane	25	4·86	—	55
Diisopentyl sulphide	n-Octane	25	5·06	—	55
Butylisopentyl sulphide	n-Octane	25	4·84	1·68	55
Diisoheptyl sulphide	n-Octane	25	4·90	—	55
Diisooctyl sulphide	n-Octane	25	4·72	—	55
Diallyl sulphide	n-Octane	25	3·64	1·46	55
Thiophen	Benzene	25	5·49	1·73	55
2,5-Dimethylthiophane	Benzene	25	5·76	1·70	55
Dicyclohexyl sulphide	n-Octane	25	3·78	—	55
Dibenzyl sulphide	n-Octane	25	3·78	—	55
Diphenyl sulphide	n-Octane	25	1·99	1·56	55
Phenylmethyl sulphide	n-Octane	25	1·88	—	55
Phenylethyl sulphide	n-Octane	25	2·63	1·48	55
Phenyl-α-naphthyl sulphide	n-Octane	25	1·75	—	55
Ethyl isothiocyanata	Carbon tetrachloride	30	5·09	3·67	57
Ethylene trithio-carbonate	Carbon tetrachloride	30	5·86	4·47	57
Tetramethylthiourea	Carbon tetrachloride	30	7·52	4·70	57
Ethyl thiocarbonate	Carbon tetrachloride	30	7·12	3·24	57
Tetramethylurea	Carbon tetrachloride	30	5·20	3·66	57
Dimethylformamide	Carbon tetrachloride	30	5·56	3·84	57

which μ_c values have been obtained are amines, ethers, sulphides, carbonyls, and thiocarbonyls. These values as well as the dipole moments of the free donors (μ_D) are given in Table 2. For the most part, discrepancies between values of μ_c obtained for the same compound by different research groups may be accounted for by some of the factors discussed in section II or by solvent effects. However, there are some rather large differences between reported values for the trialkylamines that cannot be explained by these factors. Moments between 10D and 12·4D were reported for Et_3N and Me_3N in early work on these compounds,[21, 47, 48] while in more recent publications values between 5·2D and 6·9D have been given.[49, 50] These higher values are now believed to have been caused by the formation of ionic species in secondary reactions that occur at the concentrations usually used to measure ϵ' and d. Evidence that this is the correct explanation has been obtained from several sources. Firstly, in the ultraviolet spectra of iodine–amine solutions in dioxan, at concentrations similar to those used in dielectric measurements, the characteristic ultraviolet charge–transfer band observed in dilute solutions of the complex in hexane is absent, while bands which can be attributed to the triodide ion are present.[33] Secondly, Boule isolated and identified Et_3N–HI from concentrated benzene and dioxan solutions of the complexes.[49] Furthermore, using very dilute solutions in order to avoid the formation of ionic compounds he was able to obtain values of μ_c ranging from 5·0D to 5·9D in n-heptane and cyclohexane solutions. The work of Funatsu and Toyoda in which the values of ϵ' and d were obtained over a period of time following the preparation of the solutions, also shows that there are secondary reactions leading to more polar species.[37] These authors found that as the elapsed time between mixing the components and taking the measurements increased, the calculated dipole moments also increased. Extrapolating back to the time of mixing, dipole moments between 6·5D and 6·9D were obtained. Finally, when dielectric constants were measured at temperatures low enough to decrease the formation of ionic species, low values were found for the dipole moments (Et_3N-I_2, $\mu_c = 6\cdot9D$, $T = -50°C$; Me_3N-I_2 $\mu_c = 6\cdot4D$, $T = -40°C$.[50]

Another example of an iodine complex for which large discrepancies in dipole moments occur is the 2-picoline complex ($\mu_c = 2\cdot83D$[51] and $\mu_c = 6\cdot26D$.[52] There is no reason to believe that the larger value of μ is due to ion formation. In fact, because it is smaller than the moment reported by the same authors for the iodine pyridine complex[51] the lower value may be the doubtful one. In many systems it is observed that the difference between the moment of the complex and the moment of the corresponding free donor increases as the stabilities of the complexes increase. This is particularly evident when the comparisons are between ethers and sulphides,[53–56] between carbonyls and thiocarbonyls,[55] and between amines and arsines.[56] Since the equilibrium constants obtained for the 2-picoline complex indicate that it is more stable than the pyridine complex,[57, 58] it would also be expected that the differences between μ_c and μ_D should be greater for 2-picoline than for pyridine. This is true when the higher value of μ_c is considered [($\mu_c - \mu_D) = 3\cdot2D$ for pyridine and 3·5D for 2-picoline][52] but not for the lower value [($\mu_c - \mu_D) = 2\cdot2D$ for pyridine and 0·9D for 2-picoline].[51]

B. Dipole Moments and Theories of Complex Formation

1. n-Donors

One of the major reasons for determining μ_c has often been the desire to evaluate the importance of the charge-transfer contribution to the ground state of various types of donor-acceptor complex. For most of the complexes in Tables 1 and 2, a, b and ($b^2 + abS$) have been calculated from equation (2) and the normalization condition. However, since as pointed out by Trotter and Hanna, dipole moments are secondary data,[14] values of a, b, and ($b^2 + abS$) calculated from them have large uncertainties. Another problem associated with the evaluation of these quantities concerns the values of μ_0, μ_1, and S used in the computations. From some of the values reported[55, 56] it is clear that the percentage contribution of the charge-transfer structure $[100 \ (b^2 + abS)]$ is not very sensitive to the values chosen for S, even though a and b may be affected. This is not unexpected since, in those calculations where μ_0 is taken as zero, equation (2) reduces to $(b^2 + abS) = \mu_c/\mu_1$. However, even when μ_0 is taken into consideration, varying S from 0·1 to 0·3 has almost no effect on $(b^2 + abS)$ for a series of complexes with ethers, thioethers, carbonyls, thiocarbonyls, arsines, and amines.

The values of μ_1 used in these calculations are calculated by assuming that the positive charge is on the donor atom and the negative charge is at the centre of the iodine molecule. A knowledge of the geometries, the iodine bond-distances, and the distance from the donor atom to the nearest iodine atom in the complexes enables values of μ_1 to be calculated. Using this method the following values are obtained: amine complexes $\mu_1 = 17·7D$;[55,56] arsine complexes $\mu_1 = 19·8D$,[56] oxygen complexes $\mu_1 = 18·9D$,[55] and sulphur complexes $\mu_1 = 20·4D$.[55, 56] Ratajczak and his coworkers have suggested that for the amine–iodine complexes $\mu_1 = 20·9D$ be used rather than $17·7D$.[59] They believe that this higher value is nearer the correct value since it gives results (which will be discussed later) that are in good agreement with spectroscopic data. Also, the higher value, in agreement with a suggestion given by Mulliken and Person, indicates that the positive charge is not on the nitrogen atom and that the negative charge is not centred at the middle of the iodine bond.[4] Obviously a higher value of μ_1 results in a smaller calculated value for the contribution of the charge-transfer structure. This is true whether the calculation ignores μ_0 or not, although a sample calculation for the Et_2NH complex indicates that when μ_0 is considered in equation (2), the value of $(b^2 + abS)$ is not as sensitive to a change in μ_1.

The major problem regarding these calculations arises from the methods used to treat μ_0. Until recently, the value of μ_0 was considered small enough to be ignored, even if the free donor itself had a dipole moment. However, recent calculations[52, 59] concerning the complexes of alkylamines and pyridines indicate that neglecting μ_0 can have a significant effect on $(b^2 + abS)$. In these papers μ_0 values have been calculated as the sum of μ_D and μ_{ind}, where μ_{ind} is the moment induced in iodine by the polar donor molecule. The values of μ_{ind} were estimated using a method originally developed by Frank in order to determine the effect of a nonpolar solvent on the measured moment of a polar solute.[60] This induced moment is dependent on the values of μ_D and is found to vary from 0·35D for Et_3N[59] ($\mu_D = 0·80D$) to 0·80D[59] and 0·95D[52] for pyridine ($\mu_D = 2·20D$). The corresponding values of μ_0 are 1·15D to 3·00D and 3·15D respectively. Introducing these values of μ_0 into equation (2) while keeping the same value of μ_1 lowers $(b^2 + abS)$ by approximately a third of the value obtained when μ_0 is ignored, e.g., for Et_2NH, $(b^2 + abS)$ decreases from 0·34 to 0·24 and for pyridine, $(b^2 + abS)$ decreases from 0·25 to 0·14. Some authors have carried out these calculations ignoring

μ_{ind} and taking $\mu_0 = \mu_D$.[55, 56] This may not be a bad approximation when the donors do not have large moments themselves, since μ_{ind} increases as μ_D increases. However, the values of $(b^2 + abS)$ for the complexes of compounds such as tetramethylthiourea ($\mu_D = 4\cdot70D$, $\mu_c = 7\cdot52D$) and tetramethylurea ($\mu_D = 3\cdot66D$ $\mu_c = 5\cdot20D$) are likely to be overestimated $[(b^2 + abS) = 0\cdot18$ and $0\cdot09$ respectively].

Keeping in mind the deficiencies in these calculations, several conclusions can still be drawn from the values of $(b^2 + abS)$ that have been obtained. Comparisons of those results computed on the same basis, indicate that as the strength of the interaction increases the contribution of the charge–transfer structure also increases. As the donor atom is changed from oxygen to sulphur or nitrogen to arsenic it is found that the values of $(b^2 + abS)$ increase.[55, 56] It is also clear that even in the most stable of the complexes, charge–transfer contributions although real are not the dominant factors.

Further attempts to prove that the charge-transfer structure contributes to the properties of these complexes have been based upon several empirical correlations of the values of $\Delta\mu = (\mu_c - \mu_D)$ or a and b calculated from μ_c with properties of the free donors or the complexes. Two of these involve the values of $\Delta H^\ominus$ (enthalpy of association) and $\Delta\mu$. In the first, a linear relationship between $\Delta H^\ominus$ and $(\Delta\mu^{\frac{1}{2}})$ was observed with a small number of amines,[61] while in the second, a linear correlation between $\Delta H^\ominus$ and $\Delta\mu/r$ was found for a large number of compounds (r is the distance between the donor atom and the nearest iodine atom).[48, 62, 63, 64] Among the compounds used in this second correlation are sulphides, ethers, and the same amines used in the first correlation as well as other amines. It is interesting to note that the relationship involving $(\Delta\mu)^{\frac{1}{2}}$ can also be extended to include the ethers and sulphides if $(\Delta\mu/r)^{\frac{1}{2}}$ is used in the correlation. That both $\Delta\mu/r$ and $(\Delta\mu/r)^{\frac{1}{2}}$ are linear functions of $\Delta H^\ominus$ indicates that the experimental data, although extensive, do not cover a large enough range of $\Delta H^\ominus$ or $\Delta\mu$ values to distinguish between the two. Another factor to be considered is that the correlations do not appear to be sensitive to large errors in μ_c. For example in the correlation with $\Delta\mu/r$, the high values of μ_c proposed for the iodine complexes of the trialkylamines ($10\cdot5D$ for Et_3N, $8\cdot5D$ for $(C_5H_{11})_3N$ and $8\cdot2D$ for $(C_8H_{15})_3N$) are probably in error because of secondary reactions. Also if the values of $\Delta\mu$ are corrected for μ_{ind}, in order to account for electrostatic interactions, both of the linear relations still have good correlation coefficients. Gur'yanova and his coworkers have extended the linear relation with $\Delta\mu/r$ to include complexes other than those with iodine.[64] These include ether, thioether and amine complexes with metal halides such as $SnCl_4$, $TiCl_4$, $SnBr_4$, $TiBr_4$, and BF_3.

Attempts to justify both of these relations have been based on the charge–transfer theory. In the treatment given by Ratajczka and Orville-Thomas[61] it is assumed that for weak complexes $a \simeq 1 > bS$. With this assumption the value of $\Delta H^{\ominus}$ given by application of the variation method to equation (1) is:

$$-\Delta H^{\ominus} = \frac{aW_0 + bH_{01}}{a + bS} = W_0 + bH_{01} \tag{26}$$

where $W_0 = \langle \psi_0 | \mathsf{H} | \psi_0 \rangle$ and $H_{01} = \langle \psi_0 | \mathsf{H} | \psi_1 \rangle$. Under these conditions equation (2) reduces to $b^2 = \Delta\mu/\mu_1$, and eliminating b results in a linear relation of ΔH^0 with $(\Delta\mu)^{\frac{1}{2}}$ or $(\Delta\mu/r)^{\frac{1}{2}}$.

$$-\Delta H^{\ominus} = H_{01}(\Delta\mu/\mu_1)^{\frac{1}{2}} + W_0 \tag{27}$$

For the series of complexes with a constant donor atom it is easy to assume that H_{01} and W_0 should each be constant, but it does not seem reasonable to assume that they would remain constant for the whole series of amines, ethers and sulphides. This latter assumption is necessary in order to have all the different groups fall on the same line. On the other hand, Gur'yanova and his coworkers use relationships involving the excited state ($\psi_E = a^*\psi_0 - b^*\psi_1$) and the charge-transfer excitation energy (hv) to derive the equation, $\Delta H^{\ominus} = \Delta\mu/Arhv$, where A is a function of a, b, a^*, b^*, and S.[48, 62, 63, 64] The observed linearity of $\Delta H^{\ominus}$ vs $\Delta\mu/r$ was taken as proof that hv/A was constant for the complete series of complexes with both iodine and metal halides. Since neither of these relations is exact and the dipole moment data have a limited range these equations can only give a poor indication of the relevance of the charge-transfer theory.

Using second-order perturbation theory another relationship that can be derived from equation (1) is

$$b/a = \frac{-H_{01} - SW_0}{I_D - EA_A - C} \tag{28}$$

where I_D is the vertical ionization potential of the donor, EA_A is the vertical electron-affinity of the acceptor, and C is the coulomb energy term of the charge-transfer structure.[59] If for a series of similar complexes it is assumed that H_{01}, S, and C are constants then a/b can be written as a linear function of the ionization potential of the donors.

$$a/b = -\left(\frac{1}{H_{01} - SW_0}\right)I_D + \frac{EA_A + C}{H_{01} - SW_0} \tag{29}$$

The equation was tested with a group of amine complexes for which values of a/b were calculated from equation (2) using $\mu_0 > 0$. Straight lines were obtained for the plots. However, the gradients of the lines were sensitive to the values of μ_1 used to obtain a/b. When $\mu_1 = 20.93$D, the slope obtained for the curve gives $(H_{01} - SW_0) = 1.8$ eV.* This value is in good agreement

* $1\,\mathrm{eV} \approx 1.6021 \times 10^{-19}\,\mathrm{J}$.

with values suggested by Mulliken and Person on the basis of spectroscopic data.[4] If $C = 3.3$ eV, then from the intercept of this line the electron affinity of iodine is 0·85 eV. This agrees very well with the value suggested by Briegleb and Czekalla (0·81 eV)[65] but not with the value given by Person (1·7 eV).[66] Since μ_{ind} has already been taken into account in the calculation of a/b, it is not likely that the parallelism between the ionization potential and polarizability of donor can be responsible for this correlation.

Another observation that can be rationalized by considering the contributions of the charge-transfer structure to the ground-state complexes is the effect of the dielectric constant of the solvent on the dipole moment.[35] It was found that the moments of some amine complexes in ternary solutions changed as the concentrations of the amine, and consequently the dielectric constant of the media (ϵ'_{SD}) varied. This change in moment was interpreted as arising because the higher dielectric constant of the solvent increased the stability of the charge-transfer structure which in turn increased its contribution to the ground state and the moment of the complex. The differences observed were as much as 1D, higher than what would have been expected if μ_c were due purely to an electrostatically induced dipole.

2. π-Donors

Since neither of the components in the benzene–iodine complex possess a dipole moment, it has been customary to assume that μ_0 is negligible and that the charge-transfer structure is entirely responsible for the observed dipole moment (0·58D–0·70D in benzene,[19–20] 1·8D in cyclohexane).[21] Using this assumption, the fractional contribution of the dative structure to the benzene complex has been found[3] to be 0·03 or 0·08. Attempts to obtain further confirmation that charge-transfer causes the dipole moment have been based on molecular-orbital calculations of the dipole moments. Fukui and his coworkers using the LCAO approximation for the components and a semi-empirical perturbation method that included charge-transfer found a dipole moment of 1·31D for the complex.[67] Dobrescu using a second-order perturbation treatment and starting from SCF MO's for the components found that including charge-transfer gives a moment of 0·61D.[68, 69] Considering the uncertainties of the experimental values of μ_c and of the MO calculations the agreement between the two is reasonable. However, since both of these calculations ignore any contributions from electrostatic induction terms, they do not show that it is necessary to introduce the concept of charge transfer in order to explain the existence of the moment. Dobrescu and Sahini have extended these calculations to predict the moments for the iodine complexes with pyrrole, thiophen and furan, although there are no experimental values available for comparison.[70]

Charge-transfer theory has also been employed by Gerbier to discuss the values of μ_c found for the iodine complexes of other nonpolar and polar aromatic hydrocarbons.[20, 71] For complexes of the nonpolar hydrocarbons it was assumed that μ_c was equal to the dipole moment due to charge transfer (μ_{CT}), while for polar donors $\mu_c = \mu_0 + \mu_{CT}$. According to the charge-transfer theory the two components of μ_c should be perpendicular to each other. If it is assumed that each of the isomeric xylenes would have the same μ_{CT}, the value of $\mu_{CT} = \mu_c$ for the p-xylene complex can be used to calculate μ_c for the o- and m-xylene complexes. As the values of μ_c calculated in this way are in good agreement with the experimental values, the assumption appears to be valid. Since o- and m-xylene are polar while p-xylene is not, the apparent consistency of μ_{CT} indicates that any moments induced by the dipole or quadrupole moments of the donor cannot be responsible for the appearance of μ_{CT}.

Hanna and his co-workers[11,12] and Mantione[13] have demonstrated that μ_0 is significant for the iodine–benzene complex and may indeed be large enough to account for a considerable portion of the observed dipole moment. A value of $\mu_0 > 0$ is thought to arise because the electrostatic field due to the quadrupole moment of benzene can induce a dipole moment in an iodine molecule placed perpendicular to the ring on the hexagonal axis. Both Hanna[11] and Manitone[13] assume that the strength of the electrostatic field can be computed for each CH group of benzene as the sum of the fields due to contributions from the σ-charges in the CH bonds and π-electrons. They differ, however, in the procedures used to obtain the value of the latter term. Hanna's calculations is based on a quadrupole moment originating from an electron in a Slater $2p$ orbital, while Mantione has shown that such a moment will be cancelled by the combined quadrupole moments of three identical sp^2 orbitals. However, since the $p\sigma$- and $p\pi$-orbitals are not exactly identical, the π-electrons do contribute to the field strength, although with less of an effect than claimed by Hanna. In spite of this defect, Hanna's procedure may result in a more accurate measure of the field strength, since the quadrupole moment of benzene calculated from it agrees with the moment obtained from the second virial coefficients. Once the field strength as a function of distance along the hexagonal axis was obtained, two methods were used to find μ_{ind}. In the first μ_{ind} was calculated from $\mu_{ind} = \alpha_\parallel E$, where $\alpha_\parallel$ is the parallel component of the polarizability and E is the field strength at the centre of the I_2 bond. This procedure was used by both Hanna and Mantione, but Hanna also used $\mu_{ind} = \alpha_\parallel (E_1 + E_2)/2$ where E_1 and E_2 are the field strengths at the nearest and farthest iodine atoms, respectively. Examining the shape of a plot of field strength vs distance reveals that the first procedure tends to underestimate μ_{ind}, while the second tends to over-estimate it. The results of these calculations are that μ_{ind} is between $0.97D$ and $1.37D$ from Hanna's computation and $0.66D$ from Mantione's. From these results it is clear that μ_{ind} accounts for a large portion of the experimental dipole moment.

These calculations have been extended by Hanna and his coworkers to include the benzene and p-xylene complexes with chlorine and bromine.[13] In this work, moments due to both charge transfer and electrostatic induction were obtained, with approximately one- to two-thirds of the total moments due to the electrostatic interactions. Unfortunately there are no experimental values available for comparison with these calculated moments. However, if the results given for these complexes are extrapolated to the iodine–p-xylene complex, an electrostatically induced moment of 0·5D is found, while experimental values of μ_c are 0·9D[19] and 0·79D.[20] Hanna and his coworkers point out that the calculated values of μ for the chlorine complexes decrease when the donor is changed from benzene to p-xylene, although such methylation of the ring should add to the moments.[3] Experimental values of μ for the iodine complexes indicate that the moments do increase on changing from the benzene to p-xylene complex (Table 1). This discrepancy was blamed on errors in the molecular orbital functions used to calculate the field strength, particularly that of p-xylene. Since in all of these treatments, other possible sources of moments induced by electrostatic interactions (e.g. the effect of the halogen quadrupole on the aromatic donor) have been ignored, these calculated moments are presumably lower limits.

Consideration of electrostatic interactions have also been used in these papers to obtain values of the energies of complex formation. These results indicate that a good part of the thermodynamic stability of the complexes can be accounted for by electrostatic interactions. We shall not, however, discuss these calculations in any detail. Finally, it should be noted that the short lifetimes observed for the halogen complexes are also consistent with a model based upon electrostatic interactions.

IV. COMPLEXES WITH π^*-ACCEPTORS

A. Experimental Results

Tables III, IV, and V, contain dipole moments determined for π^*-acceptors in environments where intermolecular interactions can take place. For some of these compounds, as with some halogen complexes, there are large discrepancies between the moments reported by different groups (e.g. for 2,6-dinitrophenol in benzene the reported values of μ are 3·89D[72] and 3·2D[73]). Thus, comparisons of the moments of a compound in different solvents or of different compounds in the same solvents should only be made when the results were obtained by the same workers, using the same procedures.

TABLE III. Dipole moments and (μ) and relaxation times (τ) of nonpolar π^*-acceptors in donor solvents

Acceptor	Solvent	μ/D	τ/ps	Ref.
Tetracyanoethylene	Mesitylene	0·80	13·3	74
Tetracyanoethylene	p-Dioxan	0·78–0·74	15·0–15·6	74
p-Chloranil	Mesitylene	0·82	10·5	74
p-Chloranil	p-Dioxan	0·75	10·7	74
2,5-Dichloro-p-benzoquinone	Mesitylene	0·42	9·3	74
2,5-Dichloro-p-benzoquinone	p-Dioxan	0·4	7·0	74
p-Benzoquinone	Mesitylene	0·38	7·7	74
p-Benzoquinone	p-Dioxan	0·3	5·7	74
1,3,5-Trinitrobenzene	p-Dioxan	0·4	3·0	74
1,3,5-Trinitrobenzene	Benzene	0·10	1·45	75
p-Dinitrobenzene	Benzene	0·06	1·10	75
Octafluorobenzidine	Benzene	0·16	1·59	75
Hexafluorobenzene	Benzene	0·3	—	31
Hexafluorobenzene	Benzene	0·83	—	30
Hexafluorobenzene	Benzene	0·1	—	28
Hexafluorobenzene	Benzene	0	0	7
Hexafluorobenzene	p-Xylene	0·4	—	31
Hexafluorobenzene	p-Xylene	0·2	—	29
Hexafluorobenzene	Mesitylene	0·3	—	31
Hexafluorobenzene	Mesitylene	0·2	—	29

TABLE IV. Dipole moments of π^*-acceptor–π-donor complexes (μ_c) in ternary solutions and of the component donor and acceptor molecules and (μ_A) respectively

Acceptor [a]	Donor	Solvent	μ_c/D	μ_D/D	μ_A/D	Ref.
TNB	Hexamethylbenzene	CCl_4	0·87	0	0	b
TNB	Stilbene	CCl_4	0·82	0	0	b
TNB	Naphthalene	CCl_4	0·69	0	0	b
TNB	Durene	CCl_4	0·55	0	0	b
TNB	Benzene	CCl_4	2·31	0	0	78
p-CA	Durene	CCl_4	0·90	0	0	34
p-CA	Naphthalene	CCl_4	0·90	0	0	34
TCNE	Durene	CCl_4	1·26	0	0	34
TCNE	Durene	CCl_4	1·32	0	0	36
TCNE	Naphthalene	CCl_4	1·28	0	0	34
TCNE	Hexamethylbenzene	CCl_4	1·35	0	0	34
TCNE	Hexamtthylbenzene	CCl_4	1·64	0	0	36
TCNE	Hexamethylbenzene	CCl_4	2·8	0	0	44
TCNE	Pyrene	CCl_4	2·0	0	0	93
TCNE	Benzene	CCl_4	0·75	0	0	36
TCNE	Toluene	CCl_4	0·90	0·37	0	36

[continued overleaf]

Table IV.—*continued*

Acceptor[a]	Donor	Solvent	μ_c/D	μ_D/D	μ_A/D	Ref.
TCNE	o-Xylene	CCl_4	1·06	0·54	0	36
TCNE	m-Xylene	CCl_4	1·01	0·37	0	36
TCNE	p-Xylene	CCl_4	0·96	0	0	36
TCNE	1,2,3-Trimethylbenzene	CCl_4	1·23	0·56	0	36
TCNE	1,2,4-Trimethylbenzene	CCl_4	1·16	0·37	0	36
TCNE	1,3,5-Trimethylbenzene	CCl_4	1·12	0	0	36
TCNE	1,2,3,4-Tetramethylbenzene	CCl_4	1·38	0·58	0	36
TCNE	1,2,3,5-Tetramethylbenzene	CCl_4	1·32	0·37	0	36
TCNE	Pentamethylbenzene	CCl_4	1·51	0·37	0	36
TCNE	Thiazole	CCl_4	2·3	1·72	0	92
TCNE	Phenyl-4-thiazole	CCl_4	2·4	1·33	0	92
TCNE	Phenyl-2-thiazole	CCl_4	2·3	1·21	0	92
TCNE	Phenyl-5-thiazole	CCl_4	2·6	1·89	0	92
NB	Benzene	CCl_4	4·86	0	3·92	78
NB	Aniline	Cyclohexane	4·52	1·53	3·92	86
NB	o-Chloroaniline	Cyclohexane	5·17	1·75	3·92	86
NB	o-Bromoaniline	Cyclohexane	5·31	1·75	3·92	86
NB	o-Iodoaniline	Cyclohexane	5·59	1·65	3·92	86
o-DNB	Aniline	Cyclohexane	7·10	1·53	5·66	87
o-DNB	o-Chloroaniline	Cyclohexane	10·18	1·75	5·66	87
o-DNB	m-Chloroaniline	Cyclohexane	7·60	2·64	5·66	87
o-DNB	o-Bromoaniline	Cyclohexane	10·34	1·75	5·66	87
o-DNB	o-Iodoaniline	Cyclohexane	8·24	1·65	5·66	87
m-DNB	Aniline	Cyclohexane	5·84	1·53	3·75	87
m-DNB	o-Chloroaniline	Cyclohexane	6·06	1·75	3·75	87
m-DNB	m-Chloroaniline	Cyclohexane	6·05	2·64	3·75	87
m-DNB	o-Bromoaniline	Cyclohexane	6·12	1·75	3·75	87
m-DNB	o-Iodoaniline	Cyclohexane	6·62	1·65	3·75	87
m-DNB	Naphthalene	Benzene	3·47	0	3·75	16
m-DNB	α-Naphthol	Benzene	4·24	1·49	3·75	16
m-DNB	β-Naphthol	Benzene	4·53	1·58	3·75	16
m-DNB	α-Naphthylamine	Benzene	4·31	1·48	3·75	16
m-DNB	β-Naphthylamine	Benzene	4·35	1·78	3·75	16
p-DNB	Aniline	Cyclohexane	3·11	1·53	0	87
p-DNB	o-Chloroaniline	Cyclohexane	3·91	1·75	0	87
p-DNB	m-Chloroaniline	Cyclohexane	4·15	2·64	0	87
p-DNB	o-Bromoaniline	Cyclohexane	3·52	1·75	0	87
p-DNB	o-Iodoaniline	Cyclohexane	3·72	1·65	0	87
PA	Naphthalene	p-Dioxan	2·53	0	2·61	16
PA	α-Naphthol	p-Dioxan	3·35	1·49	2·61	16
PA	β-Naphthol	p-Dioxan	3·11	1·57	2·61	16
PA	α-Naphthylamine	p-Dioxan	3·99	1·47	2·61	16
PA	α-Naphthylamine	p-Dioxan	5·08	1·47	2·61	85
PA	Aniline	p-Dioxan	6·74	1·53	2·61	85
PA	N,N-Dimethylaniline	p-Dioxan	6·06	1·58	2·61	85
PA	β-Naphthylamine	p-Dioxan	6·96	1·75	2·61	85

[a] TNB = 1,3,5-trinitrobenzene; p-CA = p-chloranil; TCNE = tetracyanoethylene; NB = nitrobenzene; DNB = dinitrobenzene; PA = picric acid.

[b] G. Briegleb and J. Czekalla. Z. *Elektrochem. Ber. Bunsen. Physik. Chem.,* **59**, 184 (1955).

TABLE V. Dipole moments (μ) of polar acceptors in various solvents.

Acceptor	Solvent	μ/D	Ref.
1. Maleic anhydride	Carbon tetrachloride	3·86	17
2. Maleic anhydride	Benzene	3·20–3·55	17
3. Maleic anhydride	p-Xylene	3·69	17
4. Maleic anhydride	Mesitylene	3·59	17
5. Maleic anhydride	p-Dioxan	3·93	17
6. Phthalic anhydride	Carbon tetrachloride	5·87	17
7. Phthalic anhydride	Benzene	4·71	17
8. Phthalic anhydride	p-Xylene	4·75	17
9. Phthalic anhydride	Mesitylene	4·85	17
10. Picric acid	Benzene	1·65	79
11. Picric acid	p-Dioxan	2·40	79
12. Styphnic acid	Benzene	2·33	79
13. Styphnic acid	p-Dioxan	2·54	79
14. 2,4,6-Trinitroanisole	Benzene	2·25	79
15. 2,4,6-Trinitroanisole	p-Dioxan	2·22	79
16. Phthalic anhydride	Benzene	5·20	79
17. Phthalic anhydride	p-Dioxan	5·30	79
18. Pyromellitic dianhydride	Benzene	2·41	79
19. Pyromellitic dianhydride	p-Dioxan	1·74	79
20. Nitroindandione	Benzene	6·20	79
21. Nitroindandione	p-Dioxan	6·26	79
22. Picric acid	Benzene	1·51	72
23. Picric acid	p-Dioxan	2·13	72
24. 2-Nitrophenol	Benzene	3·22	72
25. 2-Nitrophenol	p-Dioxan	3·07	72
26. 2,4-Dinitrophenol	Benzene	3·02	72
27. 2,4-Dinitrophenol	p-Dioxan	3·51	72
28. 2,6-Dinitrophenol	Benzene	3·9	72
29. 2,6-Dinitrophenol	p-Dioxan	3·47	72
30. 1-Nitro-2-naphthol	Benzene	3·68	72
31. 1-Nitro-2-naphthol	p-Dioxan	3·79	72
32. 2-Nitro-1-naphthol	Benzene	3·79	72
33. 2-Nitro-1-naphthol	p-Dioxan	3·62	72
34. 2,4-Dinitro-1-naphthol	Benzene	4·19	72
35. 2,4-Dinitro-1-naphthol	p-Dioxan	4·40	72

1. Nonpolar Acceptors

It is difficult to establish that $\mu_c > 0$ for these molecules in nonpolar donor solvents solely from the measurement of static dielectric constants of binary solutions because of the large uncertainties in the values of P^a (Section IIB2). However, the results of dielectric loss measurements on a wide variety of such acceptors in donor solvents such as p-dioxan, benzene, and mesitylene indicate that the dipole moments vary from 0·1D to 0·8D (first 12 rows of

Table III).[74, 75] Since these values of μ_c were calculated using the formal concentration of the acceptor, it is likely that they represent the lower limits of μ_c. Relaxation times found for these complexes are also given in Table III. As in the case of the iodine complexes (Section IIIA), the observed relaxation times would have been longer if the relaxation process involved only the rotation of a stable complex. This can be explained by assuming that the moment occurs only during complex formation and that dissociation of the complex is a faster process than molecular rotation.[74] As further support for this idea, Crump and Price point out that as the acceptor strength decreases so does τ. Additionally there is a linear relationship between $\log \tau$ and the electron affinity of the acceptors. Presumably as the complexes become less stable the rates of dissociation increase.

A number of conflucting results have been reported for hexafluorobenzene when it is dissolved in aromatic solvents. Values of μ ranging from $0.33D^{(31)}$ to $0.83D^{(30)}$ have been given for hexafluorobenzene. However, careful measurements in benzene, p-xylene, and mesitylene solutions over the entire mole-fraction range indicate that the molar polarizations are additive at all concentrations.[28, 29] Nevertheless, the molar refractions show negative deviations from additivity in each solvent. Such results are consistent with the formation of slightly polar weak complexes. The values of K are not large enough nor the data accurate enough to allow equation (20) to be used to calculate μ_c. However, a reasonable estimate of K values from calorimetric data indicates that at best $0 < \mu_c < 0.2D$. Dielectric relaxation data also indicate that a slightly polar species is formed between hexafluorobenzene and hexamethylbenzene when the two are dissolved in carbon tetrachloride.[5, 7]

Dielectric loss measurements have been used to show that p-xylene can also act as an electron acceptor towards heterocyclic amines and sulphides.[76, 77] This conclusion derives from the observation that relaxation times of these polar donor molecules are longer when measured in p-xylene that in cyclohexane, even though p-xylene has the lower viscosity. Presumably because the p-xylene–donor complex is a larger entity than a free donor molecule in cyclohexane it has a longer relaxation time. Compounds included in these studies and their observed relaxation times at 25°C in the two solvents are: pyridine ($\tau_{C_8H_{10}} = 4.3$ ps, $\tau_{C_6H_{12}} = 3.0$ ps), 4-methylpyridine ($\tau_{C_8H_{10}} = 9.8$ ps, $\tau_{C_6H_{12}} = 7.2$ ps), quinoline ($\tau_{C_8H_{10}} = 10.4$ ps, $\tau_{C_6H_{12}} = 10$ ps), isoquinoline ($\tau_{C_8H_{10}} = 15.4$ ps, $\tau_{C_6H_{12}} = 10.5$ ps), phthalazine ($\tau_{C_8H_{10}} = 12.9$ ps, $\tau_{C_6H_{12}} = 6.7$ ps at 50°C), acridine ($\tau_{C_8H_{10}} = 14.7$ ps, $\tau_{C_6H_{12}} = 15.1$ ps at 50°C), and thiophen ($\tau_{C_8H_{10}} = 2.8$ ps, $\tau_{C_6H_{12}} = 2.1$ ps). Since the relaxation times are nearly identical in the two solvents for both acridine and quinoline, it was concluded that these two donors have virtually no interaction with p-xylene. Presumably, steric factors in these two molecules would prevent the p-xylene ring from coming close enough to the nitrogen for the lone-pair electrons to interact with the π-molecular orbitals. Further evidence that the nitrogen lone-pair is the donor site comes from the observation that p-xylene has a larger effect on the relaxation time of phthalazine than on isoquinoline, even though both molecules have similar donor abilities. It is likely that interaction of p-xylene with phthalazine would be favoured because there are two nitrogen donor sites. Apparently, these interactions are not strong enough to affect the dipole moments of the donors in any consistent way.

Since Briegleb's initial reports of the dipole moments of TCNE and TNB complexes in ternary solutions,[3] only a few reports have appeared on this type of study (Table IV). However, some of the values reported in these studies conflict with the earlier values. Thus, the moment reported for the TNB–benzene complex in carbon tetrachloride $(2·31D)^{[78]}$ using the Few and Smith method differs a great deal from that reported by Briegleb $(0·80D)^{[3]}$ while the value obtained from dielectric relaxation measurements on binary solutions are even smaller $(0·10D).^{[75]}$ Chan and Liao have reported the moments for TCNE complexes with a large number of methylbenzenes. They have given the TCNE–hexamethylbenzene (HMB) complex moment as $1·64D$ while Briegleb has reported $1·35D.^{[34]}$ These discrepancies as well as others that occur for many polar acceptors in binary solutions may be due to unsuspected presence of termolecular complexes. Numerous studies[43] have indicated that such complexes (AD_2) are formed in solutions where $C_D > C_A^0$. Since both complexes may be present in the solutions used in the measurements while the moments are calculated on the assumption that only one complex is present they would not only be incorrect but dependent on the concentration ranges used in the experiments. In order to show that the formation of AD_2 could influence dielectric measurements, the original dielectric data given by Briegleb and his coworkers for TCNE–HMB[34] have been recalculated using association constants for AD and AD_2 formation obtained from more recent spectroscopic measurements (Section II.C.2).[44] The moments found by this procedure are $2·8D$ for the 1:1 complex and zero for the 1:2 complex. The zero moment is what would be expected on the basis of a DAD structure for a termolecular species. For other systems such as TCNE–durene the association constants are smaller and the data are not exact enough to enable values of μ for both complexes to be calculated. However, even with the lower association constants the amounts of AD_2 present can be large enough to have a considerable effect on the calculated moment of the 1:1 complex.

2. Polar Acceptors

(a) *Relaxation times*. For these acceptors, interactions with donor molecules to form complexes lead to species whose relaxation times and dipole moments differ from those of the free acceptors in inert solvents. Generally, relaxation times are longer for the complexes than for the free acceptors, because formation of a complex gives a polar species that has a larger molecular volume than the free acceptor. Results of dielectric loss measurements for maleic and phthalic anhydrides in various donor solvents illustrate this effect very well. Crump and Price found that the relaxation times are much longer in benzene, p-xylene, mesitylene or p-dioxan than in carbon tetrachloride (by a factor of two or three), even though the aromatic donors are less viscous than carbon tetrachloride.[17] Furthermore, these relaxation times are found to have a narrow distribution which can be attributed to rapid exchange between two polar species, the free acceptor and its complex. The rapid exchange is also shown by the results obtained when τ is measured as a function of the concentrations of anhydride and donor in ternary solutions with carbon tetrachloride. Lowering the ratio of the concentration of free acceptor to complex, increases the relaxation times. Relaxation times of picric acid (PA) and 2,6-dinitrophenol (DNP) in benzene also indicate complex formation.[72] The observed relaxation times (for PA $\tau = 50$ ps; for DNP $\tau = 25$ ps) are larger than the values estimated for the free acceptor on the basis of their molecular size and the microscopic viscosity of the solvent (for PA $\tau_{cal} = 18$ ps and for DNP $\tau_{cal} = 15$ ps).

(b) *Dipole moments.* The moments for these complexes are either greater or smaller than the moment of the free acceptor in an inert solvent. To explain this, one must consider the moment induced in the donor species by the permanent dipole of the acceptor rather than any charge-transfer contributions.[17] Depending on the geometry of the complex the induced moment may either add to or subtract from the permanent moment of the acceptor. In the interactions of π-donors and π^*-acceptors the total moments of the complexes will be less than that of the free acceptors because the two components are stacked over each other, and the moments induced in the donors will oppose the permanent moments. Calculations based upon standard electrostatic methods indicate that at a separation of 3·5 Å a molecule with a permanent moment of 4D induces a moment of 0·9D in a molecule whose polarizability is 10^{-23} cm^3. A contribution from a moment due to charge transfer would have the opposite effect on the complex moment since the two vector components would be perpendicular to each other. On the other hand, for a n-donor such as dioxan the arrangement of the components and of the induced and permanent moments is most likely to be linear. In this case the moments have the same directions and the dipole moment of the complex will be greater than that of the free acceptor. Comparisons of the dipole moments for phthalic and maleic anhydrides in several aromatic donors, p-dioxan and carbon tetrachloride show these trends (Table V lines 1 to 9).[17] For some other polar acceptors, dipole moments have been measured in aromatic solvents and p-dioxan but not in inert solvents. In such cases, the differences between the moments in benzene and p-dioxan may be explained by the notion that the acceptor induces a dipole in the donor that can in the case of benzene decrease the total moment or in the case of p-dioxan increase the total moment. The moments of phthalic anhydride, pyromellitic dianhydride (PMDA), 2,4,6-trinitroanisole, nitroindandione, picric acid (PA), and styphnic acid in benzene and p-dioxan have been determined by Lutskii and his coworkers (Table V, lines 10–21).[79] While Richards and Walker have also obtained values of μ for PA in the same solvents (Table V, lines 22 and 23).[72] Although the results given by these different groups are not in satisfactory agreement, the differences between the two solvents are the same in both reports. Except for 2,4,6-trinitroanisole and PMDA the moment of each of these compounds is larger in dioxan than in benzene. For 2,4,6-trinitroanisole the moment in both solvents is virtually the same, while for PMDA the moment is larger in benzene. Consideration of the geometry of PMDA and its complexes with aromatic donors can account for this reversal. In the solid state these molecules have a nonplanar skewed chair structure.[80, 81] This means that the permanent moment would be in the mean plane of the aromatic ring and make an angle of approximately 15° with the five-membered anhydride ring. Since aromatic donors interact with the anhydride rings, it is likely that the induced and permanent moments will not be in parallel planes and the resulting dipole will be higher than that of the isolated acceptor. Dipole moments obtained for some other nitrophenols and nitronaphthols in these two solvents (Table V, lines 24–35) have been interpreted on the basis of an interaction between dioxan and the hydrogen-bonded chelate ring.[72] For some of these compounds (three out of seven) the moments are smaller in dioxan than in benzene. However, calculations based upon changes in the angle between the moment induced in dioxan and the permanent moment were used to rationalize these variations. A weakness of this treatment is that the moments induced in benzene have been ignored.

Additional evidence regarding the effects of induced dipoles, comes from the data for naphthalene complexes with m-dinitrobenze (m-DNB) and PA.[16] The moments of these compounds were obtained from experiments in which solid complexes were prepared and dielectric constants measured for benzene solutions of the PA complex and dioxan solutions of the m-DNB complex. In both cases the moment of the complex (Table IV) is less than that of the free acceptor.[16]

Similar arguments may also be used to discuss the dipole moments reported for molecules that are usually not thought of as electron acceptors. From nmr studies it has been established that N-methylamides that cannot hydrogen-bond may still form complexes with aromatic hydrocarbons.[82, 83] It might, therefore, be expected that ureas and thioureas could also form complexes with benzene. Lien and Kumler have measured the moments of cyclic ureas and thioureas in both benzene and p-dioxan.[84] The moments are smaller in benzene than in p-dioxan for all of the compounds examined. For the nonmethylated compounds the differences in moments between the two solvents are as large as 1D and are attributed to interactions with the hydrogens on the nitrogen atoms. However, the N-methylated compounds have differences ranging from 0·01D to 0·26D. In these cases the interactions would most likely be at the positive end of the dipole, i.e. at the nitrogens. It is noteworthy that thioureas which have the highest positive charges at the nitrogens appear to experience the largest solvent effect as indicated by the differences in dipole moment.

Several studies have also been made in which dipole moments were obtained for complexes composed of two polar constituents. In these investigations, dipole moments were calculated from the dielectric constants measured for binary solutions of the solid complexes[16, 85] or for ternary solutions using the Few and Smith method (Section II C1).[86, 87] Since both of the free components are polar, interpretation of these data is more difficult.

In the first of these studies the picrates of aniline, N,N-dimethylaniline, α- and β-naphthylamine and ethylamine were examined[85] while in the second, moments of PA and m-DNB complexes of α-, β-naphthols and α-, β-naphthylamine were studied (Table IV).[16] The values given for the picrate of α-naphthylamine are in poor agreement (the values of μ_c differ by 1.09D). This disagreement might be accounted for by the failure of either group of workers to take into account the dissociation of the complex (Section IIB1). Raman and Soundararajan, however, did note that for each of these picrates, plots of ϵ' and d vs X_2 were nonlinear at very low concentrations and linear at slightly higher concentrations.[85] The $P_{2\infty}$ values used to obtain μ_c were calculated from the linear portion of the curves. These authors have interpreted their results in terms of a linear arrangement of the components, with the nitrogen of the amine acting as a lone-pair donor and PA acting as a hydrogen-bond donor. This interpretation may be discounted since it is likely that the planes of the interacting molecules lie parallel one above the other.[88] The results themselves are interesting, since the experimental values of μ_c are larger than the scalar sum of the component moments, whereas the data given by Lutskii and coworkers for both the PA and m-DNB complexes indicate that the observed moment is less than the scalar sum of the component moments.[16] These latter authors argue that since the planes of the two components lie parallel, either a parallel or antiparallel arrangement of the two permanent dipoles would be anticipated. Using the moments of the components and their angles of inclination to the perpendicular of the molecular plane, they have calculated values of μ for these two configurations. For all of the examples cited, the observed moments are intermediate to the moments calculated for the two extreme configurations, although for the m-DNB complexes the observed moments are only slightly less than those calculated for the parallel configuration. The authors conclude that the differences between the observed and calculated moments arise from moments due to intermolecular charge-transfer. However, it is also possible that the observed moment is influenced by mutually induced dipoles as well as an averaging process due to the rotation of one molecule relative to another.

Ternary solutions have also been used to obtain values of μ and K for the complexes of nitrobenzene (NB) and the isomeric dinitrobenzenes (o-, m-, p-DNB) with aniline and halogen substituted anilines.[86, 87] From the variation of K with temperature, values of $\Delta H^{\ominus}$ have also been found. Since there are several aspects of these thermodynamic results which appear to conflict with previous studies,[89] the observed dipole moments should be viewed with caution. These thermodynamic results indicate that with some of the donors, NB forms more stable complexes than do the various DNB isomers, and that the relative stabilities of the isomeric DNB complexes change as the donor molecule is changed. Also, in many cases relative stabilities indicated by the values of K are opposite the trends in stabilities based on $\Delta H^{\ominus}$ values. This last observation is, of course, due to apparently large changes in $\Delta S^{\ominus}$ for the different complexes. It is possible, however, that the results are in error because termolecular complexes are present in the solutions. That termolecular species may be present is indicated by the earlier spectroscopic studies on DNB complexes with aniline in which the value of K were wavelength dependent.[90, 91] For most but not all of the complexes, the observed dipoles are found to be less than the scalar sum of the moments of the components. The differences between this sum and the observed moments decrease as $-\Delta H^{\ominus}$ decreases and K increases. Polle and his coworkers explain these trends on the basis of a 2:2 association involving both hydrogen-bond formation and charge-transfer interaction. They argue that one aniline molecule is almost coplanar with a nitrobenzene molecule and that the two are held together by a hydrogen bond between the amine and the nitro group. Two of these complexes are the held together by charge-transfer from the aniline ring of one pair to the nitrobenzene ring of the other pair. Presumably, the vector sum of the moment due to the amino group and the charge-transfer moment is directed at an angle to the moment from the nitro groups. Steric repulsions between the ortho groups of the two molecules in the hydrogen-bonded pair increase the angle between the charge-transfer moment and the moment of the amino groups which in turn lead to an increased moment of the complex. The validity of these ideas is questionable since it is unlikely that an equilibrium reaction to form 2:2 complexes from the components could take place without the formation of termolecular complexes as well. This treatment also ignores any effects of moments induced by the permanent dipoles.

Complexes between polar components in binary solutions have also been studied by Voronkov and his coworkers using only refractive indices (Section IIB[2]).[26, 27] This technique cannot be used to find the dipole moment of the complexes, but can be employed to find the stoichiometries and relative strengths of the interactions. Systems that have been examined by this technique have contained as acceptors mono-substituted benzenes in which the substituent is an electron-withdrawing group such as NO_2, CN, NCS, $COCH_3$, or halogen atom. The donors used were lone-pair donors such as ethers, sulphides, amines, esters, alkyl chlorides, and ethoxysilanes. For many of the systems examined, the conclusions regarding complexation are in agreement with the results obtained from other sources such as spectroscopy, cryoscopy, and density isotherms. In the majority of cases the deviations of the refractive indices from ideal behaviour reach their maximum values at concentrations that indicate 1:1 mole ratios for the complexes. However, for some of the systems, specifically those containing furan and piperidine, termolecular complexes are indicated by the position of the maximum deviations. From the relative sizes of the maximum deviations, these authors indicate that the order of donor ability towards most of the acceptor species is tetrahydrofuran > furan > piperidine > pyrrole > thiophen.[27] Also for a given donor the acceptor ability of the substituted benzenes increases as the electron withdrawing ability of the substituent increases. One drawback of this method is that it is not specific for donor–acceptor complexes since deviations from ideal behaviour occur for systems that are not donor–acceptor pairs, i.e. N,N-dimethylaniline interacting with ethers.

B. Influence of Charge Transfer on Dipole Moments

In the previous section, the moments of many of the complexes with polar acceptors were rationalized in terms of moments induced in the donor by the acceptors. This factor is not applicable to the complexes between two non-polar components (e.g. TCNE–benzene) or a complex with only a slightly polar donor (e.g. TCNE–toluene). However, the polarization induced in a molecule by the atomic charges on a neighbouring nonpolar molecule can result in an appreciable dipole moment for the complex.[5, 7, 9, 10, 12, 92] Calculations of this induced moment are similar to those previously discussed for the iodine aromatic hydrocarbon complexes (Section IIIB2). The interaction energies for these complexes have been considered to arise from the van der Waals (London) forces alone[5, 7, 9, 10, 13, 92] or in combination with charge-transfer forces.[12] Agreement between calculated and experimental energies have been considered to be satisfactory in most cases. We shall not discuss the calculation of stabilization energies in any detail since we are primarily interested in the dipole moments. The procedures used in these papers give values of μ that account for widely differing fractions of the experimental moments. However, the important point of these calculations is that the induced moments are real and can account for a significant portion of the experimental moments, although it is difficult to obtain accurate values for them.

To calculate these induced moments, the electrostatic fields due to the nonpolar molecules must be found. These fields have been calculated from standard equations assuming point-charge distributions. Le Fèvre has used three methods to calculate these distributions from bond moments.[7] In the first method the bond moment divided by the bond-length was taken as the point charge at the two atoms of a bond while in the second the bond moment was assumed to act as a point charge at the centre of the bond. In the last method it was assumed that half the bond dipole acts as a point charge at the midpoint of the bond and half at the more electronegative atom. Using HMB as the donor, Le Fèvre calculated the induced moment for a series of complexes with hexafluorobenzene, p-CA, TNB, and TCNE. These values range from 0·3D for the hexafluorobenzene complex to 1·7D for the TNB complex. These results were only slightly dependent upon the values chosen for the field strength, the intercomponent distance or the polarizabilities. If these calculated values are used as μ_0 in equation (2) then the percentage contributions of the charge transfer are less than half those given by Briegleb.[3] As this treatment ignores the polarization of the acceptor by the donor these calculated values are the lower limits of the induced moments. Mantione[9, 10] and Hanna and his coworkers[12] have obtained the field strengths from MO calculations of σ- and π-electron densities at the atoms of both donors and acceptors. These charges were used to find the mutually induced dipoles in the respective components. The charge distribution for TCNE given by these two groups differ a great deal. Mantione's values for the induced moments of the TCNE complexes with pyrene, naphthalene, and durene and the nitrobenzene complex with naphthalene are in almost perfect agreement with the experimentally observed moments for the complexes. This leads her to claim that there is no need to consider charge-transfer structures at all. On the other hand, the moments calculated by Hanna and his coworkers for TCNE complexes with benzene and p-xylene account for only about a quarter of the experimental values of μ_c. This leads them to the conclusion that charge transfer is still necessary to account for the experimental values of μ_c. Calculations have also been carried out on a series of TCNE–thiazole complexes using Mantione's procedure.[90] The results of these calculations indicate that approximately 50 percent of the experimental moments may be accounted for by the induced moment.

Chan and Liao have recently interpreted the experimental dipole moments of TCNE complexes with a series of methylated benzenes solely on the basis of charge-transfer theory.[36] They observed that the moments of the complexes increased as the number of methyl groups on the benzene ring increased. If the moments of the free donors are used as μ_0 in equation (2), then the calculated fractional contributions of the charge-transfer structures increase as the ionization potentials of the donors decrease. These observations are consistent with charge-transfer theory.[4] However, they cannot be taken as proof of this theory since electrostatically induced moments will also depend on the polarizabilities of the donors which in turn will depend on the ionization potentials.

Molecular orbital calculations have also been made on π–π^* complexes. In one of these studies, the pyrene and naphthalene complexes with TCNE were examined using a second-order perturbation theory.[93] In this treatment two charge-transfer states, D^+A^- and D^-A^+, are considered. The empirical parameters involved were chosen so as to give a good agreement between the theoretical and observed dipole moments. The agreement of calculated and observed values of transition moments and stabilization energies was used as a test of the validity of the calculations. Kuroda and his coworkers regard the agreement as satisfactory. However, this treatment does not indicate that charge transfer is necessary to rationalize the properties of the complex. Yoshida and Kobayashi used the Pariser–Parr–Pople MO method to make calculations on the p-BQ–benzene complex.[94] In this method the complex was treated as a single conjugated π-electron system. The calculations give good values for the electronic spectra but are poor at predicting the energy and intermolecular distance. The calculated dipole moment is small with very little charge transfer found.

V. INTERACTIONS OF HALOGENATED ALKANES

The existence of weak interactions between a variety of lone-pair donors and polyhalogenated alkanes have been inferred from the results of measurements of both static dielectric constants and dielectric loss. In these studies, the polyhalogenated hydrocarbons have been used either as solvents or solutes. We shall exclude from our discussion those interactions where formation of hydrogen bonds is a major factor (e.g. those involving chloroform and pentachloroethane). Sharpe and Walker have published several papers concerning the interactions of carbon tetrachloride with alkyl and heterocyclic amines, ethers, and sulphides.[95, 96, 97] According to these authors, complex formation between a polar solute and carbon tetrachloride occurs if the difference between the moment of the solute in this solvent and in benzene is greater than 0·04D. This criterion is supported by observations on the dipole moments in these two solvents for a group of polar solutes which are not expected to have any specific interactions with either solvent. The solvent effect on the moments of these compounds (11 compounds for which the moments in the two solvents were measured by the same authors) is very small, the differences in moment between the two solvents vary from −0·04D to +0·03D. On the other hand differences ranging from slightly negative to +0·17D were found for amines, ethers, and sulphides while the difference is +0·47 for 4-methylpyridine N-oxide. It is likely that the increased dipole moment of the solute in carbon tetrachloride is caused by one of the chlorine atoms in the solvent acting as an acceptor for the lone-pair electrons of the solute. The alternative explanation that the amines could interact with the π-orbitals of benzene was rejected because for 4-methylpyridine N-oxide, 4-methylpyridine, and other amines no change in moment is observed when measurements are made in benzene or cyclohexane. However it should be pointed out, that data on the relaxation times of pyridine and other nitrogen heterocycles have been interpreted as indicating weak acceptor properties for p-xylene, even though there is only a small effect on the observed dipole moments (Section IV.A.1). Le Fèvre and his coworkers had earlier suggested that the large difference in μ for pyridine in the two solvents (0·12D) could be caused by an increase in the atomic polarization (P^a) arising from solvation effects rather than formation of a slightly more polar adduct.[98] However, this is not likely, since it would be necessary for the value of P^a to change from 1·5 cc to ca 12·5 cc in order to account for the change of μ in the two solvents. In 4-methylpyridine N-oxide, the change in $P_{2\infty}$ is nearly three times the value of $R_{2\infty}$ itself. Assuming these arguments are correct, one can reasonably expect the differences in the moments to be a function of the pK_a values of the solutes. If donor molecules with large steric hindrance are excluded, a linear correlation is found. Very small or negative values for the differences in dipole moments were explained on the basis of steric factors (2.6-diphenylpyridine), electronic factors (aniline), intramolecular hydrogen-bonding tying up the lone pair (8-hydroxyquinoline), and the directional nature of the dipole moment (4-nitropyridine). In this last example, the interaction of the lone-pair electrons with the chlorine atom can decrease the observed dipole moment since the total dipole vector is opposite the lone-pair dipole vector. Dielectric measurements made on ternary solutions of pyridine in benzene plus carbon tetrachloride, carbon tetrabromide, or hexachloroethane[96] and ethyl ether in carbon tetra-bromide–benzene mixtures[97] also give evidence for an interaction of the lone-pair donors with a halogen atom of the polyhalogenated hydrocarbons. In these experiments, values of P_2 calculated at fixed mole-fractions of the lone-pair donor were dependent on the concentration of the nonpolar halogenated compound. Plots of P_2 vs $X_{halogen}$ are linear with positive slopes for all of the solutions except those with carbon tetrabromide where there is a large positive nonlinear slope. Similar plots for solutions containing halogens such as p-dibromo and p-dichlorobenzene, which are not expected to be good electron-acceptors, had zero gradients. Therefore the size of the gradient apparently is a measure of the strength of the interaction. On this basis the trend of acceptor strength is carbon tetrabromide > hexachloro-ethane > carbon tetrachloride. The higher acceptor strength of the bromide is in accord with the higher polarizability of the bromine atom. Comparison of the relaxation times for pyridine in carbon tetrachloride ($\tau = 5\cdot0$ ps) and cyclohexane ($\tau = 2\cdot0$ ps–$3\cdot1$ ps) also indicate that there is an interaction between pyridine and carbon tetrachloride.[76] This dielectric evidence is supplemented by calorimetric studies on mixtures of pyridine and carbon tetrachloride.[99]

A recent determination of the dipole moment of pyridine in carbon tetra-chloride over the entire concentration range and at several temperatures gave results which may be inconsistent with a 1:1 interaction between the halogen and pyridine.[100] It was observed that μ increases as the concentra-tion of pyridine increases, reaching a maximum at a mole fraction close to $0\cdot5$. However, if only a 1:1 complex is formed then as the concentration of pyridine increases both the fraction complexed and the apparent dipole moment would decrease. It may be, however, that this anomaly appears because the Onsager equation was used to calculate μ. This equation is most appropriate to calculations in dilute solutions or pure liquids,[2] and might give false results at the higher concentrations.

Broadus and Vaughan have reported the dipole moments of N-alkyl-pyrazoles in a variety of solvents.[101] These results are consistent with the results given by Sharpe and Walker[95, 96, 97] since carbon tetrachloride produces a larger effect on the dipole moment than would be predicted by a consideration of a general solvent effect. Again it is found that the difference between the moments in carbon tetrachloride and benzene correlate with the relative pKa values of the donors. Several hydrocarbon solvents that gave a "classical" solvent effect on μ were used while the solvents capable of hydrogen bonding and carbon tetrachloride had anomalously high values.

Dielectric relaxation data obtained for the polar chlorinated ethanes dis-solved in p-dioxan, aromatic hydrocarbons, and cyclohexane have also been used as evidence of intermolecular interactions.[102] The observed relaxation times indicate that in *p*-dioxan and aromatic hydrocarbons complexes between solute and solvent are present. The relaxation times for most of the halogen compounds lengthens in the order cyclohexane < benzene < *p*-xylene < mesitylene < *p*-dioxan. As only *p*-dioxan has a higher viscosity than cyclohexane, this order can be interpreted in terms of complex formation. This trend in relaxation times parallels the increased basicity and decreased ionization potential of the hydrocarbons. *p*-Dioxan does not conform well with the data for the hydrocarbon systems although a viscosity correction would improve this. However, the systems are not completely analogous since *p*-dioxan is a lone-pair donor while the hydrocarbons are π-electron donors. For most of the halogenated ethanes the main source of this inter-action can be attributed to the formation of a hydrogen bond. However, from nmr work on chloroethanes, Kuntz and his coworkers concluded that 1,1,1-trichloroethane and 1,2-dichloroethane cannot form hydrogen bonds.[103] Therefore these two molecules must be associated through the halogen atoms. The differences between the free energy of activation for molecular reorientation in cyclohexane and a basic solvent were calculated from the relative relaxation times in the two solvents. These values agree with those obtained from nmr work.

REFERENCES

1. C. P. Smyth. "Dielectric Behavior and Structure", McGraw Hill, New York (1955).
2. C. J. F. Botcher. "Electric Polarization", Elsevier, Amsterdam (1952).
3. G. Briegleb. "Electronen-Donator-Acceptor-Komplexe", Springer Verlag, Berlin (1961).
4. W. B. Person and R. S. Mulliken. "Molecular Complexes: A Lecture and Reprint Volume", Wiley, New York (1969).
5. R. J. W. Le Fèvre, D. V. Radford and P. J. Stiles. *J. Chem. Soc. (B)*, 1297 (1968).
6. M. J. S. Dewar and C. C. Thompson, Jr. *Tetr. Suppl.*, **7**, 97 (1966).
7. R. J. W. Le Fèvre, D. V. Radford, G. L. D. Ritchie and P. J. Stiles. *Chem. Comm.*, 1221 (1967).
8. J. P. Malrieu and P. Claverie. *J. Chim. Phys.*, **65**, 785 (1968).
9. M. J. Mantione. *Theor. Chim. Acta*, **11**, 119 (1968).
10. M. J. Mantione. *Theor. Chim. Acta*, **15**, 141 (1969).
11. M. W. Hanna. *J. Amer. Chem. Soc.*, **90**, 285 (1968).
12. J. L. Lippert, M. W. Hanna and P. J. Trotter. *J. Amer. Chem. Soc.*, **91**, 4035 (1969).
13. M. J. Mantione. *Int. J. Quant. Chem.*, **3**, 185 (1969).
14. M. T. Rogers and W. K. Meger. *J. Phys. Chem.*, **66**, 1397 (1962).
15. G. Briegleb, J. Czekalla and A. Hauser. *Z. Phys. Chem. (Frankfurt am Main)*, **21**, 99 (1959).
16. A. E. Lutskii, V. V. Dorofeev and B. P. Kondratenko. *J. Gen. Chem. USSR (Eng. Transl.)*, **33**, 1961 (1963).

F

17. R. A. Crump and A. H. Price. *Trans. Faraday Soc.*, **65**, 3195, (1969).
18. V. L. Brownsell and A. H. Price. *Chem. Soc. Spec. Publ.*, **20**, 83 (1966).
19. F. Fairbrother. *J. Chem. Soc.*, 1051, (1948).
20. J. Gerbier. *Compt. Rend. Acad. Sci.*, **262B**, 685 (1966).
21. Y. K. Syrkin and K. Anisimova. *Dokl. Akad. Nauk, SSSR*, **59**, 1457 (1948).
22. E. Charney and E. D. Becker. *J. Amer. Chem. Soc.*, **83**, 4468 (1961).
23. S. Soundarajan and M. J. Vold. *Trans. Faraday Soc.*, **54**, 1155 (1958).
24. R. C. Cass, H. Spedding and H. D. Springall. *J. Chem. Soc.*, 3451 (1957).
25. C. C. Meridith, L. Westland and G. F. Wright. *J. Amer. Chem. Soc.*, **79**, 2385 (1957).
26. M. G. Voronkov and A. Y. Deich. *Teor. i Eksperim. Khim.*, *Akad. Nauk UKr. SSR (Eng. Transl.)*, **1**, 443 (1965).
27. M. G. Voronkov, A. Y. Deich and E. V. Akatova. *Khim. Geterotsikl. Soedin.*, *Akad. Nauk Lat. SSR, (Eng. Transl.)*, **2**, 5 (1966).
28. M. E. Baur, D. A. Horsma, C. M. Knobler and P. Perez. *J. Phys. Chem.*, **73**, 641 (1969).
29. M. E. Baur, C. M. Knobler, D. A. Horsma and P. Perez. *J. Phys. Chem.*, **74**, 4594 (1970).
30. C. C. Meridith and G. F. Wright. *Can. J. Chem.*, **38**, 1177 (1960).
31. W. A. Duncan, J. P. Sheridan and F. L. Swinton. *Trans. Faraday Soc.*, **62**, 1090 (1966).
32. G. Kortum and H. Walz. *Z. Electrochem.*, **57**, 73 (1953).
33. K. Toyoda and W. B. Person. *J. Amer. Chem. Soc.*, **88**, 1629 (1966).
34. G. Briegleb, J. Czekalla and G. Reuss. *Z. Phys. Chem. (Frankfurt am Main)*, **30**, 333 (1961).
35. S. Kobinata and S. Nagakura. *J. Amer. Chem. Soc.*, **88**, 3905 (1966).
36. R. K. Chan and S. C. Liao. *Can. J. Chem.*, **48**, 299 (1970).
37. A. Funatsu and K. Toyoda. *Bull. Chem. Soc. Jap.*, **43**, 279 (1970).
38. E. N. Gur'yanova and I. P. Goldshtein. *J. Gen. Chem. USSR (Eng. Transl.)*, **32**, 13 (1962).
39. I. G. Arzamanova, E. N. Gur'yanova and I. P. Goldshtein. *Proc. Acad. Sci. USSR (Eng. Trans.)*, **155**, 403 (1964).
40. A. V. Few and J. W. Smith. *J. Chem. Soc.*, 2781 (1949).
41. W. Waclawek. *Bull. Acad. Polon. Sci., Ser., Math., Astron, Phys.*, **19**, 875 (1970).
42. R. K. Chan and S. C. Liao. *Can. J. Chem.*, **48**, 2988 (1970).
43. R. Foster, in "Molecular Complexes". Vol 2, ed. R. Foster, Elek, London; Crane Russak, New York (1974), p. 107.
44. R. Foster and N. Kulevsky. *J. Chem. Soc., Faraday Trans. I*, **69**, 1427 (1973).
45. G. W. Nederbragt and J. Pelle. *Mole, Phys.*, **1**, 97 (1959).
46. J. E. Anderson and C. P. Smyth. *J. Amer. Chem. Soc.*, **85**, 2903 (1963).
47. H. Tsubomura and S. Nagakura. *J. Chem. Phys.*, **27**, 819 (1957).
48. I. G. Arzamanova and E. N. Gur'yanova. *J. Gen. Chem. USSR (Eng. Transl.)*, **36**, 1171 (1966).
49. P. Boule. *J. Amer. Chem. Soc.*, **90**, 517 (1968).
50. A. J. Hamilton and L. E. Sutton. *Chem. Commun.*, 460 (1968).
51. S. S. Singh and G. P. Saxena. *Ind. J. Chem.*, **8**, 1116 (1970).
52. L. Sobczyk and L. Budziszewski. *Rocz. Chem. Ann. Soc. Chim. Pol.*, **40**, 901 (1966).
53. I. G. Arzamanova and E. Gur'yanova. *J. Gen. Chem. USSR (Eng. Transl.)*, **33**, 3412 (1963).
54. I. G. Arzamanova and E. Gur'yanova. *Proc. Acad. Sci. USSR (Eng. Transl.)*,

157, 683 (1964).
55. S. N. Bhat and C. N. R. Rao. *J. Amer. Chem. Soc.*, **90**, 6008 (1968).
56. P. C. Dwivedi. *Ind. J. Chem.*, **9**, 1408 (1971).
57. A. I. Popov and R. H. Rygg. *J. Amer. Chem. Soc.*, **79**, 4624 (1957).
58. K. R. Bhakar and S. Singh. *Spectrochim. Acta,* **23A**, 1155 (1967).
59. H. Ratajczak, Z. Mielke and W. J. Orville-Thomas. *J. Mole. Struct.*, **14**, 165 (1972).
60. F. C. Frank. *Proc. Roy. Soc., Ser. A*, **152**, 171 (1936).
61. H. Ratajczak and W. J. Orville-Thomas. *J. Mole. Struct.*, **14**, 149 (1972).
62. I. G. Arzamanova and E. N. Gur'yanova. *Proc. Acad. Sci. USSR (Eng. Transl.),* **166**, 79 (1966).
63. I. P. Goldshtein, E. N. Kharlamova and E. N. Gur'yanova, *J. Gen. Chem. USSR (Eng. Transl.),* **38**, 1925 (1968).
64. E. N. Gur'yanova. *Russ. Chem. Rev. (Eng. Transl.),* **37**, 863 (1968).
65. G. Briegleb and J. Czekalla. *Angew. Chem.*, **72**, 401 (1960).
66. W. B. Person. *J. Chem. Phys.*, **38**, 109 (1963).
67. K. Fukui, A. Imamura, T. Yonezawa and C. Nagata. *Bull. Chem. Soc. Jap.*, **35**, 33 (1962).
68. M. Dobrescu. *Rev. Roum. Chim.*, **16**, 1825 (1971).
69. M. Dobrescu. *Rev. Roum. Chim.*, **16**, 1841 (1971).
70. M. Dobrescu-Civreanu and V. E. Sahlni. *Rev. Roum. Chim.*, **18**, 921 (1973).
71. J. Gerbier. *Compt. Rend. Acad. Sci.*, **261**, 5037 (1965).
72. J. H. Richard and S. Walker. *Trans. Faraday Soc.*, **57**, 406 (1960).
73. A. H. Price. *J. Phys. Chem.*, **64**, 1442 (1960).
74. R. A. Crump and A. H. Price. *Trans. Faraday Soc.*, **66**, 92 (1970).
75. F. F. Hanna and K. N. Abd-El-Nour. *Z. Phys. Chem.* (Leipzig), **246**, 292 (1971).
76. J. Crossley, W. F. Hassell and S. Walker. *Can. J. Chem.* **46**, 2181 (1968)..
77. J. Crossley and S. Walker. *Can. J. Chem.*, **46**, 2369 (1968).
78. V. N. Nagareva, N. N. Polle and E. G. Polle. *J. Gen. Chem. USSR (Eng. Transl.),* **40**, 2306 (1970).
79. A. E. Lutskii, V. V. Prezhdo and M. I. Kolesnik. *Russ. J. Phys. Chem. (Eng. Transl.),* **45**, 955 (1971).
80. J. Boyens and F. H. Herbstein. *J. Phys. Chem.*, **69**, 2153 (1965).
81. J. Boyens and F. H. Herbstein. *J. Phys. Chem.*, **69**, 2160 (1965).
82. R. M. Moriarity. *J. Org. Chem.*, **28**, 1296 (1963).
83. A. A. Sandoval and M. W. Hanna. *J. Phys. Chem.*, **70**, 1203 (1966).
84. E. J. Lien and W. D. Kumler. *J. Med. Chem.*, **11**, 214 (1968).
85. R. Raman and S. Soudararajan. *Can. J. Chem.*, **39**, 1247 (1961).
86. N. N. Polle, E. G. Polle and L. N. Pushnikova. *J. Gen. Chem. USSR (Eng. Transl.),* **42**, 202 (1972).
87. N. N. Polle and E. G. Polle. *J. Gen. Chem. USSR (Eng. Transl.),* **42**, 1430 (1972).
88. S. C. Wallwork. *J. Chem. Soc.*, 494 (1961).
89. R. Foster. "Organic Charge–Transfer Complexes," Academic Press, London and New York (1969), Chapter 7.
90. J. Landauer and H. M. McConnell. *J. Amer. Chem. Soc.*, **74**, 1221 (1952).
91. V. S. Anikeev, E. G. Polle and B. V. Tronov. *Teor. i Eksperim. Khim., Akad. Nauk Ukr. SSR (Eng. Transl.),* **4**, 432 (1968).
92. J. N. Bonnier and R. Arnaud. *J. Chim. Phys.*, **68**, 423 (1971).
93. H. Kuroda, T. Amano, I. Ikemoto and H. A. Kamatu. *J. Am. Chem. Soc.*, **89**, 6056 (1967).
94. Z. Yoshida and T. Kobayashi. *Theor. Chim. Acta*, **23**, 67 (1971).

95. A. N. Sharpe and S. Walker. *J. Chem. Soc.*, 2974 (1961).
96. A. N. Sharpe and S. Walker. *J. Chem. Soc.*, 157 (1962).
97. A. N. Sharpe and S. Walker. *J. Chem. Soc.*, 2340 (1964).
98. A. D. Buckingham, J. Y. H. Chau, H. C. Freeman, R. J. W. Le Fèvre, D. A. A. S. Narayana Rao and J. Tardif, *J. Chem. Soc.*, 1405 (1956).
99. K. W. Morcom and D. N. Travers. *Trans. Faraday Soc.*, 62, 2063 (1963).
100. H. A. Rizk and H. G. Shinouda. *Z. Phys. Chem.* (Frankfurt am Main), 88, 264 (1974).
101. J. D. Broadus and J. D. Vaughan. *J. Phys. Chem.*, 72, 1005 (1968).
102. J. Crossley and C. P. Smyth. *J. Amer. Chem. Soc.*, 91, 2482 (1969).
103. M. D. Johnston, Jr., F. P. Gasparro and I. D. Kuntz, Jr., *J. Amer. Chem. Soc.*, 91, 5715 (1969).

3 Solvent Effects on Charge-transfer Complexes

K. M. C. Davis

K. M. C. DAVIS

*Chemistry Department, University of Leicester, Leicester LE1 7RH,
England*

I. INTRODUCTION

Since the vast majority of published data on molecular complexes concerns measurements of complex properties in the solution phase, solvent effects on molecular complex properties are obviously of major importance. Yet most workers in the field have not been primarily concerned with the investigation of solvent effects. The solvent has been considered to be present merely as a host to the components and their complexes and in many cases has been ignored in the discussion of properties of complexes. Often the solvent is chosen on the basis of its supposed inertness to avoid the complications arising from solute–solvent interactions. It has been increasingly obvious, however, that there is no solvent which may be regarded as being inert in the sense that solute properties in solution are identical with the same properties measured in the gas or vapour phase. During the last few years data on vapour-phase complex formation have been appearing in the literature and there has been a recent comprehensive review[1] of the results. Reliable results on complex formation in the vapour phase are essential as a basis for understanding the effect of solvent on solution equilibria particularly since the theory[2] of charge-transfer complexes is based on the concept of the isolated donor, acceptor and complex, a situation only approached in the low pressure gas phase. There have been several detailed reviews and compilations of data on molecular complexes[3-8] and all of these discuss some aspects of solvent effects on such complexes. The intention of the present review is to discuss the effect of solvent on the spectroscopic and thermodynamic properties of that class of complexes in which charge-transfer forces contribute to the stabilisation of the complex. Much of the discussion will centre on the charge-transfer (CT) properties such as the typical CT absorption band. At the same time it must be recognised that in the interaction of an electron donor and acceptor to form a complex, other forces, such as electrostatic interactions via dipolar and quadrupolar forces, play an important role. The participation of such forces in the interaction has been discussed by Hanna and Lippert.[9] Hydrogen-bonded complexes will not be considered in detail although they may be considered to be explicable in terms of charge-transfer interactions.[6] Intramolecular complexes will be discussed as will charge-transfer-to-solvent spectra although the latter are not strictly due to intermolecular complexes. The review will concentrate on the interpretation of solvent effects, it is not the intention to present exhaustive lists of data although representative compilations of data will be given where necessary.

II. CHARGE-TRANSFER BAND POSITION

A. Theoretical Considerations

1. *Nature of the Charge-Transfer Absorption Band*
A description of the charge-transfer absorption band may be given using a

simple resonance structure description. Here we will merely outline the nature of the ground and excited state. Detailed accounts of the theory have been given elsewhere.[2, 6] On this theory the ground state of a 1:1 complex may be formulated as

$$\psi_N = a\psi_0(D, A) + b\psi_1(A^- D^+) \tag{1}$$

where ψ_0 and ψ_1 are the wavefunctions of the no-bond and ionic structures corresponding to states in which D and A are held by classical van der Waals forces and in which an electron has been transferred from the electron donor, D, to the electron acceptor, A. The function is normalized

$$\int \psi_N \psi_N \, dt = a^2 + b^2 + 2abS_{01} = 1 \tag{2}$$

where

$$S_{01} = \int \psi_0 \psi_1 \, dt \tag{3}$$

For weak complexes, S_{01} is small and $a^2 \gg b^2$. b^2 represents the weight of the dative structure or fraction of the electron transferred from D to A in the ground state and

$$a^2 + b^2 \approx 1 \tag{4}$$

Quantum theory demands that there also be an excited state ψ_V given by

$$\psi_V = a^*\psi_1(D^+ - A^-) - b^*\psi_0(D, A) \tag{5}$$

This may be termed a charge-transfer state. ψ_V is normalized by

$$\int \psi_V \psi_V \, dt = a^{*2} + b^{*2} - 2a^*b^*S_{01} = 1 \tag{6}$$

The CT transition is then the transition $V \leftarrow N$ and since $a^{*2} \gg b^{*2}$, the transition is essentially a transfer of an electron from D to A. The theory may also be used[6] to predict a moderately high intensity for the transition $V \leftarrow N$ in accord with the values of ϵ, the molar absorption coefficient (extinction coefficient), and f, the oscillator strength generally observed.[5] The CT transition thus involves a ground and excited state with very different dipole moments which suggests that solvent polarity effects should be marked on the position of the CT absorption band.

2. Theory of Solvent Shifts of Spectral Absorption Bands

Before discussing the observed solvent shifts of CT bands, it will be instructive to consider the theoretical background to such shifts.

(a) *Spectral shifts between vapour and solution* The transition energies hv_{CT} in vapour and solution phases may be related by the cycle[1, 10]

$$
\begin{array}{ccc}
\text{DA (g)} & \xrightarrow{\ hv_{CT}\,(g)\ } & \text{D}^+\!\!-\!\!\text{A}^-\ (g) \\[2pt]
\Big\downarrow {\scriptstyle \Delta X_{DA}(solv)} & & \Big\downarrow {\scriptstyle \Delta X_{D^+\!-A^-}\,(solv)} \\[2pt]
\text{DA (soln)} & \xrightarrow{\ hv_{CT}(soln)\ } & \text{D}^+\!\!-\!\!\text{A}^-\ (soln)
\end{array}
\qquad (7)
$$

where ΔX_{solv} includes all solvation contributions to the stabilization of the complex in ground and excited states. Thus

$$hv_{CT}(g) = hv_{CT}(soln) - [\Delta X_{D^+ -A^-}(solv) - \Delta X_{DA}(solv)]. \qquad (8)$$

Thus, if solvent stabilization of the excited state is greater than that for the ground state, we might expect a shift to lower energies of the CT band on moving from vapour to solution. In practice the cycle (7) is not easily accessible to calculation since it is the Franck–Condon excited state[11] to which the transition occurs. The Franck–Condon principle states that electronic transitions occur in a time, short compared with that required for a change in position of the nuclei. Thus the excited state is formed with the geometry of the ground state. The Franck–Condon restriction also applies to movement of the solvent molecules and hence the Franck–Condon excited state will be surrounded by the equilibrium ground-state solvation shell. In a time, short compared with the fluorescence lifetime, the solvent shell will relax to form the equilibrium excited state, lower in energy than the Franck–Condon state. Any fluorescence will then occur from this equilibrium excited state to the Franck–Condon ground state which will still be surrounded by the equilibrium excited-state solvation shell. This is the source of the Stokes shift separation of absorption and fluorescence bands in solution. Figure 1 illustrates the various energy levels. All the above assumes that ground and excited states possess different electronic distributions. The difference between the Franck–Condon and equilibrium states will depend on the change in electron distribution on excitation. This should be large for CT complexes and we may predict at least, that Stokes shifts will be large for CT spectra.

(b) *Solute–solvent interactions* The important types of interaction contributing to the $\Delta X_{(solv)}$ terms in (7) may be classified[12] as
  1. Dipole–dipole interaction where both solute and solvent are polar.
  2. Solute permanent dipole–solvent-induced-dipole interaction where the former is polar and the latter is not.
  3. Solvent permanent dipole–solute-induced-dipole interactions for the reverse case.
  4. Dispersion interaction between the solute transition-dipole and the dipole induced in the solvent.

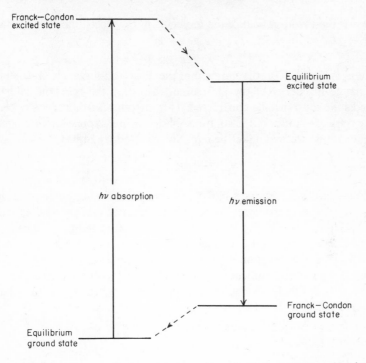

FIG. 1. Schematic energy level diagram for absorption and emission transitions in solution.

5. Specific interactions such as hydrogen-bonding, CT-complex formation, etc.

6. Solvent-cage compression effects on the solute where the solute occupies a volume larger than the site in the solvent quasi-lattice.

1–4 may be classified as general electrostatic solvent effects and much of the theoretical work on solvent shifts involves these interactions. However, where present, 5 and 6 are likely to dominate the solvent–solute interaction.

(c) *Theoretical expressions for solvent shifts*   Most frequently, theoretical discussions of solvent shifts are based on the bulk properties of the solvent medium such as dielectric constant, ϵ', and refractive index, n. Basu[13] has reviewed some of the theoretical approaches. Dielectric theory derives the solvation energies in terms of a reaction field, F, induced by the solute in an Onsager[14] cavity of radius a. The reaction field is difficult to determine accurately but useful approximations have been developed.[15–27] Representing the dipole moment as a point dipole located in the centre of the cavity

F*

yields for the reaction field of the molecule in the electronic ground state

$$F_R = \frac{2(\epsilon' - 1)}{a^3(2\epsilon' + 1)}\mu_g \tag{9}$$

where μ_g is the total dipole moment of the dissolved molecule. To determine the change in energy of the electronic transition, the reaction field in the excited state must also be considered. This introduces other terms depending on the refractive index of the solution. The resulting expressions are complex. As an example we may take the equation derived by Liptay.[23, 25]

$$\bar{\nu}_{soln} = \bar{\nu}_g - \frac{(\mu_e - \mu_g)F_{RM}}{hc} - \frac{2(n^2 - 1)}{a^3(2n^2 + 1)}D \tag{10}$$

$\bar{\nu}_{soln}$ and $\bar{\nu}_g$ are wavenumbers of the absorption maximum in the solution and the gas phase respectively, μ_e and μ_g are the excited and ground state dipole moments respectively, F_{RM} is the effective reaction field, the mean of the reaction fields in ground and excited states, and D is a term describing the dependence of the position of the absorption band on the dispersion inter- actions between the molecule and surrounding solvent molecules. Since usually $D > 0$, the dispersion term always causes a red shift with increasing refractive index of the medium.

$$F_{RM} = \frac{1}{a^3}\left[\frac{\epsilon' - 1}{2\epsilon' + 1}\left(1 - \frac{2a_g(\epsilon' - 1)}{a^3(2\epsilon' + 1)}\right)^{-1}2\mu_g \right.$$
$$\left. + \frac{(n^2 - 1)}{(2n^2 + 1)}\left(1 - \frac{2a_g(n^2 - 1)}{a^3(2n^2 + 1)}\right)^{-1}(\mu_e - \mu_g)\right] \tag{11}$$

a_g is the polarizability tensor in the ground state. The above equation applies when ground and excited state dipole moments are parallel. Analogous equations may be derived for the case where the two moments are not parallel. The equation underlines the complexity of the interaction and has been reasonably successful in predicting shifts. The equation also shows the fundamental impossibility of deriving a single-parameter function to cover solvent effects over the whole solvent range. However, for a non-polar solute, F_{RM} disappears giving

$$\nu \propto \frac{(n^2 - 1)}{(2n^2 + 1)} \tag{12}$$

while for molecules with a large ground-state dipole moment, the refractive index term is small compared with that due to the reaction field and

$$\nu \propto \frac{2(\epsilon' - 1)}{(2\epsilon' + 1)} \tag{13}$$

The important predictions to be made from these equations are firstly that non-polar solutes should invariably suffer a red (bathochromic) shift on moving from vapour to solution and further red shifts as the refractive index of the medium increases. These shifts should be small owing to the small range of solvent refractive indices. Secondly, for polar molecules, the solvent dependence will be determined by the solute dipole moment in the ground state and the change in dipole moment on excitation. An increase in dipole moment on excitation, $\mu_e > \mu_g$, should lead to a red shift. Complexes with neutral ground states fall into this group and should thus exhibit red shifts. If the dipole moment decreases on excitation, $\mu_e < \mu_g$, then a blue (hypsochromic) shift should occur with increasing dielectric constant. Complexes with a contribution of the dative state in the ground state of greater than 50 per cent ($b^2 > 0.5$) should therefore exhibit blue shifts.

This dielectric constant dependence is superimposed on the red shift due to dispersion interactions and the change in dipole on excitation. Thus if the dipole moment increases on excitation there should always be a red shift from vapour to solution. Blue shifts may occur if the dipole moment decreases on excitation depending on the balance between the two terms in equation (10). The largest solvent shifts are to be expected for highly polar solutes which undergo a large reduction in dipole moment on excitation. This accounts for the very large hypsochromic shifts observed for solvatochromic molecules and complexes.[28, 29] These molecules are ionic in the ground state, the charge being annihilated on excitation. They thus fulfil the conditions of large μ_g and large values of $\mu_e - \mu_g$ required for large shifts of the absorption band.

(d) *Specific effects*   The most obvious specific interactions of solute and solvent will be through hydrogen-bonding interactions. In protic solvents, hydrogen bonding of solute to solvent may produce anomalous shifts giving deviations from the predictions of the dielectric theory. Suppan[30] has calculated the shifts for the intramolecular CT transition in aniline between hydrogen bonding and hydrogen neutral solvents with the same dielectric constant function. These are shown in Table I.

TABLE I. Hydrogen-bonding effects (cm^{-1}) in the solvent shift of aniline[a]

Solvents	(^1L_b)		(^1L_a)	
Dimethylformamide	34 000 ⎫			
Ethanol	34 300 ⎬ 300		42,300 ⎫	
Water	35 700	1400	43,480 ⎬ 1180	
Δq_{g-e} (electrons)	0.20		0.18	
$\Delta \mu_{g-e}$ (for 2.7 Å) (D)	2.7		2.4	

From $\Delta E = \Delta q_{g-e} q_{OH} [1/d - (1/d + r)]$ with $q_{OH} = 0.4e$; $d = 2 \times 10^{-8}$ cm; and $d + r = 3 \times 10^{-8}$ cm.

[a] Reproduced with permission from P. Suppan, *J. Chem. Soc.* (A), 3125, (1968).

The calculations involve interaction of the O—H solvent dipole with the change in charge distribution on going from ground to excited state. A blue shift is predicted in accord with observations on similar intramolecular transitions.[30] Similar anomalies would be expected for solvents capable of acting as electron donors or acceptors particularly for CT transitions where competitive donor–acceptor interaction by the solvent will modify the ground and excited state electron distributions.

(e) *Solvent cage effects* On the lattice model of liquids, for a solute with larger dimensions than the site in the quasi-crystalline structure of the solvent, the intermolecular forces will tend to reduce the volume occupied by the solute.[31] This compression effect of the solvent cage will be small for intramolecular transitions. For CT complexes, however, the intermolecular bond is weak and bond compression will occur more easily.[31,32] Both Prochorow and Tramer[31] and Kroll[32] have discussed this effect for molecular complexes. R_x, the average ionic radius of the ion pair, is usually less than R_e, the equilibrium intermolecular separation in the complex ground-state. Since the potential energy curve for the ground state is shallow, important modifications of R_e may take place under the compressive effects of the solvent cage. This should result in red shifts from vapour to solution since the effect of bond compression will be to displace the Franck–Condon transition along the R-axis as shown in Fig. 2.

Trotter[33,34] has discussed the bond compression model and has explained vapour-to-solution red shifts as resulting from mechanical pressure by the solvent. Offen[35] has suggested that the bond compressibilities assumed by Trotter are unreasonably high but that the observed red shifts of CT bands on applying mechanical pressure may be explained by assuming smaller compressibilities. These pressure shifts[35–37] add support to the participation of solvent cage compression shifts in the observed vapour–solution shifts. Kroll[32] has derived an expression for the shift from gas to solution:

$$h\nu_{\text{CT(gas)}} - h\nu_{\text{CT(soln)}} = e^2\left(1 - \frac{1}{\epsilon'}\right)\left(\frac{1}{R_x} - \frac{1}{R_e}\right) \tag{14}$$

where ϵ' is the dielectric constant. This predicts a red shift if $R_e > R_x$, the shift increasing with ϵ'.

B. Experimental Results

1. Vapour–Solution Shifts

Much of the general theory of band shifts discussed above, refers to calculations based on single molecules. It is to be expected that the situation for

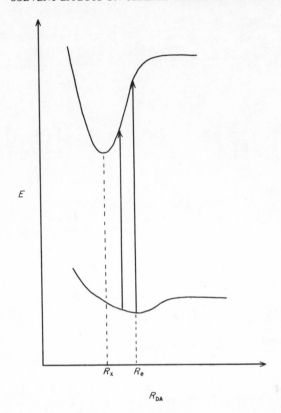

R_x R_e

R_{DA}

FIG. 2. Potential energy surfaces for ground and excited states of a weak complex showing the effect of bond compression on the absorption frequency.

complexes will be more complicated since interaction of the solvent can occur not only with the complex dipole but also with local dipoles on the component donor and acceptor moieties. These local dipoles may be similar to or even greater than the magnitude of the complex dipole moment and are unlikely to be in the same direction.[38] Murrell[39] has argued that much of the intensity of CT transitions comes from intensity borrowing from local excitations of donor and/or acceptor. Such borrowing, involving as it does, quantum mechanical mixing of CT and component wavefunctions will depend on the energy separation of CT and component excited states which may well vary with solvent.

The majority of complexes have a small contribution of the dative structure in the ground state and the excited state is virtually ionic. The vapour-to-solution red shift as predicted by the dielectric theory should thus be substantial as the change in dipole moment on excitation will be large. Table II

TABLE II. Charge-transfer band maxima (cm⁻¹) in the vapour phase and in solution in a solvent of low polarity[a]

Complex	$\bar{\nu}_{max}$ (vap.)/cm⁻¹	Ref.	$\bar{\nu}_{max}$ (sol.)/cm⁻¹	Ref.	Solvent	$\Delta\bar{\nu}_{max}$/cm⁻¹
Benzene–iodine	~37 310	41, 42	34 720	43	n-heptane	2590
Mesitylene–iodine	33 220	40	30 580	43	n-heptane	2640
Diethyl ether–iodine	42 740	44	40 000	43	n-heptane	2740
Furan–iodine	35 210	45	31 750	45	n-heptane	3460
Diethyl sulphide–iodine	34 480	46	33 000	32	n-heptane	1489
Thiocyclopentane–iodine	34 130	32	32 680	32, 47	n-heptane	1450
Iodine–iodine	~40 820	59	~34 070	60	n-heptane	6750
Benzene–TCNE	28 990	48	26 530	32	n-heptane	2460
Toluene–TCNE	27 030	32	25 000	32	n-heptane	2030
o-Xylene–TCNE	25 380	32	23 870	32	n-heptane	1510
p-Xylene–TCNE[b]	{23 530, 26 880}	32, 49	{21 880, 25 130}	32, 49	n-heptane	{1650, 1750}
Mesitylene–TCNE	23 640	32	22 370	32	n-heptane	1270
Durene–TCNE[b]	{21 140, 23 260}	32	{20 000, 21 980}	32	n-heptane	{1140, 1280}
Pentamethylbenzene–TCNE	20 960	32	20 000	32	n-heptane	960
Hexamethylbenzene–TCNE	19 690	32	18 940	32	n-heptane	750
Naphthalene–TCNE[b]	{21 100, 26 320}	50, 51	{17 990, 22 990}	50, 51	CCl₄	{3110, 3330}
Biphenyl–TCNE[b]	{22 220, 28 570}	50, 51	{20 000, 25 710}	50, 51	CCl₄	{2220, 2860}
Acenaphthene–TCNE	18 100	52	15 300	52	CCl₄	2800
Pyrene–TCNE[b]	{13 500, 16 530, 22 990}	50, 51	{14 010, 20 490, 25 770}	53	CCl₄	{510, 3960, 2780}
Diethyl ether–CO(CN)₂	38 200	31, 54	34 800	31, 54	n-hexane	3400
Dioxan–CO(CN)₂	37 700	31, 54	34 600	31, 54	n-hexane	3100
Benzene–CO(CN)₂	34 600	31, 54	32 400	31, 54	n-hexane	2200
Toluene–CO(CN)₂	33 100	31, 54	30 300	31, 54	n-hexane	2800
p-Xylene–CO(CN)₂[a]	{33 000, 28 500}	31, 54	{29 800, 26 000}	31, 54	n-hexane	{3200, 2500}
Trimethylamine–SO₂	36 200	55, 56	36 600	55, 56	n-heptane	−400
trans-But-2-ene–SO₂	40 000–41 700	48	38 200	87	n-hexane	1800–3500
Diethylamine–I₂	33 300	58	38 500	3	n-heptane	−5200
Triethylamine–I₂	31 750	58	36 000	3	n-heptane	−4250

[a] Data taken, with permission, from the compilation by M. Tamres, "Molecular Complexes", Vol. 1. (Ed. Roy Foster), Elek. Science, 1973, p. 49.
[b] Multiple CT bands.

shows a selection of data on CT transitions in the vapour phase and in solution in a non-polar solvent, taken from the review by Tamres.[1] The list of complexes for which both vapour and solution data is available is somewhat restricted, but it is clear from Table II that in the majority of cases, the predicted red shift is observed. The predictions of both the dielectric and bond compression theories are validated. At present it is not possible to separate the contributions due to the two effects which awaits the development of a more sophisticated theoretical treatment.

From Table II it may be seen that all types of complex studied exhibit red shifts. For iodine complexes, the shifts vary with strength of complexing.[1] The very weak iodine dimer[1, 59, 60] exhibits a very large shift of ca 6000 cm^{-1}, although the experimental results are subject to some uncertainty. The aromatic hydrocarbon–iodine complexes[1, 40–43] show shifts of the order of 3000 cm^{-1} or more. The n-donor complex of diethyl ether[1, 43, 44] with iodine which is more stable has a shift of less than 3000 cm^{-1} while in the stronger alkyl sulphide–iodine complexes[1, 32, 46, 47] the shifts have been reduced to the order of 1500 cm^{-1}. The very strong complexes of aliphatic amines with iodine appear to give blue shifts[3, 58, 166] as does the SO$_2$– trimethylamine complex[55, 56] which gives a blue shift of ca 400 cm^{-1}. The π-complexes of tetracyanoethylene[32, 48, 50, 51] show red shifts of 1000–2000 cm^{-1} again decreasing with donor strength. This trend to lower red shifts with increasing donor strength is not followed for the complexes of carbonyl cyanide with ethers[31, 54] and aromatic hydrocarbons.[31, 54] Here exactly the reverse trend is observed, the shift increasing with donor strength. Whereas the trend in shifts is fairly systematic for a series of closely related donors with one acceptor, there appears to be no simple pattern emerging for widely different donors. The dependence of the red shift on strength of complexing may be rationalised in terms of the dielectric theory predictions. The b coefficient in equation (1) should increase with increasing donor strength. This should cause an increase in the complex ground-state dipole moment and a decrease in $(\mu_e - \mu_g)$. The shifts should then decrease in line with this trend. The results for the carbonyl cyanide complexes are hard to rationalise on this basis. The bond compression theory also would predict a larger red shift as the intermolecular bond becomes weaker. Other factors such as solvation of the components may well be involved. It is tempting to explain the blue shift for the trimethylamine–SO$_2$ complex by ascribing more than 50 per cent dative character to the ground state. Further blue shifts are found for this complex on increasing the solvent polarity as are also observed for other strong complexes such as the iodine complexes of N,N-dimethylthioformamide,[61] triphenylarsine,[62] and tetramethylthiourea.[63] Christian and Grundnes[56] calculate, however, that dipole moment data for the trimethylamine–SO$_2$ complex do not support a ground state of more than 50 per cent ionic character. They propose that the observed blue shift is the result of a polarisation red shift as predicted and a larger blue shift caused by solvent enhancement of the contribution of the dative structure to the ground-state wave-function. This would cause a separation of the ground and excited states with a resultant blue shift. It is of interest here to mention that the position of the CT band for the complex SO_2-I^- appears to be insensitive to change of solvent.[64] Here the ground and excited states presumably have equal dative character with a virtually zero resultant shift. The explanation of blue shifts in which the ground-state dative character is modified by the solvent is also favoured by Offen and Abidi[65] to explain such shifts in highly polar solvents. While solvents of high polarity may well enhance the charge separation in the ground state, it is difficult to see sufficient stabilisation to cause large blue shifts arising from interaction with a solvent such as n-heptane.

2. Solvent Dependence of Absorption Band Shifts

(a) *Correlation with dielectric theory*   Theoretically the dielectric theory predicts for the CT transition, a correlation with the dielectric constant function or, for non-polar solvents, with the refractive index function. Mason[66] and Murrell[67] in reviews of CT spectra, predicted increasing bathochromic shifts with increasing solvent polarity, although early attempts failed to correlate the shift with the dielectric constant.[3] Rosenberg and co-workers[68-70] pointed out that for non-polar complexes, the Franck–Condon restriction would mean that the orientation of solvent around the ground state would not stabilize the highly polar excited Franck–Condon state.[72] The major correlation might therefore be expected with a measure of the polarizability interactions, i.e. the refractive index function. They found that the red shift for a number of π-complexes increased with decreasing refractive index. A similar effect was observed by Ham[71] in the course of pressure studies on iodine complexes. Hydrogen bonding and electron-donor solvents gave anomalous shifts. Suppan and coworkers[30, 73] found reasonably good correlations of the band frequency of intramolecular CT transitions with the dielectric constant function, $(\epsilon' - 1)(2\epsilon' + 1)^{-1}$ as shown in Fig. 3.

The polarizability function $(n^2 - 1)(2n^2 + 1)^{-1}$ correlated well with the shifts in the non-polar group of solvents.[30, 73] The data were scattered, the hydrogen-bonding solvents again being anomalous, and correlations within a particular group of solvents such as the alcohols were poor.

Voigt[74] and Inokuchi and coworkers[51] found very good correlations between the polarizability function and both the first and second CT band positions for tetracyanoethylene (TCNE) complexes. The group of solvents studied consisted of hydrocarbons and halogenated hydrocarbons. When

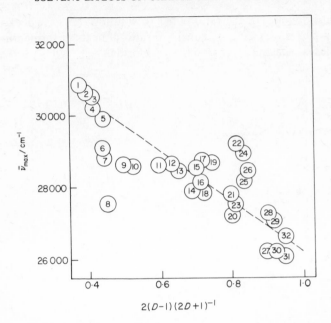

FIG. 3. $\bar{\nu}_{max}$ (cm^{-1}) for the first absorption band of 4-nitro-aniline plotted against $2(D - 1)$ $(2D + 1)^{-1}$ where D is the dielectric constant. Reference numbers of solvents: (1) iso-octane; (2) n-heptane; (3) cyclohexane; (4) carbon tetrachloride; (5) tetrachloroethane; (6) mesitylene; (7) benzene; (8) dioxan; (9) triethylamine; (10) valeric acid; (11) propionic acid; (12) iso-propyl ether; (13) diethyl ether; (14) bromoform; (15) chloroform; (16) n-butyl acetate; (17) m-dichloro-benzene; (18) iso-butyl acetate; (19) chlorobenzene; (20) methyl benzoate; (21) methyl acetate; (22) 1-chlorobutane; (23) ethyl benzoate; (24) iso-propyl bromide; (25) dichloromethane; (26) o-dichlorobenzene; (27) benzyl alcohol; (28) acetone; (29) benzonitrile; (30) pentan-1-ol; (31) propan-2-ol; (32) ethanol. [Reproduced with permission from M. B. Ledger and P. Suppan, *Spectrochim. Acta*, **23A**, 641 (1967)].

extrapolated to zero both groups found good agreement with the observed gas phase transition energies. Although good linear plots were obtained this was only so for a particular group of solvents. Voigt[74] found different slopes for the hydrocarbon and halogenated hydrocarbon groups of solvents. The linearity of the plot was good for any one group and both plots extra-polated to the same predicted gas-phase value (Fig. 4).

As for the shifts from vapour to solution, the solvent shifts decreased with increasing donor strength as measured by the donor ionization potential. The results accord with the predictions of Bilot and Kawski[19] who stated that the ratios of shifts should be in the ratio of the differences between the square of the excited and ground state dipole moments.

$$\frac{\Delta E_{\text{I}}}{\Delta E_{\text{II}}} = \frac{\mu_{e_{\text{I}}}^2 - \mu_{g_{\text{I}}}^2}{\mu_{e_{\text{II}}}^2 - \mu_{g_{\text{II}}}^2} \qquad (15)$$

Voigt[74] also found that the ratio of gas to solution shifts for benzene and durene complexes of TCNE equalled the ratio of heats of formation of the two complexes measured in carbon tetrachloride.

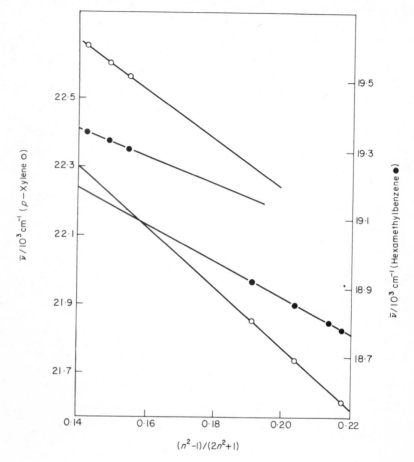

FIG. 4. Charge-transfer energies (cm^{-1}) of tetracyanoethylene complexes with p-xylene and hexamethylbenzene vs the refractive index function $(n^2 - 1)(2n^2 + 1)^{-1}$ for perfluoro and hydrocarbon nonpolar solvents. [Reproduced with permission from E. M. Voigt, *J. Phys. Chem.*, **70**, 598 (1966)].

Although for these limited groups of solvent, a reasonable agreement is found with the dielectric theory predictions, the correlations found over the whole solvent range are poor. Even for intramolecular CT transitions where reasonably good overall correlations with both dielectric and polarizability functions occur, considerable anomalies occur in the case of hydrogen-bonding solvents.[30, 73] For intermolecular complexes[51, 65, 68–70, 74, 75, 76] the generally observed red shift in non-polar solvents becomes a blue shift for the more highly polar solvents. This leads, for instance, to the value of $\bar{\nu}_{CT}$ being the same in the vapour and in acetone solution for the complex of hexamethylbenzene–tetracyanoethylene.[65] This reversal in the trend from that predicted is yet to be explained satisfactorily. Several hypotheses have been advanced in explanation and some or all of these undoubtedly contribute. Czekalla and Meyer[77] have proposed that the complex changes in nature on interaction with a polar solvent. The smaller $\Delta H^{\ominus}$ values for complex formation in these solvents would decrease the binding energy leading to a looser complex and consequently to a lower coulomb energy in the excited state. Offen and Abidi[65] favour an explanation involving a larger splitting of the ground and excited states. Here the polar solvent is held to increase the ease of formation of the dative structure in the ground state again leading to larger splitting of the two states. This is analogous to the explanation advanced for the blue vapour-to-solution shifts for strong complexes.[56] Davis and Symons[38] have pointed out that solvent shifts of complexes are often of comparable size and sign to shifts of intramolecular transitions of the constituent donor and acceptor. They conclude from this that solvation may be controlled by local polar centres on D and A rather than by the usually small ground state complex dipole. Emslie and Foster[78] have advanced similar arguments and also point out that polar solvents will form a more highly organized structure around the complex whereas solvents of low polarity will form a more disorganized structure. Beaumont and Davis[79] have shown that complexes with a neutral ground-state formed from components with large local dipoles often exhibit larger shifts between a given pair of solvents than the blue shifts in the same solvents found for ionic ground-state complexes whose component ions lack strong local dipoles. Strong complexes with a large complex dipole, presumably able to dominate the solvent organization, also give large shifts in polar solvents.[79]

Solvent-shift data have been used in conjunction with the dielectric theory to obtain excited-state dipole moments and the contributions of the dative structure to the two states. The dipole moments of ground and excited states are given by

$$\mu_g = b^2\mu_1 + abS\mu_1 \tag{16}$$

and

$$\mu_e = a^{*2}\mu_1 - a^*b^*S\mu_1 \tag{17}$$

where μ_1 is the dipole moment of the completely charge-separated state.

Chakrabarti and Basu[80] used solvent shift data and the equation[13,81]

$$h\nu_{\text{gas}} - h\nu_{\text{soln}} = \frac{1}{a^3}\frac{(2n^2 + 1)}{(n^2 + 2)}\left[2\mu_g(\mu_g - \mu_e)\left(\frac{\epsilon' - 1}{\epsilon' + 2} - \frac{n^2 - 1}{n^2 + 2}\right)\right] +$$
$$\left(\frac{n^2 - 1}{n^2 + 2}\right)(\mu_{g2} - \mu_{e2}) \qquad (18)$$

Since $\mu_e > \mu_g$ for non-polar complexes, a red shift should be obtained with increasing polarity. They used the observed blue shift from carbon tetra-chloride to chloroform to methyl acetate and by reversing the sign of the shift obtained reasonable values of 8–15D for tetrahalogenated quinone–aromatic hydrocarbon complexes.† Hence they obtained a^* and b^* co-efficients and estimated 45–75 per cent ionic character for the excited state. Offen and Abidi[65] used the same method of sign reversal and again obtained reasonable values for excited-state dipole moments for tetrachlorophthalic anhydride–hexamethylbenzene. As they point out, there is no theoretical justification for this method, the signs of the shifts being opposite to those theoretically predicted. Dipole moments of excited states may also be obtained from fluorescence measurements and are broadly in agreement with those obtained from absorption studies.

The general failure of the macroscopic solvent polarity parameters to correlate with solvent shifts over the whole solvent range may be attributed to two main causes. Firstly, the specific interactions between solute and solvent are ignored and these are likely to become more important as the solvent polarity increases. Secondly, the model used is of necessity, simplified using a model of the solute and solvent as point dipoles surrounded by a homogeneous, isotropic, dielectric continuum. What is required is a para-meter which represents the solvent situation in the microscopic region sur-rounding the solute. To this end, many workers have sought an empirical probe for solvent polarity.

(b) *Empirical solvent polarity scales* A solvent dependent property is sought which should mirror the microscopic solvent environment including specific interactions. Thus, if we assume that the contributions from the various intermolecular forces are similar in the standard process and in the process under investigation, a good correlation should emerge. Solvent polarity parameters have been reviewed by Reichardt[82] and have been critically discussed by Katritzky[83] and by Nagy[84] and their coworkers.

---

† The values of dipole moment are quoted as in the original literature in debye units (D). This unit is related to the *SI* unit, the coulomb metre (C m) by:

$$D = 3\cdot335640 \times 10^{-30}\,\text{C m}$$

The first attempts used the solvent-dependent reaction rates of solvolysis occurring by an S_N1 process. Since the rate determining step in this process involves ionisation, the rate is markedly solvent dependent. Grunwald and Winstein[85–87] defined a Y value as

$$\log \frac{k}{k_0} = mY \tag{19}$$

where k_0 and k are the rate constants in the standard solvent, 80 per cent ethanol in water, and the solvent under investigation. Y values gave better correlations with rate data than did dielectric constant. The Y values were limited to a few polar solvents and their mixtures with water. Using other reactions, values were obtained for a wider range of solvents.[88] Other parameters derived from rate data include X[89] values and Ω[90] values*. These parameters correlate well with each other although discrepancies do occur particularly for protic solvents.

The first suggestion as to the use of solvatochromism was by Brooker and his coworkers.[91] The first comprehensive scale of solvent polarity values was derived by Kosower[28, 92] who used the charge-transfer transitions in 1-alkylpyridinium halides. Using 1-ethyl-4-carbomethoxypyridinium iodide (1) as a standard, the CT transition energies in a large number of solvents could be obtained. Since this is an ionic ground-state complex in which the charge is annihilated during the transition; the hypsochromic shifts are large due to the stabilisation of the ground-state ion-pair as the solvent polarity increases. Z values are the transition energies in kilocalories and a selection is shown in Table III. Since 1 is insoluble in non-polar

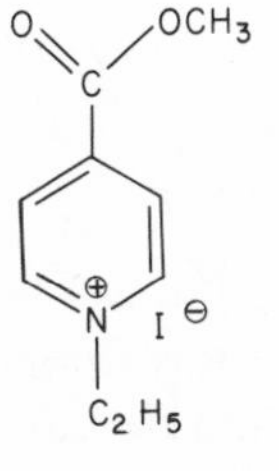

1

---

* Ω is defined by $\log(k_n/k_x) = \log(N/X) = \Omega$ where N/X is the ratio of endo-product to exo-product in the solvent-dependent Diels–Alder addition of cyclopentadiene to methylacrylate, k_n and k_x being the specific rate constants.

X is defined by $\log(k/k_0 = pX$. k and k_0 are the rate constants of an electrophilic aliphatic substitution on an organometallic compound in a given solvent or in glacial acetic acid which is used as the standard solvent ($X = 0$), p is a constant which depends on the electrophile and on the organometallic compound.

TABLE III. A selection of solvent polarity parameters, Z, E_T and ϵ', the dielectric constant at 25°C.

Solvent	Z/kcal mol^{-1}	E_T/kcal mol^{-1}	ϵ
Water	94·6	63·1	78·5
Formamide	83·3	56·6	109·5
Ethylene glycol	85·1	56·3	37·7
Methanol	83·6	55·5	32·6
N-Methylformamide		54·1	182·4
Ethanol	79·6	51·9	24·3
Acetic acid	79·2	51·9	6·2
Propan-1-ol	78·3	50·7	20·1
Butan-1-ol	77·7	50·2	17·1
Propan-2-ol	76·3	48·6	18·3
Propylene carbonate		46·6	65·1
Nitromethane		46·3	38·6
Acetonitrile	71·3	46·0	37·5
Dimethyl sulphoxide	71·1	45·0	48·9
t-Butanol	71·3	43·9	12·2
Dimethylformamide	68·5	43·8	36·7
Acetone	65·7	42·2	20·7
Dichloromethane	64·2	41·1	8·9
Chloroform	63·2	39·1	4·7
Ethyl acetate	59·4	38·1	1·85
Tetrahydrofuran		37·4	7·4
1,4-Dioxan		36·0	2·2
Diethyl ether		34·6	4·2
Benzene		34·5	2·3
p-Xylene		33·2	2·4
Carbon disulphide		32·6	2·6
Carbon tetrachloride		32·5	2·2
Cyclohexane		31·2	2·0
n-Hexane		30·9	1·9

solvents, Z for these solvents had to be obtained using a secondary standard, pyridine-N-oxide. The shifts for this compound are small and the correlation with Z only moderately good and hence the Z values for non-polar solvents are less reliable. In highly polar solvents, the CT band moves under the intramolecular absorption band of the pyridinium ion and Z values are obtained from the good correlation of Z with Y.. For the pyridinium halide complexes, the large ground-state dipole will be solvated by the solvent dipoles in an orientation to minimize the ground-state energy. The region of solvent around the complex was termed by Kosower[28] the cybotactic region. The complex dipole in the ground state is at right angles to the pyridinium ring but will lie in the plane of the ring in the excited state. Since there is no solvent reorganization on excitation, the excited state should be de-stabilized by the ground-state solvent shell. It is doubtful, however, that ground-state stabilization and excited-state destabilization will be equal in magnitude as Kosower[28] suggested. A similar polarity scale is given by the E_T values of Reichardt and coworkers.[29, 82, 93] These values are the CT transition energies in kilocalories of the intramolecular pyridinium phenol-betaines 2 and 3. These compounds have the largest solvatochromic shifts of any compound yet investigated, the CT band moving from 810 nm in diphenyl ether to 453 nm in water for 2. Using 3 which is soluble in hydrocarbon solvents and has shifts correlating well with those of 2, a comprehensive range

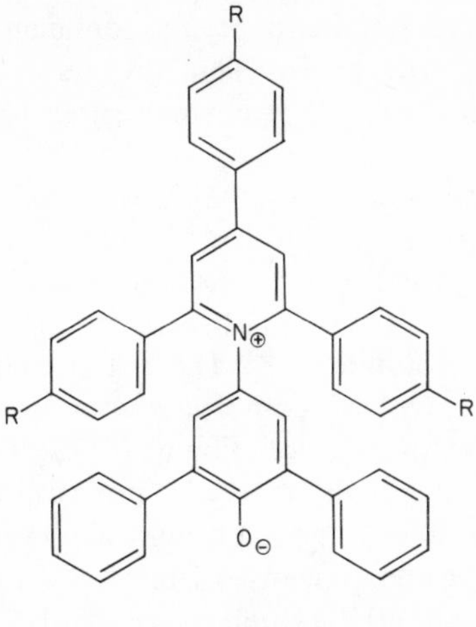

2 R = H
3 R = CH$_3$

of E_T values may be obtained over the whole solvent range. A selection of values is shown in Table III. E_T values are particularly useful in studies in non-polar solvents for which Z is suspect. Z and E_T have an excellent linear correlation over the middle range of solvent polarities although they disagree in highly polar and non-polar solvents. It is interesting to note[29] that solvents with very high dielectric constants like formamide and N-methylformamide do not have equally high E_T values which casts light on the failures of dielectric theory to predict shifts in these solvents.

One of the most general solvent polarity parameters is provided by the S and R values of Brownstein.[94] He uses a linear free enthalpy relationship analogous to the Hammett[95] relationship. Brownstein's equation is

$$\log \frac{k_s}{k_E} = S \cdot R \qquad (20)$$

where k_s is the rate constant, equilibrium constant or spectral shift in a solvent at 298 K and k_E is the corresponding value in ethanol. S is a solvent dependent parameter ($S = 0.00$ for ethanol) and R is a measure of the sensitivity of the system towards a change of solvent. Brownstein uses compound **1** as a standard with an R value of 1.00. With this standard solvent and process one hundred and fifty-eight S and seventy-eight R values were determined yielding a comprehensive generalization of empirical data. Two other parameters worthy of mention in the context of this review are the S_M values[96] and $\Delta\bar{v}_D$ values.[97] S_M values are based on the quaternization rates of tri-n-propylamine by methyl iodide while $\Delta\bar{v}_D$ values in cm^{-1} measure the electron-donor or electron-acceptor properties of the solvent and are the values of the O—D stretching frequency in monodeuterated methanol using benzene as a standard solvent. Excellent reviews of solvent polarity parameters have been published[82, 96, 98] in which other parameters are included.

(c) *Correlation of charge–transfer band shifts with empirical solvent parameters*
Both Z and E_T values correlate well with shifts of other solvent sensitive absorptions. Z gives a good linear correlation with the n–σ^* absorption in alkyl halides,[99] the n–π^* and π–π^* absorptions of mesityl oxide[28, 92] and π–π^* absorption in phenol blue.[28, 92] The first attempts to seek correlations with CT band shifts involved complexes with ionic ground-states. Kosower[100] found a good linear correlation with a slope close to unity between Z and E_{CT} ($=h\nu_{CT}$) for tropylium iodide, not surprisingly in view of the similarity of the transitions. A good straight line of unit slope is found for Z vs E_T, where again the charge transfer from O^- of the phenolate ion to the pyridinium ring in **2** is essentially similar to that in **1**. Similar plots with slope again close to unity are obtained for E_{CT} of halide-ion complexes of other planar cations such as N-methylquinolinium and N-methylacridinium.[79] Davis and Beaumont[101] have argued that the solvent sensitivity of such transitions, involving electron transfer from a highly charge-localized anion to the cation within an ion pair should be close to the maximum observable. They suggest that transitions involving larger charge-delocalized anions should exhibit less solvent sensitivity since the solvent organization in complexes such as **1** should be dominated by the small anion. The lower charge density on the cation will have a much smaller solute–solvent interaction. This is borne out by results on iodide ion–neutral-acceptor complexes which again give slopes close to unity for E_T vs E_{CT}.[102, 103] Kosower and Ramsay[104] found the slope of Z vs E_{CT} for the intramolecular transition in pyridinium cyclopentadienylide to be much greater than unity although a good straight line was obtained. A similar high value for the slope was found for the tropylium–pentamethoxycarbonylcyclopentadienylide (PCCD) complex.[79] For complexes of pyrylium and thiopyrylium cations with

1,1,3,3-tetracyanopropenide and tricyanomethanide anions[105] intermediate values of slope 1·5–2·0 were found for the Z vs E_{CT} plots. Complexes of tropylium cation with aromatic hydrocarbon donors[79, 106] and of the PCCD anion[79] with tetracyanoethylene show an almost complete insensitivity to solvent polarity of $\bar{\nu}_{CT}$, although large shifts are obtained for solvents like dichloromethane and nitrobenzene, presumably due to specific complex formation.[79] This is in marked contrast with results for halide ion–neutral acceptor complexes[102] which have a solvent sensitivity of $\bar{\nu}_{CT}$ equal to that of complexes with a fully ionic ground-state, although the SO_2-iodide ion complex also has $\bar{\nu}_{CT}$ insensitive to solvent.[64] Although all ion–neutral-molecule complexes should have small dipole moment differences between ground and excited states, it appears that $\bar{\nu}_{CT}$ is very sensitive to solvation of the components of the complex in the ground state.

For complexes of halide ions with neutral acceptors the plots of E_T vs E_{CT} yield two straight lines of different slope. One of slope close to unity is obtained for the polar solvent group, the other of higher slope is found for the non-polar and ion-pairing group of solvents.[102] This has been interpreted[102] as due to the raising of the energy of the CT transition by the extra stabilization of the ground state by the cation within a contact or solvent-separated ion-pair. There should thus be two complexes present in solvent mixtures owing to specific solvation of the two forms, I^-A and M^+I^-A. In solvent mixtures of methyl cyanide and t-butanol, CT bands due to both species may be observed. In ion-pairing solvents, $\bar{\nu}_{CT}$ is strongly dependent on the cation radius whereas in highly polar solvents it is independent of cation radius.[102, 103]

If we now turn to neutral ground-state complexes, a bathochromic shift is expected compared with the hypsochromic shifts found for ionic and ion–neutral-molecule complexes. There is good correlation of the red shifts of $\bar{\nu}_{CT}$ with Z and E_T for solvents of low polarity.[38, 78] Data obtained by Emslie and Foster[78] are shown in Fig. 5. For polar solvents and particularly the protic solvents, hypsochromic shifts are obtained[38, 78] a good correlation with Z again being obtained. Burgess[107, 108] found good linear correlations of $\bar{\nu}_{CT}$ for transition metal complexes with E_T on two separate lines for protic and aprotic solvents. It should be emphasised that solvents suspected of undergoing specific interactions with the complex components have been excluded from the correlation shown in Figure 5.[78] Benzenoid and aza-aromatic solvents form complexes with electron acceptors as do amine impurities in formamide and dimethylformamide.[78] These solvents always appear anomalous in correlations of $\bar{\nu}_{CT}$ with solvent polarity.[38, 78, 102] Apart from this group, anomalous shifts occur[78] for other solvents shown to act as electron donors such as dioxan,[109] ethers,[110] alkyl cyanides,[111] sulphoxides[112] and ketones.[78] All these solvents act as lone pair n-donors.

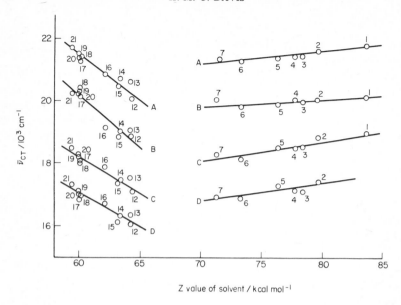

FIG. 5. Plots of $\bar{\nu}_{CT}$ vs Z value of solvent for the complexes (A) 1,3,5-trinitrobenzene–N,N-dimethylaniline; (B) 1,2,4,5-tetracyanobenzene–N,N-dimethylaniline; (C) chloranil–acenaphthene; (D) chloranil–pyrene in solvents (1) methanol; (2) ethanol; (3) propan-1-ol; (4) butan-1-ol; (5) propan-2-ol; (6) decan-1-ol; (7) 2-methylpropan-2-ol; (12) 1,1,2,2-tetrachlorethane; (13) dichloromethane; (14) 1,2-dichloroethane; (15) chloroform; (16) 1,1-dichloroethane; (17) cyclohexane; (18) 2,2,4-trimethylpentane; (19) n-hexane; (20) n-heptane; (21) ethyl acetate. [Reproduced with permission from P. H. Emslie and R. Foster, *Rec. Trav. Chim. Pays-Bas*, **84**, 255 (1965)].

Again the difficulty of obtaining a single parameter to cover shifts in all solvents is underlined. Z and E_T appear to cover general polarity interactions plus hydrogen bonding but give separate correlations for polar and non-polar solvents whereas dielectric constant and refractive index functions ignore specific interactions. Nagy and coworkers[84] have shown that where a group of solvents have a specific interaction with the solute then this interaction dominates the overall solvent effect. They examined the CT absorption for the complex acenaphthene–tetrachlorophthalic anhydride in 32 solvents belonging to four classes: (*a*) halogenated solvents, (*b*) aromatics, (*c*) aromatics with one functional group and (*d*) *n*-donor solvents. (*a*) and (*b*) should represent general solvent interactions, (*d*) specific (CT) interactions and (*c*) mixed interactions. No correlation with dielectric constant was found but the polarizability function, while giving no simple correlation, gave a better general trend. Using S_M values[96] to represent both polar and polarizability effects, they found an excellent correlation with $\bar{\nu}_{CT}$ for classes (*a*) and (*b*) on two separate parallel straight lines. Those class (*c*) solvents, having both π- and *n*-donor properties such as pyridine, did not correlate while simple halogenated benzenes fell on the line for aromatic solvents. $\bar{v}_{CT}$ in n-donor solvents gave a complete scatter when plotted against S_M. If for the n-donor solvents, $\Delta\bar{v}_D$ values were used, an excellent straight-line correlation emerged showing a blue shift with increasing $\Delta\bar{v}_D$. This is interpreted in terms of the orientation of the solvent dipoles with respect to the solute dipole. A perpendicular orientation favours the n-donor interaction while the dipolar interaction is a maximum for a parallel orientation. For ionic complexes the dipolar interaction is strong enough to overcome the n-donor interaction and the n-donors fall into the general correlation of $\bar{v}_{CT}$ with Z and E_T. Nagy's results provide good evidence that specific interactions are largely responsible for the anomalous blue shifts. One other correlation of interest is that found by Huong, Platzer and Josien.[113] They found a good correlation between $\bar{v}_{CT}$ and the solubility parameter, δ,[114] of the solvent for the pyridine–iodine complex red shift in a number of solvents of low polarity. They conclude that the solubility parameter represents both dielectric and cage effects of the solvent. It remains to be seen whether δ will also explain shifts in polar solvents.

To sum up one may quote Offen and Abidi[65] who state that the complexity of the microscopic situation prevents a complete assessment of the change after complex solvent interaction and of the large variation in shifts between closely related complexes or solvents. Only for groups of solvents with similar interaction with the complex and components, can good correlations with an appropriate solvent polarity parameter be obtained. One possible hope for a general parameter lies in the use of multi-parameter equations. Katritzky and coworkers[83] have proposed an equation

$$\Delta F_m - \Delta F_1 = \rho_A(A_m - A_1) + \rho_B(B_m - B_1) + \cdots \qquad (21)$$

where ΔF_m and ΔF_1 are measures of a solvent dependent phenomenon in a solvent S_m and a standard solvent S_1; A_m, B_m and A_1, B_1 are relative magnitudes of the interaction mechanisms A, B in solvents S_m and S_1; ρ_A, ρ_B are the sensitivity of ΔF to the interaction mechanisms A, B, i.e. ρ_A, ρ_B depend only on the solute. Correlations of $\bar{v}_{CT}$ were found to improve when E_T values were combined with dielectric and refractive index functions. The idea is formally equivalent to the S and R values of Brownstein[94] which have also been found to give good correlations with CT band shifts.[51]

Finally one may mention the effect of solvent on the various terms in the theoretical expression for the energy of the CT band, namely:[1, 2, 6, 115]

$$hv_{CT} = I_D - (EA_A - G_1 + G_0) + X_1 - X_0 \qquad (22)$$

where I_D and EA_A are the vertical ionization energy of the donor and the vertical electron affinity of the acceptor respectively, G_1 and G_0 contain attractive and repulsive terms, X_1 and X_0 are resonance terms. Kroll[32] has discussed the effect of solvent on the various terms for TCNE complexes, Rao and coworkers for iodine complexes[58] and Strong and coworkers for halogen atom complexes.[116] The terms in (22) are found to vary considerably in vapour and solution phases.

(d) *The Blue-Shifted iodine band*   In iodine complexes, the iodine visible absorption band undergoes a blue shift, attributed by Mulliken[117] to greater exchange repulsion between the σ_μ excited MO of iodine and the adjacent donor. The blue shift increases with increasing strength of complexing, an isosbestic point normally being observed between free and complexed iodine bands.[43, 118] In the vapour phase there is difficulty in identifying the blue shifted band.[32, 41, 45, 119] One suggested explanation for this is that removal of the solvent cage raises the intermolecular distance and hence decreases the orbital overlap. Consequently the absorption of the complexed iodine would decrease.[41, 120] Other suggestions have involved differences in the geometry of the complex in the two phases[6, 32] and freer rotation of the complex in the vapour phase leading to fewer favourable orientations. Using high concentrations of diethyl sulphide in the vapour phase, Tamres and Bhat[121] found an increase in intensity of the iodine visible band in the region expected for the blue shifted band. The increase was greater at higher donor concentrations and decreased with increasing temperature as expected for an equilibrium between free and complexed iodine. No absorption maximum or isosbestic point was observed. Resolution of the band was carried out using the value for the equilibrium constant determined for the complex.[46] The results showed that the free iodine band was shifted to longer wavelengths in solution while the blue shift of the complexed iodine band was greater in solution than in the vapour phase. The two bands are sufficiently separated in solution to enable resolution and observation of the isosbestic point whereas they are overlapped in the vapour phase. The extra blue shift in solution has been attributed to the solvent cage effect,[31–34] the reduction of the intermolecular distance leading to greater repulsion between the excited state of iodine and the electron donor.[1, 121] Orientation effects may also contribute to the change in blue shift.

(e) *Contact charge-transfer absorption bands*   Contact charge-transfer was first suggested by Orgel and Mulliken[122] as a possible explanation of the breakdown of the expected relationship between intensity and stability in series of related complexes.[2, 122] Absorption of light to populate the excited CT state is held to occur during collision between donor and acceptor. The complex has no stability in the ground state, i.e. the equilibrium constant is zero. Many weakly interacting species give a characteristic absorption held to be due to contact CT absorption. Oxygen[123–131] and iodine[132–136] are the best studied systems but nitric oxide,[137–139] tetranitromethane,[140] bromine,[141] metal fluorides[142] and halogenomethanes[143–146] have also been studied. The main feature of contact CT spectra is the extension to longer wavelengths of a donor or acceptor absorption band on addition of the second component. Seldom is an absorption maximum observed. Some of the donors involved in contact complexing are solvents commonly used in studies of CT complex formation such as n-heptane and carbon tetrachloride. One effect of contact complexing is the distortion of the donor band. Davis and Farmer[146] found a broadening of aromatic hydrocarbon absorption bands by contact complex formation with carbon tetrachloride, the integrated intensity of the band being preserved. Where component and complex absorption bands overlap it is the practice to deduct the component absorption from that due to the complex. In any subsequent calculation of molar absorption coefficient and equilibrium constant from the CT band intensity, it is often assumed that the component absorption is unaffected by change of medium.

Very little investigation of solvent dependence of contact CT spectra has been carried out. Tamres and Grundnes[147] observed iodine–hydrocarbon contact CT bands in the gas phase and found that the band was blue shifted compared with the solution spectra.[133, 134] From collision theory they estimated a value for K of ca $0.13\ 1\ mol^{-1}$ which is similar to that calculated by Prue[148] for contact complexing in solution. The distinction between contact complexing and very weak complexes is really not possible.[1, 149] The shifts observed are very similar to those observed for weak complexes going from vapour to solution. Davis and coworkers[150] have found that the intensity of the contact CT band of oxygen complexes of benzene is markedly dependent on solvent viscosity in ethanol–ethylene glycol mixtures whereas the intensity of conventional complex CT bands are not. This is analogous to the observation that contact CT bands disappear on freezing the solution[151] and underlines the collisional nature of the interaction.

III. THE INTENSITY OF THE CHARGE-TRANSFER BAND

A. Theoretical Considerations

The molar absorption coefficient (extinction coefficient), ϵ, is defined by Beer's Law

$$I = I_0 \, 10^{-\epsilon cl} \qquad (23)$$

where I and I_0 are the transmitted and incident intensities of the light at a given frequency, l is the path length in centimetres and c is the concentration of the absorber in moles litre^{-1}. The oscillator strength, $\mathbf{f}$, for the electronic transition is defined by

$$\mathbf{f} = 4{\cdot}318 \times 10^{-9} \int \epsilon_v \, dv \qquad (24)$$

the integration being made over the whole absorption band. The oscillator strength is related to the magnitude of the electronic transition dipole μ_{VN} for a CT transition between states V and N.[152]

$$\mu_{VN} = 0{\cdot}0958 \left[\frac{\int \epsilon v \, dv}{\bar{v}} \right]^{\frac{1}{2}} \qquad (25)$$

where $\bar{v}$ is approximately the wavenumber of the band maximum. μ_{VN} is given by

$$\mu_{VN} = \int \psi_V \mu_{OP} \psi_N \, d\tau \qquad (26)$$

where μ_{OP} is the dipole moment operator. Approximately[152]

$$\mu_{VN} = (a^*b\mu_1 - ab^*\mu_0) + (aa^* - bb^*)\mu_{01} \qquad (27)$$

where

$$\mu_0 = \int \psi_0 \mu_{OP} \psi_0 \, d\tau \quad \text{etc.}$$

Alternatively

$$\mu_{VN} = a^*b(\mu_1 - \mu_0) + (aa^* - bb^*)(\mu_{01} - S_{01}\mu_{01}) \qquad (28)$$

where S_{01} is the overlap integral between the two states. We see that the intensity depends on the dipole-moment difference between the dative and no-bond structures and the fraction of electron transfer in ground and excited states. The first term in (28) is usually held to be more important but for weak complexes, the second term may be important.[66, 152] The above equation shows that the intensity of absorption is proportional to the square of the transition dipole and applies to gas-phase complexes. An increase in intensity might be expected in solution due to the change of the effective electric field of the light in a polarizable medium. Theoretical treatments of this effect have been discussed by several workers.[22, 23, 25, 153–156] The expression of Chako[153] predicts, for the absorption band of a single molecular species, an intensification from gas to solution related to a function of the refractive index.

$$\frac{\epsilon_{soln}}{\epsilon_{gas}} = \frac{(n^2 + 2)^2}{9n} \qquad (29)$$

where ϵ_{soln} and ϵ_{gas} are the integrated intensities and n the refractive index. This predicts an intensification of up to 30 percent and a ratio $\epsilon_{soln}/\epsilon_{gas}$ and hence f_{soln}/f_{gas} of greater than one. Unfortunately most authors do not publish f values for CT complexes, only quoting ϵ_{max}, the molar absorption coefficient at the band maximum. ϵ_{max} may be used to make comparisons of intensities since the band width for CT absorptions is strikingly indepen-dent of changes of donor, acceptor and medium. Again for single molecular absorbers, Abe[155] has found that f_{soln}/f_{gas} is sometimes less than one, contrary to the predictions of the Chako[153] model. He derives an expression from dielectric and field effects of a solvent on the solute which gives a better indication of the experimental results for the π–π^* transition in nitrobenzene and the n–π^* transition in acetone. Liptay[23, 25] has developed a theory based on dielectric effects on the transition moment which deals successfully with transitions in some dye molecules.[157]

$$\mu_{ga}^{sol} = \mu_{ga} + \alpha_{ga} \, \text{FRM} + \frac{2(n^2 - 1)}{a^3(2n^2 + 1)} \, W_{ga} \qquad (30)$$

where μ_{ga} and μ_{ga}^{sol} are transition moments in the gas phase and in solution; α_{ga} is the transition polarizability tensor, FRM the effective electric field, the mean of F in the ground and excited states and W_{ga} is a vector measuring the effect of the dispersion interaction on the transition moment. This predicts that ϵ should increase with increasing polarity of the solvent for cases where the dipole moment increases on excitation and should decrease for the reverse case where $(\mu_e - \mu_g)$ is negative. Equation (30) is similar to the equation for band shifts and Liptay[25] points out that the scatter on plots of $\bar{v}_{CT}$ using equation (10) should be mirrored on similar plots of the integral absorption using (30), since the reasons for deviations from theory should be similar in the two cases. Plots of band shifts as a function of integral absorption should thus eliminate deviations and this has been confirmed in some cases.[157] As for solvent shifts, no theory at the present time is able to deal with intensity changes with solvent of all bands in all solvents. In fact the situation is less satisfactory for theories of intensity changes.

B. Experimental Results for Complexes

The situation for complexes is bedevilled by the lack of reliable values for ϵ and f. These can not be determined directly from Beer's Law as the equi-librium constant is not known independently in the majority of cases.

Inconsistencies occur in the published values of ϵ for a given complex from different groups of workers. Where K may be determined by non-optical methods, then a direct determination of ϵ may be made but again its reliability depends on the accuracy with which K is known. Insufficient data exist for intensities in a wide range of solvents for a useful discussion in terms of the theoretical relationships. However, data do exist for vapour and solution complexes and some of these are given in Table 4. Qualitatively, the data support the predictions of Chako[153] and are also consistent with Liptay's[25] model. The gas-to-solution intensification is, however, often of an order of magnitude compared to the 30 per cent or so predicted by the theoretical models. Other explanations must be sought since this intensification is much larger than found for single molecule absorbers. Two main approaches have been suggested. Firstly explanations have been sought for a real change of intensity due to solvent effects and secondly errors in the determination of ϵ have been discussed. Rice[158] has proposed that gas-phase complexes are held mainly by van der Waals forces, there being a larger number of orientations of donor and acceptor in the complex, particularly for weak complexes. In solution he suggests that orientations will be preferred which favour maximum orbital overlap between donor and acceptor orbitals, i.e. those orientations maximizing the CT stabilization. The random orientations in the gas phase would diminish ϵ compared with the solution value. Mulliken and Person[159] do not regard this as the sole factor and suggest that the situation is considerably more complicated. The bond compression model[31–34] in which the donor–acceptor distance is reduced by the solvent cage compression, would increase f_{soln}/f_{vap} due to increased orbital overlap between donor and acceptor. This explanation is supported by the observed pressure intensification[35–37] of CT bands, an effect first proposed by Mulliken.[2, 117] Offen[35] has stressed the parallelism between solvent effects on intensity and band position and the similarity of these to changes on subjecting the complex to pressure.[37] The pressure intensification is however, of too small a magnitude to explain totally the observed gas-to-solution intensification unless unreasonably high values for the pressure exerted by the solvent cage are assumed. Again it is impossible at present to separate dielectric and solvent cage effects on intensities.

As mentioned earlier, the difficulty of separation of K and ϵ particularly in the gas phase,[1] places a limit on the reliability of ϵ values. Where non-optical determinations of K have been made, these may be used to determine ϵ from the product $K\epsilon$ obtained from the Benesi–Hildebrand[160] analysis. $K\epsilon$ may be determined accurately. Christian and coworkers[55, 56] have found that for the strong trimethylamine–SO_2 complex for which separation of K and ϵ is possible in the vapour phase, the value of K obtained is in agreement with that obtained from non-spectrophotometric methods.

TABLE IV. Molar absorption coefficients, ϵ_{max}, in vapour and in solution in a solvent of low polarity[a]

Complex	ϵ_{max} (vap.) /l mol^{-1} cm^{-1}	Ref.	ϵ_{max} (sol.) /l mol^{-1} cm^{-1}	Ref.	Solvent
Benzene–iodine	1650 ± 100	41, 42	12000	43	n-heptane
Diethyl ether–iodine	2100 ± 400	44	6000	43	n-heptane
Diethyl sulphide–iodine	16800 ± 2230	46	26400 ± 1050	32	n-heptane
Benzene–TCNE	770 ± 120	48	2330	53	CCl$_4$
Mesitylene–TCNE	1250 ± 375	32	1680	343	n-heptane
Naphthalene–TCNE[b]	{ 575 ± 167 / 610 ± 177 }	48	1640 / 1660	53	CCl$_4$
Diethyl ether–CO(CN)$_2$	~ 65	31, 54	420	31, 54	n-hexane
Benzene–CO(CN)$_2$	~ 125	31, 54	950	31, 54	n-hexane
Trimethylamine–SO$_2$	5700 ± 350	55, 56	5000 ± 40	55, 56	n-heptane
trans-2-Butene–SO$_2$	120 ± 40	48	2290	57	n-hexane

[a] Data taken, with permission, from the compilation by M. Tamres, "Molecular Complexes", Vol. 1. (Ed. Roy Foster), Elek. Science, 1973, p. 49.
[b] Multiple CT bands.

G

For this complex which shows a reverse vapour-to-solution shift of $\bar{v}_{CT}$, the intensities in vapour and solution are nearly equal. There is a general trend, vapour-to-solution intensification increasing as the complex becomes weaker while the red shift becomes larger.[1, 170]

The assumption that activity coefficients can be ignored may lead to an overestimate of ϵ in solution.[159] The thermodynamic equilibrium constant K_a, is related to the observed equilibrium quotient K, by

$$K_a = K \frac{\gamma_c}{\gamma_A \gamma_D} = K\Gamma_i \qquad (31)$$

where γ is the activity coefficient of C, A or D and Γ_i is the activity coefficient quotient on the appropriate concentration scale. Using a linear function of donor mole-fraction for Γ_X

$$\Gamma_X = k_0[1 + bX_D] \qquad (32)$$

Mulliken and Person[159] show that

$$\epsilon_{app} = \epsilon \frac{K_X^0}{K_{app}} \qquad (33)$$

where ϵ_{app} and ϵ are the apparent and true molar absorption coefficients, K_X^0 and K_{app} being the "true" equilibrium constant and the equilibrium constant neglecting activity coefficient corrections. Using regular solution theory[161] to estimate Γ_X they calculated that ϵ may be overestimated by as much as a factor of two for the benzene–iodine complex if activity coefficients are neglected. Others have argued that competition by the solvent for the donor and/or acceptor by specific solvation,[162] or solvent–component complex formation[168] may lead to overestimates of ϵ. Other workers[164–167] have stressed the effect of solution non-ideality on ϵ. Mulliken and Person[159] consider that their treatment is general and includes the treatments by these other authors as special cases. One other possibility is that 2:1 and 1:2 complexes form in solution, a situation less likely in the gas phase. Ross and Labes[168] found for the complex N,N-dimethylaniline–1,3,5-trinitrobenzene in chloroform, a decrease in ϵ_{app} from 1 800 to 608 as the donor acceptor ratio varied from 10^{-4} to 1. K_{app} increased over the same range, the product $K\epsilon'$ remaining constant. They interpreted this as due to multiple complex formation. It is not easy to say whether this would lead to over-estimate of ϵ in any particular case. Foster and coworkers[169] at one stage suggested that such results could also be explained by a non-obedience by the complex of Beers Law, i.e. that ϵ varies as the solvent medium varies over the whole donor concentration range. Although such changes in ϵ would be small, large changes in ϵ_{app} may occur. Again this is analogous with other treatments based on solution non-ideality. Changes in ϵ on changing the solvent have been measured over a range of solvents of low polarity for a few complexes.[77, 171–173] The changes are small and variable although a slight trend to smaller values in ϵ may be discerned as the solvent polarity increases. For ionic complexes a fall in intensity has been found with increasing polarity of the solvent[174] as predicted by the Liptay theory[25] for cases where $\mu_e < \mu_g$. No completely satisfactory explanation has yet emerged to explain the very large intensification of the CT band from vapour to solution and until reliable methods are determined for the unequivocal determination of ϵ, very little progress seems possible in fitting the results to theoretical equations.

IV. SOLVENT EFFECTS ON CHARGE-TRANSFER FLUORESCENCE BANDS

A. Theoretical Considerations

1. Solvent Effects on Emission Bands

Reference to Fig. 1 shows that due to the difference in polarity of the ground and excited states of CT complexes, the Franck–Condon and equilibrium states will differ in energy, particularly those of the excited state. Solvent reorganization should occur in a time short compared with the fluorescence lifetime and hence fluorescence may be expected to occur from the equilibrium excited state to the Franck–Condon ground state. Normally fluorescence bands are seen as the mirror image of the corresponding absorption band and at lower wavelengths. The 0–0 transition for absorption and fluorescence may be expected to be coincident in the absence of a solvent or if ground and excited states are very similar in their electronic distribution. Usually, in solution, however, there is a separation of the two bands known as the Stokes shift. This shift should be large for CT complexes where the dative contribution is very different in ground and excited states. Furthermore it should be strongly solvent dependent as emission and absorption bands will differ in sensitivity to solvent polarity effects. Expressions may be derived for shifts of emission bands analogous to those for absorption bands. They may be applied to give expressions[17–19] for the shift in band origins, or more approximately, in band maxima for the absorption and fluorescence bands; for example

$$\Delta v_a - \Delta v_f = \frac{2(\mu_e - \mu_g)^2}{hca^3}\left[\frac{(\epsilon' - 1)}{(2\epsilon' + 1)} - \frac{(n^2 - 1)}{(2n^2 + 1)}\right] \qquad (34)$$

Here it is assumed that the dispersive interactions contributing to the shifts in absorption and emission bands, Δv_a and Δv_f, cancel to a first approximation. As before, ϵ' and n are the solvent dielectric constant and refractive index respectively and a is the solvent cavity radius. The equation provides a method of determination of the excited-state dipole moment from the solvent dependence of the Stokes shift.

The possibility also exists of intersystem crossing to the CT triplet state. It is very unlikely, however that CT phosphorescence will be observed in fluid solution due to the long lifetime of triplet states and the consequent possibility of decomposition or non-radiative transition to the ground state.

2. "Exciplex" or Heteroexcimer Formation

For systems in which no ground-state complex formation is detectable, formation of a complex is possible between an excited donor (or acceptor) and a ground-state acceptor (or donor). This may occur because of the change in size of donor or acceptor orbitals on excitation with a consequent change in ionization potential and electron affinity. Such a complex has been termed by Birks[175] an "exciplex" or excited-state complex and is similar to the complexes formed between an excited-state molecule and a ground-state molecule of the same species, termed an excimer.[176] Exciplex formation may be represented as

$$^1A^* + D \rightleftharpoons {}^1A^*D \rightleftharpoons A^-D^+$$
$$\underset{\text{encounter complex}}{} \quad \underset{\text{exciplex}}{}$$

Knibbe and Weller[177] have stated that an exciplex should always be formed if the conditions, $(I_A - I_D) \geqslant 0$ and $(EA_A - EA_D) \geqslant 0$ are fulfilled, where I and EA are the ionization energy and electron affinity respectively. If an exciplex is formed, then the possibility exists of fluorescence from this CT state to a ground state which has no stability, a situation analogous to the reverse of contact CT absorption. Solvent sensitivity of such a band would then be a good test of the CT nature of the excited state.

3. Fluorescence Quenching

Another effect of CT complexing either by ground-state or excited-state complex formation should be the quenching of the component donor and/or acceptor fluorescence. Quenching might occur, generally, either by the binding of the fluorescer in the ground state by complexing (static quenching), or by deactivation of the fluorescer excited-state by complexing or collisional de-activation (dynamic quenching). A possible scheme may be represented as[178–181]

$$A^* + D \underset{k_{-1}}{\overset{k_1}{\rightleftharpoons}} A^*D \overset{k_2}{\longrightarrow} (A^-D^+) \overset{k_5}{\longrightarrow} A^- + D^+$$

$$k_f \nearrow \quad \searrow k_i \qquad k_2' \searrow \quad \nearrow k_3 \qquad \searrow k_4$$

$$A + h\nu_f \quad A \qquad \qquad A + D \qquad A + D + h\nu_{CT_f} \qquad A + D$$

In this scheme either the encounter complex or the exciplex may decay non-radiatively or the exciplex may decay by fluorescence. It may not be necessary to involve an encounter intermediate, i.e. $A^* + D \rightleftharpoons (A^-D^+)$ may occur directly.[181] No reference has been made in the scheme to either static quenching or intersystem crossing to triplet states. The processes initiated by k_1, with the exception of k_{-1}, will all lead to a reduction in the quantum yield of the fluorescence of A^*. Solvent might be expected to have an effect particularly on k_2 and k_4.

B. Experimental Results

1. Types of Complex Exhibiting Charge-Transfer Fluorescence
The first observations of CT fluorescence were from crystalline complexes or complexes in rigid solvents.[182–188] The earlier work has been reviewed by Briegleb.[3] The fluorescence band was observed as the mirror image of the CT absorption band and was characterized by a large, solvent-dependent Stokes shift. Stokes shifts of 5000–20000 cm^{-1} have been observed both for complexes with neutral[189–191] and ionic ground states.[192–195] This confirms the large separation between the Franck–Condon and equilibrium states originating from the large differences between ground and excited state electronic structures.[196] It is also good confirmatory evidence for the dative nature of the CT excited state.

Very few intermolecular complexes exhibit CT fluorescence in liquid solution. Only the complexes of the acceptors, tetrachlorophthalic anhydride (TCPA), pyromellitic dianhydride (PMDA) and 1,2,4,5-tetracyanobenzene (TCNB) with aromatic hydrocarbon donors have so far yielded CT fluorescence in liquid solution.[191, 196–203] CT fluorescence in liquid solution has recently been observed for intramolecular complexes of the form, donor—$(CH_2)_n$—acceptor, e.g. N-[ω-(p-nitrophenyl)alkyl]-arylamines.[204–206] The general failure to observe CT fluorescence in liquid solution must be ascribed to the efficient quenching of the CT excited state by the various modes of radiationless transition to the ground state.

By far the largest number of CT fluorescence bands observed are from systems where exciplex formation occurs. For the formation of an exciplex by $^1A^* + D \rightleftharpoons (A^- \cdots\cdots D^+)$, A^* may typically be an aromatic hydrocarbon and D an aromatic[177, 178, 207–213] or aliphatic amine.[214–216] Other systems studied include aromatic amine–aromatic ester exciplexes.[217–219] Data on some 100 exciplexes have been reviewed by Rehm and Weller[220] and reviews of the early work made by Weller[221] and by Stephenson and Hammond.[222] CT fluorescence has also been observed from intramolecular exciplexes[223–226] of the form donor—$(CH_2)_n$—acceptor, e.g. N,N-dimethyl-3-(1-naphthyl)propylamine.[226]

2. Solvent Effect on Charge-Transfer Fluorescence Band Shifts

$\bar{\nu}_{CT}$ for CT fluorescence is markedly dependent on solvent polarity, undergoing a pronounced red shift as the solvent becomes more polar. Since the absorption band red-shift is quite small, the Stokes shift also increases rapidly with increasing solvent polarity. Using equation (34) for the Stokes shift, excited-state dipole moments have been estimated[17, 177, 197] using measured values of μ_g. It is found that CT fluorescence is only observed for exciplexes and for complexes in solvents of very low polarity. The intensity of the band diminishes as the polarity increases within this group. For exciplexes, no absorption band is observable and a modification of equation (34) must be used. A plot of $\bar{\nu}_f$ against

$$\left[\frac{\epsilon' - 1}{2\epsilon' + 1} - \frac{1}{2} \frac{(n^2 - 1)}{(2n^2 + 1)} \right]$$

is reasonably linear confirming the dipolar nature of the exciplex.[210, 214, 227, 228] Beens, Knibbe and Weller[228] use an expression derived by neglecting polarizability terms.

$$\bar{\nu}_c = {}_t\bar{\nu}_c(0) - \frac{2\mu_e^2}{hca^3} \left[\frac{\epsilon' - 1}{2\epsilon' + 1} - \frac{1}{2} \frac{(n^2 - 1)}{(2n^2 + 1)} \right] + \frac{(2hcD - \mu_g^2)}{hca^3} \frac{(n^2 - 1)}{2n^2 + 1}, \quad (35)$$

where $\bar{\nu}_c$ is the wavenumber of exciplex emission. The equation still contains the dispersion interaction term $2Da^{-3}$. Assuming the ground-state dipole moment to be perpendicular to μ_e, then $2hcD$ and μ_g^2 are assumed to be of the same order and the contribution of the third term in (35) may be ignored. From the plot of $\bar{\nu}_c$ against

$$\left[\frac{\epsilon' - 1}{2\epsilon' + 1} - \frac{1}{2} \frac{(n^2 - 1)}{(2n^2 + 1)} \right], \quad 2\mu_e^2 (hca^3)^{-1}$$

may be estimated which, assuming $a = 5$ Å, gives reasonable values of ca 10D for the exciplex dipole moment. The plots were scattered due to specific solute–solvent interaction but $\bar{\nu}_c$ for two complexes plotted against each other gave good straight lines and reliable relative values of $2\mu_e^2(hca^3)^{-1}$. Davis[198] has found good linear plots with slope close to one, when $\bar{\nu}_f$ for both complex and exciplex fluorescence bands are plotted against E_T,[29, 93] the empirical solvent-polarity parameter. Similar plots have also been found for $\bar{\nu}_f$ of intramolecular complexes.[229] CT fluorescence for intramolecular complexes and exciplexes is observable over the whole range of solvent polarities. The similarity of plots of $\bar{\nu}_{CT}$ the absorption band of complexes with an ionic ground-state and $\bar{\nu}_f$ for the fluorescence of neutral ground-state complexes and exciplexes formed from uncharged species against E_T, points to the ionic nature of the CT state. The two plots are of similar magnitude of slope, the slope differing, however, in sign. Conversely, the fluorescence

CT band of ground-state ionic complexes such as the pyridinium halides is relatively insensitive to solvent polarity.[192]

The contribution of solvent-cage compression[31-34] is likely to be less important for the CT state where the ionic nature of the state, leading to a much steeper potential energy curve, will make the ionic complex less compressible. This is borne out by the small pressure shifts shown by CT fluorescence bands.[35, 187, 230] However, the strong pressure quenching shows the sensitivity of quenching to the ion-pair intermolecular distance. The fluorescence lifetime decreases[178, 181, 196, 197, 217, 231] as the solvent becomes more polar. The quantum yield of fluorescence decreases at a faster rate than the fall in fluorescence lifetime. The reduction in lifetime is not, therefore, a complete explanation of the intensity reduction. It is generally agreed that the excited state must be a contact ion-pair for emission to be observed.[181, 198, 232, 233] Increase in polarity of the solvent will favour the formation of solvent-shared and solvent-separated ion-pairs and in highly polar solvents, the free, solvated ions. These species are unable to decay by CT fluorescence and will undergo some form of non-radiative decay to the ground state. There are two slightly different views of the effect of solvent polarity. Mataga and co-workers[178, 196, 208, 211, 232, 233] believe that the solvent enhances the degree of charge transfer in the excited state with consequent weakening of the binding. The solvent may also increase the intermolecular distance again weakening the binding. The principal effect of solvent is then viewed as effecting a change in the electronic distribution in the excited state, a view shared by Koizumi and coworkers.[197, 217, 219] These workers consider the radiationless transition principally as an internal conversion to the ground state, the reaction scheme being:

$$A^* + D \underset{k_{-1}}{\overset{k_1}{\rightleftharpoons}} (A^{\delta -} \cdots\cdots D^{\delta +})$$

$$k_f \swarrow \qquad \searrow k_i$$

$$A + D + h\nu_f \qquad A + D$$

The internal conversion rate constant increases and the fluorescence rate constant decreases as solvent increases the separation of charge. Weller and coworkers[177, 181, 213, 220, 221, 228] take a slightly different view. They see the solvent-shared ion-pair being formed in polar solvents as an intermediate, with the individual solvated ions being formed in very polar solvents. The solvent-shared ion-pair may then undergo spontaneous reversal of electron transfer to produce either the ground state singlet or the triplet state. They quote in support of this, flash photolysis experiments in acetonitrile and other polar solvents which yield both the free solvated ions of donor and acceptor, and triplet state molecules. Further evidence is provided by chemiluminescence experiments.[234, 235] If the ions A^- and D^+ are reacted in solution; the CT fluorescence is observed together with the fluorescence of the acceptor hydrocarbon. It may be shown that the ions have insufficient energy to populate the first excited state of the hydrocarbon but may populate the triplet. The singlet is then populated by triplet–triplet annihilation and in cases where the triplet is less than halfway in energy between A and $^1A^*$, no chemiluminescence is observed. Other workers have observed the production of A^- and D^+ on flash photolysis.[200, 201, 203, 217, 232, 233, 237, 238] Mataga and coworkers have studied the dissociation into ions of TCNB and PMDA complexes using nanosecond flash photolysis and transient photocurrent measurements.[203] For PMDA complexes they find dissociation occurs via the CT singlet in polar solvents and via the CT triplet in non-polar solvents. In the TCNB–benzene–1,2-dichloroethane system, they quote evidence to show that ionic dissociation occurs *via* the Franck–Condon state rather than *via* the equilibrium excited-state. The intermediacy of solvent-shared ion-pairs in ionic dissociation is shown by lifetime studies; the fluorescent ion-pair, presumably a contact ion-pair, having a lifetime only one thousandth of the ion pair responsible for the ionic dissociation process.[203] The relative rate constants for radiationless transition to the ground state may be deduced from the effect of solvent on the ionic dissociation. From all these results it appears that the radiationless processes may include both intersystem crossing and internal conversion. McGlynn and coworkers[239] have investigated the intersystem crossing processes in detail, including the effect of heavy atom containing solvents in spin–orbit coupling induced intersystem crossing.

The reduction in intensity of the fluorescence band and its eventual disappearance as the solvent increases in polarity is undoubtedly due to some degree of separation of the ion pair by solvent. In intramolecular complexes and exciplexes, the solvent is unable to separate the ion pair and the CT fluorescence is still observable in polar solvents.[204–206, 223–226] However Mataga and coworkers[225] find that the quantum yield and radiative transition probability decrease in an intramolecular complex of the form, donor—$(CH_2)_n$—acceptor, as polarity increases. For $n = 1$ or 2 only anthracene fluorescence is observable in non-polar solvents, while the CT fluorescence is seen in polar solvents. Here the polar solvent is presumably able to stabilise the charge separation in the intra-ion pair while being unable to form a solvent-separated ion pair. For $n = 0$ and 3 the CT fluorescence is observable in both polar and non-polar solvents showing a structural factor is also involved. A reasonable correlation of solvent shift of $\bar{v}_{fCT}$ was obtained[226] when plotted against

$$\left[\frac{\epsilon' - 1}{2\epsilon' + 1} - \frac{(n^2 - 1)}{(2n^2 + 1)}\right]$$

as predicted by (34).

3. Quenching of Fluorescence by Complex or Exciplex Formation

The ability of certain molecules and ions to quench the fluorescence of other molecules has been studied for the last forty years. The donor–acceptor or electron-transfer nature of the interaction was recognized early on in these studies.[240–243] CT complex formation is usually accompanied by a reduction in fluorescence intensity of the components.[244–250] The equilibrium constant is often insufficient for the quenching to be entirely described in terms of static quenching.[180, 246, 248, 251, 252] Increased complexing in the excited state is generally held to be responsible for the extra quenching. Davis and Beaumont[180] have shown that the quenching of acridinium-ion fluorescence by anions is strongly solvent dependent. The equilibrium constant for complex formation decreases markedly as solvent polarity increases due to dissociation of the ion pair. In non-polar solvents the equilibrium constant and the Stern–Volmer quenching constant, K_{sv}, as defined by

$$\frac{F_0}{F} = 1 + K_{sv}[Q] \qquad (36)$$

where F and F_0 are the intensities of fluorescence in presence and absence of the quencher of concentration $[Q]$ are nearly equal showing that the quenching is entirely static. In polar solvents, $K_{sv} > K$ showing a greater contribution from excited-state complex formation.

For exciplex formation, the growth of the new CT fluorescence band is accompanied by a decrease in the component fluorescence, an isoemissive point being observed at the crossing of the two bands.[212, 217, 219, 227] The quenching, like the CT fluorescence intensity, is highly sensitive to solvent polarity. Within the non-polar group of solvents in which CT fluorescence is observable, the quenching decreases in step with the reduction of intensity. In polar solvents where the CT fluorescence does not occur, quenching increases with polarity of the solvent.[178, 179, 212, 217, 218, 221] The quenching is interpreted as being due to formation of the contact ion-pair exciplex in non-polar solvents, and by electron transfer in polar solvents.[178, 179, 181, 217] Weller and coworkers [177, 181, 213, 220, 221, 228, 253, 254] have shown from kinetic studies that quenching in polar solvents can occur by electron transfer over encounter distances of more than 7 Å, a distance more than twice as great as the interplanar distance in a contact ion-pair. In polar solvents they formulate the quenching scheme as:

$$A^* + D \xrightarrow{k_q} (A_s^- ------ D_s^+) \longrightarrow A + D$$

and in non-polar solvents as:

$$A^* + D \xrightarrow{k_c} (A^-D^+) \underset{k^i}{\overset{k_f}{\rlap{\diagup}\diagdown}} \begin{array}{l} A + D + h\bar{\nu}_f \\ A + D \end{array}$$

G*

The quenching occurs *via* electron transfer without the intermediacy of contact ion-pairs in polar solvents and *via* the contact ion-pair in non-polar solvents. The probability of fluorescent exciplex formation, γ', is given by

$$\gamma' = \frac{k_c}{k_c + k_q} \qquad (37)$$

and the fall of γ' with increasing solvent polarity is rationalized in terms of significantly different solvent effects on k_c and k_q. Mataga and Koizumi and their coworkers[178, 211, 217, 219] point out that the solvent could form a solvated ion-pair or the separated ions *via* the contact complex, the solvent influencing the electronic structure and binding energy of the exciplex. Recent results of Mataga and coworkers[203] tend to support the intermediacy of the contact ion-pair in the formation of solvent-shared or separated ions.

In many examples of fluorescence quenching by donors and acceptors, no exciplex fluorescence is observed. It is difficult to state in such cases whether exciplex formation or simply electron transfer is involved. Solomon, Steel and Weller[255] have discussed the energetics of exciplex formation. They conclude that for a given donor and acceptor in a range of solvents, the quenching rate should vary with the difference in solvation energies between the components and the complex. Using the solvent E_T value[29, 82] as a measure of this difference on the assumption that the solvation energy of the iodide ion will dominate the solvation energy difference, Brooks and Davis[256] have obtained a linear correlation between $\log k_q$ and E_T for the quenching of anthracene fluorescence by iodide ions. Such a plot tends to confirm the electron transfer nature of the quenching.

The formation of triplet states and radical ions *via* exciplex formation has considerable implications for the photochemistry of these systems. It is not intended to review these here although the solvent may play a considerable role in the chemistry of the excited state. Elsewhere in this volume, Davidson[257] has reviewed the photochemistry of excited CT states.

V. CHARGE-TRANSFER-TO-SOLVENT (CTTS) SPECTRA

A. Nature of the Transition

In CTTS spectra, the transition is from an anion, typically a halide or pseudohalide, to an excited state defined by the solvent environment:

$$X^-_{solv} \xrightarrow{h\nu_{CTTS}} (X + e^-)_{solv}$$

Various models have been proposed for theoretical treatments of the transition and these are discussed critically in the extensive review of CTTS spectra by Blandamer and Fox.[258] In most of the theoretical models,[259-263] the anion is considered to be in a solvent potential energy-well and the energy of the transition is defined in terms of the ionization energy plus a term dependent on the solvent cavity radius.[259-261] A recent treatment by Fox and Hunter,[264] however, considers the solvated anion as a donor–acceptor CT complex. The anion is the electron donor and the acceptor orbital is a shell-type orbital centred on the site of the anion and bounded by solvent. An expression for the transition energy may then be derived[264] analogous to that of Mulliken[2] for intermolecular donor–acceptor complexes. If this model is accepted, the transition energy should depend on solvent in a similar manner to the anion–acceptor complexes.[102, 103] Applying the Franck–Condon restriction to the transition would suggest that as for D^-A complexes, the transition energy will be largely determined by solvent influence on the ground-state energy of the anion. This is also predicted by the diffuse potential energy-well model[259] whereas the confined model of Smith and Symons[261] attributes changes in transition energy mainly to solvent effects on the excited state. Further discussion on the effect of solvent on the various models is contained in the review by Blandamer and Fox.[258]

B. Effect of Solvent on CTTS Transitions

$\bar{v}_{max}$ for CTTS transitions is highly sensitive to changes in solvent, exhibiting larger shifts than for any other transitions with the exception of some dyes[82] and ionic CT complexes.[28, 82, 92] There is no clear correlation with solvent dielectric constant, $\bar{v}_{max}$ first increasing as dielectric constant decreases, followed by a decrease. Simple correlations of $\bar{v}_{max}$ or E_{max} with solvation free energies and enthalpies based on the Born equation,[265] are unlikely to be successful[258] as the Born equation requires a similar trend of solvation energies to the Kirkwood[266] expression for free energy of solvation of a dipole. As for the intermolecular complexes, this failure has led to attempts to correlate E_{max} with the microscopic measures of solvent polarity, Z^{28} and E_T.[82] Kosower[28] found a good linear trend of $\bar{v}_{max}$ with Z for hydroxylic solvents although polar, aprotic solvents showed large departures from this line. Kosower and coworkers[267] argued that solvents showing anomalies in the correlation, for example, methyl cyanide, form discrete 1:1 complexes with the anion. Griffiths and Symons[268] pointed out that no correlation should be expected with solvent parameters which depend on changes of dipole on excitation as there is no dipole change in CTTS transitions. Furthermore they stressed[268] that there is no evidence for specific complex formation indeed the anomalous solvents behave identically to conforming solvents in properties such as band shifts on adding a second solvent. For specific complex formation, a band change would be expected on adding the second solvent.[258] The behaviour of $\bar{v}_{max}$ in mixed solvents is interesting in that a band shift is observed without any major change in intensity.[258] The plot of $\bar{v}_{max}$ vs mole-fraction of added co-solvent is not linear, showing that the anion is surrounded by a solvent environment of different composition to that of the bulk solvent. Similar specific solvation has also been demonstrated for complexes of the tropylium ion with aromatic hydrocarbons in solvent mixtures containing dichloromethane.[79] In t-butanol–water mixtures there is also a marked cation effect on E_{max} at high t-butanol mole fractions,[269] indicating an effect of ion pairing similar to that observed in iodide-1,3,5-trinitrobenzene complexes.[102, 103] In water–alcohol mixtures the CTTS transition has been used as a probe for microscopic changes in the solvent structure.[258, 270, 271] Marked changes are observed in the region $0.05 > X_2 > 0$ where X_2 is the mole fraction of added alcohol. The high energy shifts in this region have been used to support the confined model.[261] If it is assumed that the iodide ion follows the same endothermic trend as heat of dissolution of potassium iodide with added alcohol then an increase in E_{max} may not be attributed to changes in the ground-state stabilization since these are in the wrong sense. Data on quenching of anthracene by the iodide ion in alcohol–water mixtures[256, 272] tend to confirm ground-state destabilization of the ion. The quenching rate-constant increases sharply in the 0.04 mole-fraction region in t-butanol–water. These results have been interpreted as an enhancement of the hydrogen-bonding structure of water by the added alcohol. For further details the reader is referred to the review by Blandamer and Fox.[258]

VI. SOLVENT EFFECTS ON VIBRATIONAL SPECTRA OF COMPLEXES

A. Effects of Complexing on Vibrational Spectra

Comprehensive reviews of the theory and experimental results of the vibrational spectra of complexes have appeared recently.[6, 273, 274] We may summarize the effects of complex formation on vibrational spectra as: (1) changes in frequency of the vibrations of the donor and acceptor moieties. These are usually small except where the electron is donated or received by a particular bond orbital when significant changes in force constant, and hence frequency, may occur. Thus marked changes in the stretching frequencies of halogen and interhalogen acceptors occur due to the electron donated into the anti-bonding orbital; (2) changes in intensity of acceptor and donor vibration frequencies which may be large; (3) forbidden transitions, for example the iodine stretching-mode, may be rendered infrared active due to the change in symmetry on complexing with a donor; (4) new bands at low frequency, ascribed to intermolecular stretching and perhaps bending modes may be observed, for example the $D \cdots X—Y$ stretching mode observed for n-donor–halogen and interhalogen complexes.[274–284] Raman spectra have also been investigated for these systems.[285–286]

B. Effect of Solvent

As in the determination of other CT parameters, competition of the solvent for the components may also be important in vibrational spectra. In solution, decomposition of the complex may occur giving rise to chemically different species and thus to misleading results. This will be particularly important in polar solvents.[274] Solvent effects on both intensities and positions of the bands may also be expected. Person[273] has pointed out that on complexing in solution a D, A interaction replaces a D, S and an A, S interaction. The D–A vibrations are then seen against a background of D–S and A–S vibrations. If the solvent is strongly interacting then the complex intermolecular modes may not be seen against the background. The effect of solvent on vibrational band frequencies has been predicted by Kirkwood[287] and by Bauer and Magat.[288] The equation, known as the KBM equation is:

$$\frac{\Delta \bar{v}}{\bar{v}} = \frac{\bar{v}_g - \bar{v}_s}{\bar{v}_g} = \frac{C(\epsilon' - 1)}{(\epsilon' + 2)} \approx \frac{C(n^2 - 1)}{(n^2 + 2)} \tag{38}$$

The equation has been modified by Hallam[289] and Buckingham.[290] $\bar{v}_g$ and $\bar{v}_s$ are the wavenumbers of the transition in gas and solution, ϵ' and n, the solvent dielectric constant and refractive index and C, a constant, dependent on the intensity of the band. In most cases for both single molecular species and molecular complexes, the predicted and observed shifts are quite small although large shifts may be observed for hydrogen-bonding complexes.[289] For the strong complexes of halogens and interhalogens with amine and nitrogen heterocyclic bases, the effects of solvent on the X–Y halogen stretch and the intermolecular vibrations are quite large. Examples are shown in Table V. Using the KBM equation, the shifts estimated[273] are of the order 2–3 cm^{-1}. This is considerably less than the observed shifts and other explanations must be sought. The positions of the D–A and X–Y bands move closer together as the polarity of the solvent is increased.[279, 281, 291] Similar shifts are observed for pyridine–iodine complex on changing the pyridine concentration in cyclohexane as solvent.[280] One suggestion to explain the sign and magnitude of the shifts is that the contribution of the dative, $D^+–XY^-$, state increases with the polarity of the solvent, a similar suggestion to that invoked for anomalous ultraviolet shifts.[65]

TABLE V. Halogen ($\bar{v}_{X-Y}$) and intermolecular ($\bar{v}_{D-X}$) bands in various solvents

Complex	Solvent	$\bar{v}_{X-Y}$/cm^{-1}	$\bar{v}_{D-X}$/cm^{-1}	Ref.
Pyridine–I$_2$	cyclohexane	183	93	283
	benzene	172	103	274(b)
4-Methylpyridine–I$_2$	cyclohexane	181	88	282
	n-pentane	182	85	282
	CCl$_4$	178	94	282
	benzene	171	100	282
	toluene	171	99	282
	o-xylene	172	100	282
Pyridine–ICl	benzene	290	140	275, 279
	chloroform	283	151	274(b)
	pyridine	277	160	291
Pyridine–IBr	benzene	206	132	274(b)
	pyridine	195	145	291

Tamres and Yarwood[274b] argue that this should also lead to increases in intensities of the bands in proportion to the shift, whereas observed intensity changes are very small. They also point out that the lack of intensity change could possibly be explained by increased vibrational coupling of the low frequency bands leading to intensity redistribution. If this did occur, it would not then rule out an explanation of the shifts on the basis of increase of charge-transfer in the ground state of the complex. Person[273] has advanced the idea that the observed wavenumber of the low frequency bands may be the weighted average of bands due to say the acceptor complexed by donor and of randomly, or possibly specifically, solvated acceptor. Tamres and Yarwood,[274b] while accepting this as a possible explanation, have pointed out that shifts for pyridine–ICl in benzene and chloroform are not in the correct sequence, on the basis of observed ICl frequencies in the two solvents, assuming frequency averaging to occur. At the present time, no conclusive explanation seems to have been advanced for these large shifts in the low frequency bands. Increased CT participation in the ground state appears to be the most attractive suggestion.

The effects of the electrostatic field of the solvent on intensities have been discussed by Buckingham[292, 293] who considers a vibrating molecule in a continuous solvent dielectric including effects on the dipole derivative of the vibration due to dipolar and polarizability interactions. Results expected for DA complexes have been discussed by Person.[273] The observed intensity changes are very small for even the iodine–n-donor complexes exhibiting large solvent shifts. There are difficulties in obtaining accurate intensity data in the infrared. The effects of solvent on intensity are much smaller than those due to CT complex formation. For low energy vibrations of the components or the complex, the energy of interaction with the solvent may become of the same order of magnitude as the DA interaction energy. Donor and acceptor will be undergoing rapid exchange between the complex and randomly solvated states. Two peaks should be observable for complexed and randomly solvated acceptor if the rate of exchange between the two situations is less than the frequency difference. If the exchange rate is greater than this difference, then one peak with weighted average frequency and position should be observed.[273, 294] This is the usual situation for DA complexes, only one peak being observed, although "free" and complexed bands may be seen in hydrogen-bonded complexes and in pyridine–iodine.[274] Very low frequency bands observed for weak complexes[280, 295, 296] are thought to arise not from intermolecular modes of a CT complex but as librations or hindered rotations within the solvent cage. The absorptions are thought of as collision induced and may be partly translational in nature.[274] Yarwood and Tamres[274] have discussed in detail the interpretation of low frequency bands in terms of charge-transfer, electrostatic polarization and librational interactions. It appears that as argued by Hanna and Williams[9, 297] and discussed by Person,[273] a discussion of complex properties, particularly the low frequency vibrations, is unjustified in terms of purely CT effects. This must apply to the solvent effects also.

VI. NUCLEAR MAGNETIC RESONANCE SHIFTS

As an alternative to optical methods of determining the association constant, nmr shifts of acceptor or donor protons and in some cases of fluorine atoms, have been used. Such methods have been reviewed.[5, 298] Typically, an amended form of the Benesi–Hildebrand[160] or related equation[5] is used to analyse the data. For example

$$\frac{1}{\Delta_{obs}} = \frac{1}{K\Delta_{AD}} \cdot \frac{1}{[D]_0} + \frac{1}{\Delta_{AD}} \tag{39}$$

where $\Delta_{obs} = (\delta_{obs} - \delta_A)$, and $\Delta_{AD} = (\delta_{AD} - \delta_A)$, δ_{obs}; δ_A and δ_{AD} being the chemical shifts of a nucleus in the acceptor in the experimental situation in the isolated acceptor and in the pure complex respectively, and $[D]_0$ is the concentration of the donor which by comparison with the concentration of acceptor is in large excess. The observed shift will depend on the properties of the solution and will be different from that observed in the gas phase. This difference δ_{solv} is known as the solvent screening constant and is given by:[5]

$$\delta_{solv} = \delta_{gas} - \delta_{obs} = \delta_b + \delta_W + \delta_A + \delta_E + \delta_S \tag{40}$$

where δ_b is a contribution from bulk diamagnetic susceptibility differences between the solution and reference, δ_A measures the solvent magnetic aniso-tropy, δ_W the weak dispersion force interaction between solute and solvent. δ_E is a polar effect included where the solute is polar and δ_S measures the shift due to any specific complexing of solute and solvent. δ_b is eliminated by the use of an internal reference. The effect of solvent complexing on nmr shifts has been reviewed[298-300] and a recent critical appraisal published.[301] Here we will be concerned briefly only with one aspect, the solvent variation of Δ_{AD}, the shift of the acceptor proton in the pure complex. This may not be measured directly and is evaluated, together with K, from the Benesi–Hildebrand plot. It may loosely be considered as the nmr equivalent of the molar absorption coefficient (extinction coefficient) derived from optical methods. Early measurements showed Δ_{AD} to be dependent on solvent and on the concentration scale employed.[302] This was rationalized in terms of solvent interactions leading to the solution being more ideal on certain concentration scales than others.[167] Recently, Hanna and Rose[303] have investigated the benzene–caffeine complex using nmr. They find that Δ_{AD} differs by a factor of two in carbon tetrachloride as solvent when mole-fraction and molal concentration units are employed. If corrections are made for non-ideality on the basis of Henry's Law activity coefficients for benzene determined from vapour pressure measurements, then the same value for Δ_{AD} is obtained on the two concentration scales and the K values are also consistent. This result casts doubts on measurements made assuming an ideal mixture of species for the solution and emphasizes the need to take into account activity coefficients in the determination of CT parameters from Benesi–Hildebrand and similar equations. However, it should be emphasized that in most cases where there is an apparent dependence of Δ_{AD} on the concentration scale, the degree of complexation is low. In some other systems where $K_c \approx 101 \, mol^{-1}$,* Δ_{AD} has been shown to be independent of the concentration scale.[303a]

VIII. ASSOCIATION CONSTANTS AND RELATED THERMODYNAMIC PARAMETERS†

A. General Considerations

The equilibrium constant for the reaction $D + A \rightleftharpoons DA$ is given by

$$K_a = \frac{a_{DA}}{a_D a_A} = \frac{[DA]}{[A][D]} \cdot \frac{\gamma_{DA}}{\gamma_D \gamma_A},$$ (41)

---

* See next section.
† For a recent review of the determination of association constants see ref. 303b.

where a and γ are the activity and activity coefficient respectively. In the majority of determinations of association constants it has been assumed that the ratio of activity coefficients may be taken as unity and much of the published data gives K_c, the equilibrium constant in terms of concentration in molarity units. Another commonly used unit is K_x, the mole fraction equilibrium constant while occasionally K_m the molal constant has been used. Except in very dilute solution these constants are not related in a simple way. Expressions relating K in various units have been given by Christian.[304] Briegleb[3] in his review of the early work on complexes noted the paucity of information on solvent variation of K and also that published values of thermodynamic parameters by various authors for the same system, varied widely. Many of the determinations have been carried out in one or perhaps two solvents chosen for their inertness. Very little data exist on solvent variation of K and other thermodynamic parameters in a wide range of solvents and hardly any values are known for polar solvents. Data have been accumulated, however, for vapour phase complexes[1] and a number of authors have discussed the changes in thermodynamic parameters from gas to solution phase.[1, 149, 304] The majority of determinations of K have been carried out spectrophotometrically using the Benesi–Hildebrand[160] or related equations.[5] A basic form of the relationship, derived from the condition $[D]_0 \gg [A]_0$ is

$$\frac{[A]_0 l}{A} = \frac{1}{K_{BH}\epsilon_{BH}} + \frac{1}{\epsilon_{BH}} \tag{42}$$

where A is the complex absorbance, l, the path length, $[A]_0$ and $[D]_0$ the total concentrations of donor and acceptor and ϵ_{BH}, K_{BH} the molar absorption coefficient and association constant for the complex derived from the equation. There has been much discussion on the validity of values of K for weak complexes due to the difficulty of separation of K and ϵ from the product.[1, 5, 6, 149, 305a] A criterion has been proposed[305–309] that meaningful separation of K and ϵ may only be achieved under the conditions $0{\cdot}1K \leqslant [D]_0 \leqslant 9K$. In the gas phase where only low concentrations may be achieved, reliable values for K are only feasible for strong complexes. The product $K\epsilon$ may be determined accurately and from its temperature dependence, assuming constancy of ϵ, internal energies, enthalpies and entropies of complex formation may be determined. The unreliability of K values, has hindered the interpretation of changes in the thermodynamic parameters on changing the medium. Recently, methods have been evolved for the reliable determination of K for some systems.[304]

B. Changes in Parameters from Gas to Solution

Table VI shows a selection of data representative of the various classes of

TABLE VI. Association constant, K_c and the product, $K\epsilon$, measured in the gas phase and in solution in solvents of low polarity

Complex	K_c(vap)/l mol^{-1}	$K_c\epsilon$(vap)/10^3 l^2 mol^{-2} cm^{-1}	Ref.	K_c(soln)/l mol^{-1}	$K_c\epsilon$(soln)/10^3 l^2 mol^{-2} cm^{-1}	Solvent	Ref.
Benzene–iodine	4·45±0·6[a] (25°C)	7·43	41	0·246 (24°C)	2·95	*n*-heptane	318
Diethyl ether–iodine	6·4±1·2 (25°C)	13·4	41	1·3 (15°C)	7·12	*n*-heptane	118
Diethyl sulphide–iodine	11·5±1·1 (100°C)	193·6±3	46	180±7 (25°C)	4750	*n*-heptane	32, 47
Dimethyl Sulphide–iodine	191±20[a] (25°C); 9·5±1·3 (100°C); 84±15[a] (25°C)	3210±55[a]; 102±3; 1150±31[a]	46	71±2 (25°C)		CCl_4	340
Iodine–iodine	0·4±0·1 (332°C); 1·7[a] (25°C)	~1·2; ~5·1	341	0·13[b] (25°C)	2·04	CCl_4	342
Benzene–TCNE	6·6±3·6 (100°C); 49±30 (25°C)	5·0±2·7; 38±23	48	0·938[c]	0·893	*n*-heptane	343
Toluene–TCNE	8·1±1·0 (100°C); 76±11 (25°C)	9·3±1·2; 87±13	48	2·18[c]	2·63	*n*-heptane	343
p-Xylene–TCNE	23·1±6·2 (100°C); 280±100[a] (25°C)	18·5±1·7	32	5·77[c]	9·62	*n*-heptane	343
Mesitylene–TCNE	45·8±12·4 (100°C); 1020±350[a] (25°C)	57·3±3·6	32	13·6[c]	22·8	*n*-heptane	343
Durene–TCNE	348±120 (100°C); 11800±5900[a] (25°C)	153±10·0	32	19·8[c]	18·4	*n*-heptane	343
Naphthalene–TCNE	35·6±6 (100°C); 381±73 (25°C)	20±3; 219±41	48			*n*-heptane	343
Diethyl ether–CO(CN)$_2$	45[d]	2·48[d]	31, 54	4·04±0·21 (25°C); 3·6[d]	1·51[d]	CCl_4; *n*-hexane	344; 31, 54
Benzene–CO(CN)$_2$	~14[d]	1·52[d]	31, 54	0·4[d]	0·38[d]	*n*-hexane	31, 54
Trimethylamine–SO$_2$	154±12 (40°C); 340±35[a] (25°C)	878; 2070	55, 56	1035±20 (40°C); 2550±50 (25°C)	5170; 13800	*n*-heptane	55, 56
trans-2-Butene–SO$_2$	3·1±0·5 (25°C)	0·290±0·049	48	0·082±0·009 (25°C)	0·116±0·007	*n*-hexane	57

[a] Extrapolated value. [b] Estimated value. [c] Room temperature. [d] Temperature not specified.

complex. More complete collections of data are contained in recent reviews.[1, 149, 304] In general it may be seen that association constants and energies of formation are larger in the vapour phase than in solution. The only exceptions are the very strong complexes of iodine with n-donor,[61–63] and the sulphur dioxide–triethylamine complex[55, 56] where the stability increases, with further increases as the polarity of the medium is increased. There is a trend towards smaller differences in K and $\Delta E^{\ominus}$ as the strength of the complexes increases. Diethyl ether–iodine, for example, has K_c larger in the vapour phase whereas $\Delta E^{\ominus}$ is similar in the two phases. The stronger complexes dialkyl sulphide–iodine have both K_c and $\Delta E^{\ominus}$ of similar magnitude in both phases.[1]

We may compare the values of thermodynamic functions in gas and solution phases in the following cycle:

$$
\begin{array}{ccc}
\mathrm{D_{(g)}} + \mathrm{A_{(g)}} & \xrightarrow{\Delta X_g^{\ominus}} & \mathrm{DA_{(g)}} \\
\left\downarrow{\scriptstyle \Delta X^{\ominus}_{D(solv)}}\quad \left\downarrow{\scriptstyle \Delta X^{\ominus}_{(solv)}} & & \left\downarrow{\scriptstyle \Delta X^{\ominus}_{DA(solv)}} \\
\mathrm{D_{(soln)}} + \mathrm{A_{(soln)}} & \xrightarrow{\Delta X^{\ominus}_{soln}} & \mathrm{DA_{(soln)}}
\end{array}
$$

where ΔX_{soln} are the solvation energies of D, A and DA. We thus obtain:

$$
\Delta X^{\ominus}_{(g)} = \Delta X^{\ominus}_{(soln)} - [\Delta X^{\ominus}_{DA(solv)} - \Delta X^{\ominus}_{D(solv)} - \Delta X^{\ominus}_{A(solv)}]. \tag{43}
$$

The thermodynamic parameters will only be the same in the two phases if the solvation energy of the complex is equal to the sum of those for donor and acceptor. This is unlikely since solvent is extruded in the formation of the complex. Mulliken and Person[6] have used a more detailed cycle starting with pure liquid as the standard state for each species. The difference between $\Delta H_{(g)}$ and $\Delta H_{(soln)}$ then involves the heats of vapourisation of the three species as well as solvation heats. Using regular solution theory,[161] they estimate the solvation energies from the solubility parameters

$$
\Delta H^{\ominus}_{D(solv)} = \overline{V}_D \phi_1^2 (\delta_D - \delta_1)^2 \tag{44}
$$

where $\overline{V}_D$ is the molar volume of pure liquid D; ϕ is the volume fraction of the solvent and δ_D, δ_1 are the solubility parameters of donor and solvent respectively. δ_C, the solubility parameter of the complex, is estimated as the geometric mean of δ_D and δ_A. They estimate the enthalpy change in the vapour phase to be typically 20–$30\,\mathrm{J\,mol^{-1}}$ (5–$8\,\mathrm{kcal\,mol^{-1}}$) more negative than that in solution in an inert solvent. This may be viewed as the contribution from dispersion forces to the stability of the vapour phase complex. In solution the dispersion interactions should approximately cancel since one D–S and one A–S contact are replaced by one D–A and one S–S contact on complex formation. This treatment assumes the absence of any specific interaction with solvent and does not take into account dipolar interactions with the solvent. For strong complexes particularly, where the results are not in accord with this prediction, one or both of these possibilities must be considered.

Christian and co-workers[304, 308–311] have treated the problem in terms of the transfer energies from vapour to solution of the components and the complex. The thermodynamic energies and free energies may then be expressed as:

$$\Delta E^{\ominus}_{(soln)} = \Delta E^{\ominus}_{(vap)} + \Delta E^{\ominus}_{\underset{V \to S}{DA}} - \Delta E^{\ominus}_{\underset{V \to S}{D}} - \Delta E^{\ominus}_{\underset{V \to S}{A}} \tag{45}$$

and:

$$\Delta G^{\ominus}_{(soln)} = \Delta G^{\ominus}_{(vap)} + \Delta G^{\ominus}_{\underset{V \to S}{DA}} - G^{\ominus}_{\underset{V \to S}{D}} - E^{\ominus}_{\underset{V \to S}{A}} \tag{46}$$

Here $\Delta E^{\ominus}_{V \to S}$ and $\Delta G^{\ominus}_{V \to S}$ are the transfer energies and free energies of the components from the ideal gaseous state to ideal dilute solution in a solvent, S. They express coefficients α and α' as

$$\alpha = \frac{\Delta E^{\ominus}_{\underset{V \to S}{DA}}}{\Delta E^{\ominus}_{\underset{V \to S}{D}} + \Delta E^{\ominus}_{\underset{V \to S}{A}}} \tag{47}$$

and

$$\alpha' = \frac{\Delta G^{\ominus}_{\underset{V \to S}{DA}}}{\Delta G^{\ominus}_{\underset{V \to S}{D}} + \Delta G^{\ominus}_{\underset{V \to S}{A}}} \tag{48}$$

α and α' are taken to be nearly equal for a given solvent and complex[309, 311] and to be little dependent on temperature. For weak complexes, $\alpha < 1$ due to reduced solute–solvent contact while for very strong complexes $\alpha > 1$ due to strong complex dipole–solvent-induced-dipole interaction. The relationship between $K_{c(soln)}$ and $K_{c(vap)}$ may be written as

$$\log K_{c(soln)} = \log K_{c(vap)} + (\alpha' - 1) \log (K_{d, D} \cdot K_{d, A}) \tag{49}$$

where $K_{d, D}$ and $K_{d, A}$ are the dimensionless distribution coefficients of D and A between solution and vapour. Equation (49) has been used to correlate data for the pyridine–iodine complex.[309] α and α' may be calculated directly from (49) where $K_{c(vap)}$ and $K_{c(soln)}$ as well as the transfer energies, are known. Values of α and α' may also be calculated using lattice models of the solution and by calculations involving the transfer energies of non-polar analogues for polar solutes.[312, 313] The calculation of transfer energies thus permits the prediction of $K_{c(soln)}$ from values in the vapour phase. The variation in the difference between $K_{c(vap)}$ and $K_{c(soln)}$ with strength of complexation may be viewed as the competition between two opposing effects. Firstly the extrusion of solvent on complex formation will destabilize the complex with respect to the separated components and secondly the increased dipole–induced-dipole interaction will stabilize the complex more as the dative-state contribution increases in the ground state. The method of transfer energies is mainly of application for non-polar solvents where specific interactions are expected to be relatively unimportant. It represents one extreme view which seeks to interpret solvent effects purely in terms of non-specific interactions.

C. Specific Solvent Effects

Other workers have taken the opposite view of solvent effects on equilibrium and have sought to interpret changes in terms of specific interaction between the complex, or its components, and the solvent. On this view, no correlation of thermodynamic functions with general solvent polarity parameters such as dielectric constant or polarizability, may be carried out without first eliminating specific interactions.[5, 314] Carter, Murrell and Rosch[162] argued that the solvent should be incorporated into the equilibrium equation which is then considered to be between solvated species. Foster[5] has termed this specific solvation:

$$AS_n + DS_m \rightleftharpoons DAS_p + S_q \qquad (50)$$

where $q = n + m - p$. This again stresses the extrusion of solvent on complex formation. The equilibrium constant is then defined by:

$$K = \frac{[DAS_p]X_s^q}{[AS_n][DS_m]} \qquad (51)$$

where X_s is the mole fraction of free solvent defined relative to the total number of moles of D, A and S, both complexed and uncomplexed. K, defined in this way, is in the same units as the equilibrium constant derived from the Benesi–Hildebrand equation, K_{BH}. Assuming $[D_0] \gg [A_0]$ and $[S]_0 \gg [D]_0$,

$$K_{BH} = \frac{K - q(m + 1)}{[S]_0} \qquad (52)$$

and

$$\epsilon_{BH} = \epsilon\left(\frac{K}{K_{BH}}\right) \qquad (53)$$

where K and ϵ are the true equilibrium constant and absorption coefficient. K_{BH} will thus be an underestimate of the equilibrium constant by an amount $q(m + 1)[S_0]^{-1}$ and ϵ_{BH} an overestimate by the factor K/K_{BH}. Trotter and Hanna[167] have used a similar treatment, simplified by the assumption that the donor is not specifically solvated. Using an optimal value of $q(m + 1)$, Carter, Murrell and Rosch corrected values of ϵ and K such that a plot of ϵ vs K for a related series of complexes had a positive slope and passed through the origin as demanded by the Mulliken theory.[2] For the same complexes, the uncorrected values did not give this correlation. For strong complexes, $K \gg q(m + 1)[S_0]^{-1}$ and K_{BH} is almost identical with K. For contact complexes $K = q(m + 1)[S_0]^{-1}$, while it is also possible to envisage $K < q(m + 1)[S_0]^{-1}$. This corresponds to positive, zero and negative intercepts being obtained for the uncorrected Benesi–Hildebrand plot, all situations which have been observed for weak complexes.[145] Where solvation energies are of the same order as those involved in complex formation, it will obviously be very difficult to obtain meaningful values for K. The correction factor will obviously depend on the solvent used, being dependent on both $[S]_0$ and the solvation numbers of the various species. A criticism of the Carter, Murrell and Rosch treatment has been made in that it predicts that K and ϵ should be equal to K_{BH} and ϵ_{BH} in the gas phase. The $K\epsilon$ product should thus be equal in the two phases. In fact, as pointed out by Tamres,[1] there are a number of examples in which $K\epsilon$ for the same complex in solution and in the vapour phase differ considerably. Other criticisms are based on the failure to account for the difference in optical and NMR values for K,[316] and the relative constancy of ϵ_{BH} for some complexes in different solvents.[5, 316] In a later paper, Carter rebuts the first of these criticisms as being caused by higher order complexes and the second by pointing out that strong complexes do not require solvation correction. Other authors[318, 319] find no improvement in K vs ϵ plots using the corrections. Scott,[320, 321] however, has stressed the value of the approach. He considers that the solvent must be considered because it occupies lattice sites in the solution, and must therefore be taken into account

The other major approach to specific interaction views the solvent as a competing reagent for either or both the donor and the acceptor.[5, 163, 166] For such competition, it may be shown if both A and D are complexed by solvent,

$$K^{AD}_{c(corr)} = K^{AD}_{c(obs)}(1 + K^{AS}_c[S])(1 + K^{DS}_c[S]) \tag{54}$$

The competing interaction may be via charge-transfer or hydrogen-bonding forces, and will obviously be of major importance in protic and n-donor solvents. Merrifield and Phillips[163] have accounted for the large solvent variations in K for tetracyanoethylene complexes, assuming solvent–

acceptor complex formation. Attempts have been made to determine directly K for solvent–component complexing in order to correct K_{BH} values for complexes.[322] Other workers have used a mixture of interacting and non-interacting solvents.[323] Using the relationship:

$$K_{c(obs)}^{DA} = K_{c(corr)}^{DA} - K_c^{SD} K_{c(obs)}^{DA} [S]_0 \qquad (55)$$

where $[S]_0$ is the concentration of interacting solvent. A plot of $K_{c(obs)}^{AD}$ vs $K_{c(obs)}^{DA}[S]_0$ gives both $K_{c(corr)}^{DA}$ and K_c^{SD} from the intercept and slope. Tamres[149] has extended the corrections to 2:1 solvent–donor complexes. The extreme of solvent–component competition tends to the Carter, Murrell and Rosch approach when the stoichiometry of the complexes approaches the solvation numbers.

D. Solution Non-Ideality

Mulliken and Person[6] have argued that the specific interaction approaches may be generalized into a treatment involving the dependence of activity coefficients on concentration. The Benesi–Hildebrand and related equations are usually used without activity coefficient corrections and any departures from ideality, including specific solvation and specific complexing will result in the derived values of K being in error. Using equation (32), Mulliken and Person[6] find that the value of K for benzene–iodine may be in error by a factor of two by the neglect of non-ideality. Using regular solution theory,[161] however, Alley and Scott[324] have argued that the activity coefficient γ_{DA} should nearly equal $(\gamma_D \gamma_A)$.

Many workers[165, 167, 303, 325] have argued that the choice of concentration scale may influence the value of K_{BH}. If the solution is less ideal on one scale than on another, then the derived values of K may not agree when conversion to a common scale is made. Trotter and Hanna[167] have analysed data for the benzene–iodine complex and find that the lowest value for ϵ is obtained using a molal plot and argue that since the intercept on a molal scale refers to the condition of infinitesimal concentration of acceptor in pure donor, the data provides a lower limit for ϵ. They also argue that as solvent is displaced on complexing, then the mole-fraction and molar scales will lead to overestimate of ϵ and underestimate of K. Christian,[304] however, points out that even in solutions rigorously obeying Raoults Law, molality activity coefficients are strongly concentration dependent and molality plots should thus be avoided. This point is made also by Hanna and Rose[303] who nevertheless find that using independently determined activity coefficients for benzene, K_x and K_m values consistent with one another may be determined for the benzene–caffeine complex. The uncorrected values of K disagree markedly. Kuntz and coworkers[325] through the analysis of data in pairs of inert solvents, have shown that only K_c is independent of solvent properties such as molar volume. They conclude that molarity is the best scale to use, a point also made by Foster.[5] Christian and coworkers[305a, 326, 327] have developed a method to eliminate activity coefficient variations by using a constant activity method. Using the equilibrium,

$$(CH_3)_4NI_{3(S)} + I_2 \rightleftharpoons (CH_3)_4NI_{5(S)}$$

a constant activity of iodine may be maintained in solution or in vapour phase. Increase in total iodine concentration on addition of a donor may then be ascribed to complex formation and the concentration of complex may thus be determined directly. Some extra iodine solubility, however, may result from non-specific interactions. Values for K corrected for this show good agreement with K_{BH} whereas the K values determined directly from solubility data disagree. This confirms Scott's[320] view that the spectrophotometric value of K_{BH} includes only complexes formed in excess to random encounters between the components. The magnitude of the non-specific interactions as measured by the difference between uncorrected values of K from solubility measurements and K_{BH}, were found to depend only on γ_{I_2}, the activity coefficient of iodine. The contribution from non-specific interactions decreased with strength of the donor as earlier predicted by Mulliken and Orgel[122] in their consideration of contact complexes and orientational isomers. Activity coefficient variations from solvent to solvent for the same DA system are capable of having a marked effect on measured data. Thus the solubility studies of Satterfield[328] have shown that order-of-magnitude variations in activity coefficients occur for TCNE, triphenylene and their complex between n-heptane and chloroform.

As mentioned earlier, regular solution theory[161] has been used to estimate activity coefficients from solubility parameters. Using solubility parameter theory, a relation has been proposed between the molar association constant and the solubility parameter of the solvent, δ_S, by assuming the solubility parameters of donor, acceptor and complex to be invariant with solvent.[329]

$$\log K_c = a + b\delta_S \qquad (55)$$

where a and b are constants for a particular DA system. For a group of non-polar solvents, a good linear plot may be obtained of $\log K_c$ vs δ_S for the pyridine–iodine complex.[113] A similar linear relationship does not seem to hold for K_x in systems so far investigated.[329] Christian[330] has pointed out that this treatment is basically similar to the α-coefficient approach based on transfer energies from gas to solution. It is interesting to note that K values for a particular related series of complexes in one solvent, linearly correlate with the values in another solvent as shown by Foster and coworkers[5, 331] for the complexes of fluoranil with various donors in carbon tetrachloride and chloroform.

Moving to the variation of K with solvent, there appears to be no simple correlation of K with bulk polarity parameters such as dielectric constant although there is a general trend to lower association constant with increasing polarity.[5, 304] For strong complexes, the reverse trend occurs.[55, 56, 62, 63] This has been interpreted as loss of stabilization by extrusion of solvent for weak complexes where the complex dipole–solvent-induced-dipole contribution will be insignificant since the complex dipole is often of comparable magnitude to component local dipoles.[304] The strong complexes with their significantly larger dipoles will gain extra stabilization from the inductive interactions between solute and solvent. It appears that separation of specific and non-specific interaction is not possible at the present time, and as Christian,[304] has argued the best approach lies in accurately determining gas-phase data and combining this with transfer energies calculated from theoretical models of the solution. At the same time there is obviously a need to take activity coefficients into consideration in the evaluation of solution-phase data. Even then it is difficult to see a comprehensive theory emerging for any but the non-polar solvent group. In polar solvents, the range of interactions is such as to render the problem considerably more formidable.

E. Ionic Complexes

The ground state of ionic complexes is a contact ion-pair. Increase of solvent polarity will lead to an increase in solvent-shared and separated ion-pairs and in very polar solvents to the individually solvated ions. Association constants will thus decrease markedly with increasing polarity. A linear plot of K vs Z has been observed for the pyridinium halide complexes.[28] Observed values for K will also depend on the total ionic strength of the solution[180] as predicted by the Brönsted–Bjerrum theory.

IX. CHANGE IN NATURE OF THE COMPLEX BY INTERACTION WITH THE SOLVENT

It is not the intention in this review to discuss the chemical implications of the interaction of solvents with complexes. Comprehensive reviews have been published[4, 5, 257, 332–335] on the chemistry of CT complexes, in which the role of solvent is discussed. In the context of the effect of solvent on the physical properties of complexes which forms the bulk of this review, it is, however, worth briefly mentioning changes in the nature of the complex which occur as a result of interaction with the solvent. In complexes of

N,N,N',N'-tetramethyl-p-phenylenediamine with certain quinones[336, 337] the CT band ascribed to the normal $V \leftarrow N$ transition is observed only in solvents of low polarity. In more polar solvents, spectra are observed, identifiable as arising from the Würster's radical cation of the diamine and the radical anion of the quinone.[337] Similar results are found for other complexes. Mulliken and Person[338] have termed this action of the solvent, environmental co-operative action. For cases where the difference between donor ionization energy and acceptor electron affinity is small, the complex ground and excited states will be quite close in energy. Under the influence of a polar solvent, the ionic excited state may become lower in energy than the no-bond state. Formation of the separate radical ions of D and A is then viewed as proceeding via the solvated ion-pair. The two forms of the complex, DA and D^+-A^- are termed "outer" and "inner" complexes, the inner complex being formed via the outer by crossing of the two potential energy curves at a sufficiently low energy. The possibility now exists in a polar solvent of observation of the reverse CT transition $D^+A^- \rightarrow DA$. While there is no unequivocal observation of such a band in solution, studies on the spectra of crystals in polarized light have established[339] the existence of the transition. This modification of the relative energies of the dative and no-bond structures on increasing the polarity of the solvent gives some force to the idea that a polar solvent may drastically change the dative contribution to the ground state, mixing between ψ_0 and ψ_1, being facilitated by a reduction in the energy difference between the two states.

It is hoped that this review has given some indication of the major role played by the solvent in determining the properties of complexes. One may end by a plea for more detailed measurements on CT bands, with the publication of oscillator strengths, band-widths and the like, and of the thermodynamic parameters by taking due account of activity corrections. Only when reliable data are available for a wide variety of complexes in a large selection of solvents, can meaningful attempts be made to arrive at a definitive description of the role of the solvent.

REFERENCES

1. M. Tamres. Ref. 7, p. 49.
2. R. S. Mulliken. *J. Amer. Chem. Soc.*, **74**, 811 (1952).
3. G. Briegleb. "Elektronen–Donator–Acceptor Komplexe", Springer-Verlag, Berlin (1961).
4. L. J. Andrews and R. M. Keefer. "Molecular Complexes in Organic Chemistry", Holden-Day, San Francisco (1964).
5. R. Foster. "Organic Charge-Transfer Complexes", Academic Press, London and New York (1969).

6. R. S. Mulliken and W. B. Person. "Molecular Complexes: A Lecture and Reprint Volume", Wiley, New York (1969).
7. "Molecular Complexes", Vol I (ed. R. Foster), Elek Science, London; Crane Russak, New York (1973).
8. "Spectroscopy and Structure of Molecular Complexes", (ed. J. Yarwood), Plenum Press, London and New York (1973).
9. M. W. Hanna and J. L. Lippert. Ref. 7, p. 1.
10. M. Tamres. *J. Phys. Chem.*, **71**, 1988 (1967).
11. E. G. McRae. *J. Phys. Chem.*, **61**, 562 (1957).
12. F. L. Wehry. In "Fluorescence" (ed. G. G. Guilbault), Edward Arnold, London (1967), p. 76.
13. S. Basu. *Advances in Quantum Chemistry*, **1**, 145 (1964).
14. L. Onsager. *J. Amer. Chem. Soc.*, **58**, 1486 (1936).
15. N. G. Bakshiev. *Opt. Spectry.*, **10**, 379 (1961).
16. N. S. Bayliss. *J. Chem. Phys.*, **18**, 292 (1950).
17. E. Lippert. *Z. Elektrochem.*, **61**, 962 (1957).
18. E. Lippert. *Z. Naturforsch.*, **10a**, 541 (1955).
19. L. Bilot and A. Kawski. *Z. Naturforsch.*, **17a**, 621 (1962).
20. E. G. McRae. *J. Phys. Chem.*, **61**, 562 (1957).
21. Y. Ooshika. *J. Phys. Soc. Japan*, **9**, 594 (1954).
22. E. Weigang. *J. Chem. Phys.*, **41**, 1435 (1964).
23. W. Liptay. *Z. Naturforsch.*, **21a**, 1605 (1966).
24. W. W. Robinson. *J. Chem. Phys.*, **46**, 572 (1967).
25. W. Liptay. *Angew. Chem. Int. Ed. Engl.*, **8**, 177 (1969).
26. T. Abe. *Bull. Chem. Soc. Japan*, **38**, 1314 (1965).
27. H. Kuhn and A. Schweig. *Chem. Phys. Lett.*, **1**, 255 (1967).
28. E. M. Kosower. *J. Amer. Chem. Soc.*, **80**, 3253 (1958).
29. K. Dimroth, C. Reichardt, T. Siepmann and F. Bohlmann. *Liebigs Ann. Chem.*, **661**, 1 (1963).
30. P. Suppan. *J. Chem. Soc. A*, 3125 (1968).
31. J. Prochorow and A. Tramer. *J. Chem. Phys.*, **44**, 4545 (1966).
32. M. Kroll. *J. Amer. Chem. Soc.*, **90**, 1097 (1968).
33. P. J. Trotter. *J. Amer. Chem. Soc.*, **88**, 5721 (1966).
34. P. J. Trotter. *J. Chem. Phys.*, **47**, 775 (1967).
35. H. W. Offen. Ref. 7, p. 117.
36. H. W. Offen. *J. Chem. Phys.*, **42**, 430 (1965).
37. A. H. Ewald. *Trans. Faraday Soc.*, **64**, 733 (1968).
38. K. M. C. Davis and M. C. R. Symons. *J. Chem. Soc.*, 2079 (1965).
39. J. N. Murrell. "The Theory of the Electronic Spectra of Organic Molecules", Methuen, London (1963).
40. W. K. Duerkson and M. Tamres. *J. Amer. Chem. Soc.*, **90**, 1379 (1968).
41. F. T. Lang and R. L. Strong. *J. Amer. Chem. Soc.*, **87**, 2345 (1965).
42. E. M. Voigt and B. Meyer. *J. Chem. Phys.*, **49**, 852 (1968).
43. J. Ham. *J. Amer. Chem. Soc.*, **76**, 3875 (1954).
44. J. Grundnes, M. Tamres and S. N. Bhat. *J. Phys. Chem.*, **75**, 3682 (1971).
45. E. I. Ginns and R. L. Strong. *J. Phys. Chem.*, **71**, 3059 (1967).
46. M. Tamres and S. N. Bhat. *J. Amer. Chem. Soc.*, **94**, 2577 (1972).
47. M. Tamres and S. Searles Jr. *J. Phys. Chem.*, **66**, 1099 (1962).
48. I. Hanazaki. *J. Phys. Chem.*, **76**, 1982 (1972).
49. M. Kroll and M. L. Ginter. *J. Phys. Chem.*, **69**, 3671 (1965).

50. J. Aihara, M. Tsuda and H. Inokuchi. *Bull. Chem. Soc. Japan*, **40**, 2460 (1967).
51. J. Aihara, M. Tsuda and H. Inokuchi. *Bull. Chem. Soc. Japan*, **42**, 1824 (1969).
52. J. Aihara, M. Tsuda and H. Inokuchi. *Bull. Chem. Soc. Japan*, **43**, 3067 (1970).
53. G. Briegleb, J. Czekalla and G. Reuss. *Z. Phys. Chem.* (Frankfurt am Main), **30**, 316 (1961).
54. J. Prochorow. *J. Chem. Phys.*, **43**, 3394 (1965).
55. S. D. Christian and J. Grundnes. *Nature*, **214**, 1111 (1967).
56. S. D. Christian and J. Grundnes. *J. Amer. Chem. Soc.*, **90**, 2239 (1968).
57. D. Booth, F. S. Dainton and K. J. Ivin. *Trans. Faraday Soc.*, **55**, 1293 (1959). (1970).
58. C. N. R. Rao, G. C. Chaturvedi and S. N. Bhat. *J. Mol. Spectroscopy*, **33**, 554 (1970).
59. M. Tamres, W. K. Duerksen and J. M. Goodenow. *J. Phys. Chem.*, **72**, 966 (1968).
60. P. A. D. de Maine. *J. Chem. Phys.*, **24**, 1091 (1956).
61. H. Mollendal, J. Grundnes and E. Augdahl. *Acta Chem. Scand.*, **23**, 3525 (1969).
62. E. Augdahl, J. Grundnes and P. Klaeboe. *Inorg. Chem.*, **4**, 1475 (1965).
63. R. P. Lang. *J. Phys. Chem.*, **72**, 2129 (1968).
64. K. M. C. Davis, unpublished work.
65. H. W. Offen and M. S. F. A. Abidi. *J. Chem. Phys.*, **44**, 4642 (1966).
66. J. N. Murrell. *Quart. Rev. Chem. Soc.*, **15**, 191 (1961).
67. S. F. Mason. *Quart. Rev. Chem. Soc.*, **15**, 287 (1961).
68. H. M. Rosenberg. *Chem. Commun.*, 312 (1965).
69. H. M. Rosenberg and D. Hale. *J. Phys. Chem.*, **69**, 2490 (1965).
70. H. M. Rosenberg, E. Eimutis and D. Hale. *Can. J. Chem.*, **44**, 2405 (1968).
71. J. Ham. *J. Amer. Chem. Soc.*, **76**, 3881 (1954).
72. N. S. Bayliss and E. G. McRae. *J. Phys. Chem.*, **58**, 1002 (1954).
73. M. B. Ledger and P. Suppan. *Spectrochim. Acta*, **23a**, 641 (1967).
74. E. M. Voigt. *J. Phys. Chem.*, **70**, 598 (1966).
75. N. S. Isaacs. *J. Chem. Soc. B.*, 1352 (1967).
76. I. Ilmet and P. M. Rashba. *J. Phys. Chem.*, **71**, 1140 (1967).
77. J. Czekalla and K. O. Meyer. *Z. Phys. Chem.* (Frankfurt am Main), **27**, 185 (1961).
78. P. H. Emslie and R. Foster. *Rec. Trav. Chim. Pays-Bas*, **84**, 255 (1965).
79. T. G. Beaumont and K. M. C. Davis. *J. Chem. Soc. B*, 1010 (1968).
80. S. K. Chakrabarti and S. Basu. *Trans. Faraday Soc.*, **60**, 465 (1964).
81. S. Basu. *Proc. Natl. Inst. Sci. India*, **A30**, 561 (1964).
82. C. Reichardt. *Angew. Chem. Int. Ed. Engl.*, **4**, 29 (1965).
83. F. W. Fowler, A. R. Katritzky and R. J. D. Rutherford. *J. Chem. Soc. B.*, 460 (1971).
84. O. B. Nagy, J. B. Nagy and A. Bruylants. *J. Chem. Soc., Perkin Trans.*, **2**, 968 (1972).
85. E. Grunwald and S. Winstein. *J. Amer. Chem. Soc.*, **70**, 846 (1948).
86. S. Winstein, E. Grunwald and H. W. Jones. *J. Amer. Chem. Soc.*, **73**, 2700 (1951).
87. A. H. Fainberg and S. Winstein. *J. Amer. Chem. Soc.*, **78**, 2770 (1956).
88. S. Winstein, A. H. Fainberg and S. G. Smith. *J. Amer. Chem. Soc.*, **83**, 618 (1961).
89. (a) M. Gielen and J. Nasielski. *Rec. Trav. Chim. Pays-Bas*, **82**, 228 (1963).
    (b) M. Gielen and J. Nasielski. *J. Organomet. Chemistry*, **1**, 173 (1964).
90. A. Berson, Z. Hamlet and W. A. Mueller. *J. Amer. Chem. Soc.*, **84**, 297 (1962).
91. L. G. S. Brooker, G. H. Keyes and D. W. Heseltine. *J. Amer. Chem. Soc.*, **73**, 5350 (1951).
92. E. M. Kosower. *J. Amer. Chem. Soc.*, **80**, 3267 (1958).

93. K. Dimroth and C. Reichardt. *Palette* No. 11, p. 28 (1962), publication by Sandoz A.G. Basel, Switzerland.
94. S. Brownstein. *Can. J. Chem.*, **38**, 1590 (1960).
95. L. P. Hammett. "Physical Organic Chemistry", McGraw Hill, New York, London (1940), p. 184.
96. J. C. Jungers and L. Sajus. "L'Analyse cinetique de la Transformation chimique", 1re ed., Vol. II, Technip, Paris (1968).
97. T. Kagyia, Y. Sumida and T. Inoue. *Bull. Chem. Soc. Japan*, **41**, 767 (1968).
98. E. M. Arnett. *Prog. Phys. Org. Chem.*, **1**, 223 (1963).
99. E. M. Kosower and M. Ito. *Proc. Chem. Soc.*, 25 (1962).
100. E. M. Kosower. *J. Org. Chem.*, **29**, 956 (1964).
101. T. G. Beaumont and K. M. C. Davis. Ref. 79.
102. K. M. C. Davis. *J. Chem. Soc. B.*, 1128 (1967).
103. K. M. C. Davis. *J. Chem. Soc. B.*, 1020 (1969).
104. E. M. Kosower and B. G. Ramsay. *J. Amer. Chem. Soc.*, **81**, 856 (1959).
105. H. Yasuba, T. Imai, K. Okamoto, S. Kusabayishi and H. Mikawa. *Bull. Chem. Soc. Japan*, **43**, 3101 (1970).
106. M. Feldman and B. G. Graves. *J. Phys. Chem.*, **70**, 955 (1966).
107. J. Burgess. *Spectrochim. Acta*, **26A**, 1369 (1970).
108. J. Burgess. *Spectrochim. Acta*, **26A**, 1957 (1970).
109. J. A. A. Ketelaar. *J. Phys. Radium*, **15**, 197 (1954).
110. P. A. de Maine. *J. Chem. Phys.*, **26**, 1192 (1957).
111. A. I. Popov and W. A. Deskin. *J. Amer. Chem. Soc.*, **80**, 2976 (1958).
112. P. Klaeboe. *Acta Chem. Scand.*, **18**, 27 (1964).
113. P. V. Huong, N. Platzer and M. L. Josien. *J. Amer. Chem. Soc.*, **91**, 3669 (1969).
114. J. H. Hildebrand and R. L. Scott. "The Solubility of Non-electrolytes", Dover Publications, New York (1964).
115. R. S. Mulliken and W. B. Person. *Ann. Rev. Phys. Chem.*, **13**, 107 (1962).
116. V. A. Brosseau, J. R. Basila, J. F. Smalley and R. L. Strong. *J. Amer. Chem. Soc.*, **94**, 716 (1972).
117. R. S. Mulliken. *Rec. Trav. Chim. Pays-Bas*, **75**, 845 (1956).
118. M. Brandon, M. Tamres and S. Searle Jr. *J. Amer. Chem. Soc.*, **82**, 2129 (1960).
119. C. A. Goy and H. O. Pritchard. *J. Mol. Spectrosc.*, **12**, 38 (1964).
120. M. Tamres and J. M. Goodenow. *J. Phys. Chem.*, **71**, 1982 (1967).
121. M. Tamres and S. N. Bhat. *J. Phys. Chem.*, **75**, 1057 (1971).
122. L. E. Orgel and R. S. Mulliken. *J. Amer. Chem. Soc.*, **79**, 4839 (1957).
123. D. F. Evans. *J. Chem. Soc.*, 345 (1953).
124. D. F. Evans. *J. Chem. Soc.*, 2753 (1959).
125. D. F. Evans. *J. Chem. Soc.*, 1735 (1960).
126. H. Tsubomura and R. S. Mulliken. *J. Amer. Chem. Soc.*, **82**, 5966 (1960).
127. H. Bradley Jr. and A. D. King. *J. Chem. Phys.*, **47**, 1189 (1967).
128. H. Ishida, F. Takahishi, H. Sato and H. Tsubomura. *J. Amer. Chem. Soc.*, **92**, 275 (1970).
129. J. Jortner and V. Sokolov. *J. Phys. Chem.*, **65**, 1633 (1961).
130. M. A. Slifkin and A. C. Allison. *Nature*, **215**, 949 (1967).
131. H. Tsubomura and R. P. Lang. *J. Chem. Phys.*, **36**, 2155 (1962).
132. D. F. Evans. *J. Chem. Phys.*, **23**, 1424 (1955).
133. D. F. Evans. *J. Chem. Soc.*, 4229 (1957).
134. J. S. Ham, J. R. Platt and H. McConnell. *J. Chem. Phys.*, **19**, 1301 (1951).

135. S. H. Hastings, J. L. Franklin, J. C. Schiller and F. A. Matsen. *J. Amer. Chem. Soc.*, **75**, 2900 (1953).
136. L. M. Julien and W. B. Person. *J. Phys. Chem.*, **72**, 3059 (1968).
137. D. F. Evans. *J. Chem. Soc.*, 1987 (1961).
138. D. F. Evans. *J. Chem. Soc.*, 3885 (1957).
139. J. Jortner and V. Sokolov. *J. Phys. Chem.*, **65**, 1633 (1961).
140. D. F. Evans. *J. Chem. Soc.*, 4229 (1957).
141. D. F. Evans. *J. Chem. Phys.*, **23**, 1426 (1955).
142. P. R. Hammond. *J. Phys. Chem.*, **74**, 647 (1970).
143. F. Dorr and G. Buttgereit. *Ber. Bunsenges. Phys. Chem.*, **67**, 867 (1963).
144. R. F. Weimer and J. M. Prausnitz. *J. Chem. Phys.*, **42**, 3643 (1965).
145. K. M. C. Davis and M. F. Farmer. *J. Chem. Soc. B.*, 28 (1967).
146. K. M. C. Davis and M. F. Farmer. *J. Chem. Soc. B.*, 859 (1968).
147. M. Tamres and J. Grundnes. *J. Amer. Chem. Soc.*, **93**, 801 (1971).
148. J. E. Prue. *J. Chem. Soc.*, 7534 (1965).
149. M. Tamres and J. Yarwood. p. 253 ref. 8.
150. C. J. Bolter, C. A. G. Brooks, K. M. C. Davis and M. E. Delf. *J. Chem. Soc., Perkin Trans.*, **2**, 1350 (1973).
151. P. R. Hammond and R. R. Lake. *Chem. Commun.*, 987 (1968).
152. Ref. 6, p. 26.
153. N. Q. Chako. *J. Chem. Phys.*, **2**, 644 (1934).
154. A. D. Buckingham. *Proc. Roy. Soc., Ser. A*, **255**, 32 (1962).
155. T. Abe. *Bull. Chem. Soc. Japan*, **43**, 625 (1970).
156. N. S. Bayliss and G. Wills-Johnson. *Spectrochim. Acta*, **24A**, 551 (1968).
157. W. Liptay, H. J. Schlosser and B. Dumbacher. *Z. Naturforsch.*, **23a**, 1613 (1968).
158. O. K. Rice. *Int. J. Quantum Chem.*, **Symp. No 2**, 219 (1968).
159. Ref. 6, p. 205.
160. H. A. Benesi and J. H. Hildebrand. *J. Amer. Chem. Soc.*, **71**, 2703 (1949).
161. J. H. Hildebrand and R. L. Scott. "Regular Solutions", Prentice Hall, Englewood Cliff, N.J. (1962).
162. S. Carter, J. N. Murrell and E. J. Rosch. *J. Chem. Soc.*, 2048 (1965).
163. R. E. Merrifield and W. D. Phillips. *J. Amer. Chem. Soc.*, **80**, 2778 (1958).
164. M. Tamres. *J. Phys. Chem.*, **65**, 654 (1961).
165. R. L. Scott. *Rec. Trav. Chim. Pays-Bas*, **75**, 787 (1956).
166. J. M. Corkill, R. Foster and D. L. Hammick. *J. Chem. Soc.*, 1202 (1955).
167. P. J. Trotter and M. W. Hanna. *J. Amer. Chem. Soc.*, **88**, 3724 (1966).
168. S. D. Ross and M. M. Labes. *J. Amer. Chem. Soc.*, **79**, 76 (1957).
169. P. H. Emslie, R. Foster, C. A. Fyfe and I. Horman. *Tetrahedron*, **21**, 2843 (1965).
170. M. Tamres and J. Yarwood. Ref. 8, p. 217.
171. R. Foster and D. L. Hammick. *J. Chem. Soc.*, 2685 (1954).
172. C. C. Thompson Jr. and P. A. D. de Maine. *J. Amer. Chem. Soc.*, **85**, 3096 (1963).
173. C. C. Thompson and P. A. D. de Maine. *J. Phys. Chem.*, **69**, 2766 (1965).
174. A. H. Ewald and J. Scudder. *J. Phys. Chem.*, **76**, 249 (1972).
175. J. B. Birks. *Nature*, **214**, 1187 (1967).
176. T. L. Förster. *Angew. Chem., Int. Ed. Engl.*, **8**, 333 (1969).
177. H. Knibbe and A. Weller. *Z. Phys. Chem.* (Frankfurt am Main), **56**, 99 (1967).
178. T. Okada, H. Matsui, H. Oohari, H. Matsumoto and N. Mataga. *J. Chem. Phys.*, **49**, 4717 (1969).
179. T. Miwa and M. Koizumi. *Bull. Chem. Soc. Japan*, **39**, 2588 (1966).
180. T. G. Beaumont and K. M. C. Davis. *J. Chem. Soc. B.*, 456 (1970).

181. H. Knibbe, K. Röllig, F. P. Schäfer and A. Weller. *J. Chem. Phys.*, **47**, 1184 (1967).
182. J. Czekalla, A. Schmillen and K. J. Mager. *Z. Elektrochem.*, **61**, 1053 (1957).
183. J. Czekalla, A. Schmillen and K. J. Mager. *Z. Elektrochem.*, **63**, 623 (1959).
184. J. Czekalla, G. Briegleb and W. Herre. *Z. Elektrochem.*, **63**, 712 (1959).
185. J. Czekalla, G. Briegleb, W. Herre and H. J. Vahlensieck. *Z. Elektrochem.*, **63**, 715 (1959).
186. J. Prochorow and A. Tramer. *J. Chem. Phys.*, **47**, 775 (1967).
187. H. W. Offen and J. F. Studebaker. *J. Chem. Phys.*, **47**, 253 (1967).
188. S. Iwata, J. Tanaka and S. Nagakura. *J. Amer. Chem. Soc.*, **89**, 2813 (1967).
189. G. Briegleb and J. Czekalla. *Angew Chem.*, **72**, 401 (1960).
190. H. M. Rosenberg and E. C. Eimutis. *J. Phys. Chem.*, **70**, 3494 (1966).
191. C. A. Parker and G. D. Short. *Spectrochim. Acta*, **23A**, 2487 (1967).
192. J. S. Brinen, J. G. Koren, H. D. Olmstead and R. C. Hirt. *J. Phys. Chem.*, **69**, 3791 (1965).
193. G. Briegleb, J. Trencséni and W. Herre. *Chem. Phys. Lett.*, **3**, 146 (1969).
194. G. Briegleb, W. Herre, W. Jung and H. Schuster. *Z. Phys. Chem.* (Frankfurt am Main), **45**, 229 (1965).
195. G. Briegleb, G. Betz and W. Herre. *Z. Phys. Chem.* (Frankfurt am Main), **64**, 85, (1969).
196. N. Mataga and Y. Murata. *J. Amer. Chem. Soc.*, **91**, 3144 (1969).
197. J. Prochorow and R. Siegozcynski. *Chem. Phys. Lett.*, **3**, 635 (1969).
198. K. M. C. Davis. *Nature*, **223**, 728 (1969).
199. T. Kabayishi, K. Yoshihara and S. Nagakura. *Bull. Chem. Soc. Japan*, **44**, 2063 (1971).
200. M. Shimada, H. Masuhara and N. Mataga. *Bull. Chem. Soc. Japan*, **43**, 3316 (1970).
201. M. Shimada, H. Masuhara and N. Mataga. *Bull. Chem. Soc. Japan*, **44**, 3310 (1971).
202. H. Masuhara, N. Tsujino and N. Mataga. *Bull. Chem. Soc. Japan*, **46**, 1088 (1973).
203. M. Shimada, H. Masuhara and N. Mataga. *Bull. Chem. Soc. Japan*, **46**, 1903 (1973).
204. K. Mutai. *Chem. Comm.*, 1207 (1970).
204. K. Mutai. *Bull. Chem. Soc. Japan*, **45**, 2635 (1971).
205. K. Mutai. *Tetrahedron Lett.*, 1125 (1971).
206. K. Mutai and M. Oki. *Tetrahedron*, **26**, 1181 (1970).
207. H. Leonardt and A. Weller. *Ber. Bunsenges. Phys. Chem.*, **63**, 791 (1963).
208. N. Mataga, T. Okada and N. Yamamoto. *Chem. Phys. Lett.*, **1**, 119 (1967).
209. N. Mataga, T. Okada and H. Oohari. *Bull. Chem. Soc. Japan*, **39**, 2563 (1966).
210. N. Mataga, T. Okada and K. Ezumi. *Mol. Phys.*, **10**, 201 (1966).
211. N. Mataga and K. Ezumi. *Bull. Chem. Soc. Japan*, **40**, 1355 (1967).
212. H. Leonhardt and A. Weller. *Z. Phys. Chem.* (Frankfurt am Main), **29**, 277 (1961).
213. H. Knibbe, D. Rehm and A. Weller. *Z. Phys. Chem.* (Frankfurt am Main), **56**, 95 (1967).
214. A. Nakajima. *Bull. Chem. Soc. Japan*, **42**, 3409 (1969).
215. M. G. Kuzmin and L. N. Guseva. *Chem. Phys. Lett.*, **3**, 71 (1969).
216. A. Nakajima and H. Akamatu. *Bull. Chem. Soc. Japan*, **42**, 3030 (1969).
217. M. Koizumi and H. Yamashita. *Z. Phys. Chem.* (Frankfurt am Main), **57**, 107 (1968).
218. K. Kaneta and M. Koizumi. *Bull. Chem. Soc. Japan*, **40**, 2254 (1967).

219. H. Yamashita, H. Kokuburi and M. Koizumi. *Bull. Chem. Soc. Japan*, **41**, 2312 (1968).
220. D. Rehm and A. Weller. *Z. Phys. Chem.* (Frankfurt am Main), **69**, 183 (1970).
221. A. Weller. *Pure and Appl. Chem.*, **16**, 115 (1968).
222. C. M. Stephenson and G. S. Hammond. *Angew. Chem. Int. Ed. Engl.*, **8**, 261 (1969).
223. R. S. Davidson and G. Brimage. *Chem. Commun.*, 1385 (1971).
224. R. S. Davidson and A. Lewis. *Chem. Commun.*, 263 (1973).
225. T. Okada, T. Fujita, M. Kobota, S. Masaki, N. Mataga, R. Ide, Y. Sakata and S. Musumi. *Chem. Phys. Lett.*, **14**, 563 (1972).
226. E. A. Chandross and H. T. Thomas. *Chem. Phys. Lett.*, **9**, 393, 397 (1971).
227. N. Mataga, T. Okada and N. Yamamoto. *Bull. Chem. Soc. Japan*, **39**, 2562 (1966).
228. H. Beens, H. Knibbe and A. Weller. *J. Chem. Phys.*, **47**, 1183 (1967).
229. J. H. Barkent, J. W. Verhoeven and Th. J. DeBoer. *Tetrahedron Lett.*, 3363 (1972).
230. A. H. Kadhim and H. W. Offen. *J. Chem. Phys.*, **48**, 749 (1968).
231. W. R. Ware and H. P. Richter. *J. Chem. Phys.*, **48**, 1595 (1968).
232. H. Masuhara and N. Mataga. *Chem. Phys. Lett.*, **6**, 608 (1970).
233. H. Masuhara and N. Mataga. *Z. Phys. Chem.* (Frankfurt am Main), **80**, 113 (1972).
234. L. R. Faulkner and A. J. Bard. *J. Amer. Chem. Soc.*, **91**, 209 (1969).
235. A. Weller and K. Zachariasse. *J. Chem. Phys.*, **46**, 4984 (1967).
236. E. C. Lim. *Ber. Bunsenges. Phys. Chem.*, **72**, 274 (1968).
237. N. Yamamoto, Y. Nakato and H. Tsubomura. *Bull. Chem. Soc. Japan*, **40**, 451 (1967).
238. Y. Taniguchi and N. Mataga. *Chem. Phys. Lett.*, **13**, 596 (1972).
239. S. P. McGlynn, T. Azumi and M. Kinoshita. "Molecular Spectroscopy of the Triplet State", Prentice-Hall, Englewood Cliffs, New Jersey (1969).
240. J. Weiss. *Trans. Faraday Soc.*, **34**, 451 (1938).
241. R. Livingston and C. L. Ke. *J. Amer. Chem. Soc.*, **72**, 909 (1950).
242. J. Weiss. *Trans. Faraday Soc.*, **35**, 48 (1939).
243. J. Eisenbrand. *Z. Phys. Chem.*, **22B**, 145 (1933).
244. T. Komiyama, T. Miwa and M. Koizumi. *Bull. Chem. Soc. Japan*, **39**, 2597 (1966).
245. D. K. Majumdar. *Z. Phys. Chem.* (Leipzig), **217**, 200 (1961).
246. P. J. McCartin. *J. Amer. Chem. Soc.*, **85**, 2021 (1963).
247. G. H. Schenk and N. Radke. *Anal. Chem.*, **37**, 910 (1965).
248. A. Nakajima and H. Akamatu. *Bull. Chem. Soc. Japan*, **41**, 1961 (1968).
249. D. K. Majumdar and S. Basu. *J. Chem. Phys.*, **33**, 1199 (1960).
250. K. H. Grellmann, A. R. Watkins and A. Weller. *J. Phys. Chem.*, **76**, 469; 3132 (1972).
251. M. Gouterman and P. E. Stevens. *J. Chem. Phys.*, **37**, 2266 (1962).
252. R. Livingston, R. M. Go and T. G. Truscott. *J. Phys. Chem.*, **70**, 1312 (1966).
253. H. Knibbe, D. Rehm and A. Weller. *Ber. Bunsenges. Phys. Chem.*, **72**, 257 (1968).
254. H. Knibbe, D. Rehm and A. Weller. *Ber. Bunsenges. Phys. Chem.*, **73**, 839 (1970).
255. B. S. Solomon, C. Steel and A. Weller. *Chem. Commun.*, 927 (1969).
256. C. A. G. Brooks and K. M. C. Davis, unpublished work.
257. R. S. Davidson. This volume, ch. 4.
258. M. J. Blandamer and M. F. Fox. *Chem. Rev.*, **70**, 59 (1970).
259. G. Stein and A. Treinin. *Trans. Faraday Soc.*, **55**, 1086 (1959).
260. R. Platzman and J. Franck. *Z. Phys.*, **138**, 411 (1954).
261. M. Smith and M. C. R. Symons. *Trans. Faraday Soc.*, **54**, 338, 346 (1958).

262. C. K. Jorgenson. *Solid State Phys.*, **13**, 375 (1962).
263. C. K. Jorgenson. *Advances Chem. Phys.*, **5**, 33 (1963).
264. M. F. Fox and T. F. Hunter. *Nature*, **223**, 177 (1969).
265. M. Born. *Z. Phys.*, **1**, 45 (1920).
266. J. G. Kirkwood. *J. Chem. Phys.*, **2**, 351 (1934).
267. E. M. Kosower, R. L. Martin and V. M. Melochi. *J. Chem. Phys.*, **26**, 1353 (1957).
268. T. R. Griffiths and M. C. R. Symons. *Trans. Faraday Soc.*, **56**, 1125 (1960).
269. M. J. Blandamer, M. F. Fox, M. C. R. Symons, K. J. Wood and M. J. Wootten, quoted in ref. 258.
270. M. J. Blandamer, D. E. Clarke, T. A. Claxton, M. F. Fox, N. J. Hidden, J. Oakes, M. C. R. Symons, G. S. P. Verma and M. J. Wootten. *Chem. Commun.*, 273 (1967).
271. M. J. Blandamer, M. F. Fox and M. C. R. Symons. *Nature*, **214**, 163 (1967).
272. M. J. Blandamer, C. A. G. Brooks and K. M. C. Davis. *J. Solution Chem.*, in press.
273. W. B. Person, ref. 8, p. 29.
274. (a) J. Yarwood, ref. 8, p. 105, (b) M. Tamres and J. Yarwood, Ref. 8, p. 257.
275. J. Yarwood and W. B. Person. *J. Amer. Chem. Soc.*, **90**, 3930 (1968).
276. J. Yarwood and W. B. Person. *J. Amer. Chem. Soc.*, **90**, 594 (1968).
277. S. G. W. Ginn, I. Hague and J. L. Wood. *Spectrochim. Acta*, **24A**, 1531 (1968).
278. I. Hague and J. L. Wood. *Spectrochim. Acta*, **23A**, 2523 (1967).
279. J. Yarwood. *Spectrochim. Acta*, **26A**, 2099 (1970).
280. G. W. Brownson and J. Yarwood. *Adv. Mol. Relaxation Processes*, **6**, 1 (1973).
281. H. W. Thompson and R. F. Lake. *Proc. Roy. Soc. Ser. A*, **297**, 440 (1967).
282. R. F. Lake and H. W. Thompson. *Spectrochim. Acta*, **24A**, 1321 (1968).
283. G. W. Brownson and J. Yarwood. *J. Mol. Structure*, **10**, 147 (1971).
284. J. N. Gayles. *J. Chem. Phys.*, **49**, 1840 (1968).
285. P. Klaeboe. *J. Amer. Chem. Soc.*, **89**, 3667 (1967).
286. H. Stammreich, R. Forneris and Y. Tavares. *Spectrochim. Acta*, **17**, 775, 1173 (1961).
287. J. G. Kirkwood, quoted by W. West and R. T. Edwards. *J. Chem. Phys.*, **5**, 14 (1937).
288. E. Bauer and M. Magat. *J. Phys. Radium*, **9**, 319 (1938).
289. H. E. Hallam, in "Infrared Spectroscopy and Molecular Structure", (ed. M. Davis), Elsevier, Amsterdam (1963).
290. A. D. Buckingham. *Trans. Faraday Soc.*, **56**, 753 (1960).
291. S. G. W. Ginn and J. L. Wood. *Trans. Faraday Soc.*, **62**, 777 (1966).
292. A. D. Buckingham. *Proc. Roy. Soc. Ser. A*, **248**, 169 (1958).
293. A. D. Buckingham. *Proc. Roy. Soc. Ser. A*, **255**, 32 (1960).
294. I. N. Levine. "Quantum Chemistry II Molecular Spectroscopy", Allyn and Bacon, Boston (1970), p. 145.
295. J. P. Kettle and A. H. Price. *J. Chem. Soc., Faraday Trans.*, **2**, 1306, (1972).
296. G. W. Brownson and J. Yarwood. *Spectroscopy Lett.*, **5**, 185 (1972).
297. M. W. Hanna and D. E. Williams. *J. Amer. Chem. Soc.*, **90**, 5358 (1968).
298. R. Foster and C. A. Fyfe. In "Progress in Nuclear Magnetic Resonance Spectroscopy", Vol. 4 (eds J. W. Emsley, J. Feeney and L. H. Sutcliffe), Pergamon, Oxford and New York (1969), p. 1.
299. J. Ronayne and D. H. Williams. *Ann. Rev. NMR Spectrosc.*, **2**, 83 (1969).
300. P. Laszlo. *In* "Progress in Nuclear Magnetic Resonance Spectroscopy", Vol. 3 (eds J. W. Emsley, J. Feeney and L. H. Sutcliffe), Pergamon, Oxford and New York (1969).
301. E. M. Engler and P. Laszlo. *J. Amer. Chem. Soc.*, **93**, 1317 (1971).

H

302. M. W. Hanna and A. L. Ashbaugh. *J. Phys. Chem.*, **68**, 811 (1964).
303. M. W. Hanna and D. G. Rose. *J. Amer. Chem. Soc.*, **94**, 2601 (1972).
   (a) M. I. Foreman and R. Foster. *Rec. Trav. Chim. Pays-Bas*, **89**, 1149 (1970).
   (b) R. Foster. *In* "Molecular Complexes", Vol. 2 (ed. R. Foster), Elek Science, London; Crane Russak, New York (1974), ch. 3.
304. S. D. Christian. "Techniques of Chemistry, Vol 8, part 1: Solutions and Solubilities" (ed. A. Weissberger and M. R. J. Dack) Wiley-Interscience, to be published 1975.
305. W. B. Person. *J. Amer. Chem. Soc.*, **87**, 167 (1965).
   (a) J. D. Childs. S. D. Christian and J. Grundnes. *J. Amer. Chem. Soc.*, **94**, 5657 (1972).
306. R. A. La Budde and M. Tamres. *J. Phys. Chem.*, **74**, 4009 (1970).
307. D. A. Deranleau.*J. Amer. Chem. Soc.*, **91**, 4044 (1969).
308. S. D. Christian. *Office of Saline Water Research and Development Progress Report* No. 706, July (1971).
309. S. D. Christian, J. R. Johnson, H. E. Affsprung and P. J. Kilpatrick. *J. Phys. Chem.*, **70**, 3376 (1966).
310. S. D. Christian, A. A. Taka and B. W. Gash. *Quart. Rev. Chem. Soc.*, **24**, 20 (1970).
311. S. D. Christian and J. Grundnes. *Acta Chem. Scand.*, **22**, 1702 (1968).
312. S. D. Christian, K. O. Yeo and E. E. Tucker. *J. Phys. Chem.*, **75**, 2413 (1971).
313. S. D. Christian, R. Frech and K. O. Yeo. *J. Phys. Chem.*, **77**, 813 (1973).
314. W. J. McKinney and A. I. Popov. *J. Amer. Chem. Soc.*, **91**, 5215 (1969).
315. W. K. Duerksen and M. Tamres. *J. Amer. Chem. Soc.*, **90**, 1379 (1968).
316. P. H. Emslie, R. Foster, C. A. Fyfe and I. Horman. *Tetrahedron*, **21**, 2843 (1965).
317. S. Carter. *J. Chem. Soc. A.*, 404 (1968).
318. B. B. Bhowmik. *Spectrochim. Acta*, **27A**, 321 (1971).
319. T. F. Hunter and D. H. Norfolk. *Spectrochim. Acta*, **25A**, 193 (1969).
320. R. L. Scott. *J. Phys. Chem.*, **75**, 3843 (1971).
321. R. L. Scott and D. V. Fenby. *Ann. Rev. Phys. Chem.*, **20**, 111 (1969).
322. R. X. Ewall and A. J. Sonnessa. *J. Amer. Chem. Soc.*, **92**, 2845 (1970).
323. F. Takahishi, W. J. Karoly, J. B. Greenshields and N. C. Li, *Can. J. Chem.*, **45**, 2033 (1967).
324. S. K. Alley, Jr. and R. L. Scott. *J. Phys. Chem.*, **67**, 1182 (1963).
325. I. D. Kuntz Jr., F. P. Gasparro, M. D. Johnson, Jr. and R. P. Taylor. *J. Amer. Chem. Soc.*, **90**, 4778 (1968).
326. J. D. Childs, S. D. Christian, J. Grundnes and S. R. Roach. *Acta. Chem. Scand.*, **25**, 1679 (1971).
327. S. D. Christian, J. D. Childs and E. H. Lane. *J. Amer. Chem. Soc.*, **94**, 6861 (1972).
328. R. Satterfield, quoted by M. Tamres, ref. 1, p. 81.
329. H. Buchowski, J. Devaure, P. V. Huong and J. Lascombe. *Bull. Soc. Chim. France*, 2532 (1966).
330. S. D. Christian. *J. Amer. Chem. Soc.*, **91**, 6514 (1969).
331. N. M. D. Brown, R. Foster and C. A. Fyfe. *J. Chem. Soc. B.*, 406 (1967).
332. E. M. Kosower. *Progr. in Phys. Org. Chem.*, **3**, 81 (1965).
333. G. Balogh. *Ind. Chim. Belge*, **31**, 142 (1966).
334. P. A. Chopard. *Arch. Sci.*, **19**, 129 (1966).
335. A. K. Colter and M. R. J. Dack, Ref. 7, p. 301.
336. H. Kainer and A. Überle. *Chem. Ber.*, **88**, 1147 (1955).
337. R. Foster and T. J. Thomson. *Trans. Faraday Soc.*, **59**, 296 (1963).

338. R. S. Mulliken and W. B. Person. Ref. 6, p. 252.
339. B. G. Anex and E. B. Hill, Jr. *J. Amer. Chem. Soc.*, **88**, 3648 (1966).
340. N. W. Tideswell and J. D. McCullough. *J. Amer. Chem. Soc.*, **79**, 1013 (1957).
341. A. A. Passchier and N. W. Gregory. *J. Phys. Chem.*, **72**, 2697 (1968).
342. R. M. Keefer and T. L. Allen. *J. Chem. Phys.*, **25**, 1059 (1956).
343. W. E. Wentworth, G. W. Drake, W. Hirsch and E. Chen. *J. Chem. Ed.*, **41**, 373 (1964).
344. M. J. S. Dewar and C. C. Thompson, Jr., *Tetrahedron*, **Suppl. 7**, 97 (1966).

4 Photochemical Reactions Involving Charge-transfer Complexes

R. S. DAVIDSON

Chemistry Department, University of Leicester, Leicester LE1 7RH, England

DEFINITIONS

The term Exciplex (*Exci*ted Com*plex*) is used to describe an electronically excited molecular complex of definite stoichiometry. Complexes which fall into this broad classification include:

EXCIMERS (*Excited Di*mers)—electronically excited complexes formed between identical atoms or molecules.

HETEROEXCIMERS—electronically excited complexes formed between two *non*-identical atoms or molecules.

EXCITED CHARGE-TRANSFER COMPLEXES—complexes produced by excitation of ground state molecular complexes for which there is conclusive evidence for association in the ground state.

I. INTRODUCTION

Over the years, evidence has accumulated which demonstrates that many bimolecular photochemical reactions occur via complexes formed by association of the reaction partners. In the present decade it seems fashionable to postulate the intermediacy of such species, since in many cases they can be held responsible for the diverse, and on classical theory unexplainable courses, which many of these reactions take. It is therefore imperative that the implications and factors influencing association processes by recognized and appreciated in order that excited complexes do not become the universal scapegoat of photochemistry.

II. FACTORS AFFECTING, AND IMPLICATIONS OF, COMPLEX FORMATION

A. Ground-state Complex Formation

1. Electron donor–acceptor complex formation

(a) *Effective production of excited states* There is an abundance of evidence for the association of many varied types of molecules in the ground state.[1] In a critique by Dewar and Thompson,[2] it is pointed out that molecules may associate in solution for a variety of reasons, e.g. van der Waals forces, and if the molecules are close enough together it may be possible to observe a new electronic absorption band—a charge-transfer (CT) absorption band. Excitation into this band will produce excited CT complexes in which the two reaction partners have not had to diffuse together after the excitation process. This is therefore an extremely efficient way of producing excited complexes since competition from decay of an excited state of either reaction partner is minimal. In many cases discrete CT bands are not observed and instead the association process causes a broadening of an absorption band of one of the components. Although relatively few of the reactant molecules are correctly oriented to give a good strong new absorption band this does not mean that quite a large proportion of the reactant molecules are not complexed. On excitation of one of the reaction partners, the other reaction partner probably has only to undergo a small amount of movement in order for excited-complex formation to occur. The ready formation of excited CT complexes between reaction partners will of course have a drastic effect upon the course of the photochemical reaction. As will be seen in Section IIIa these complexes may dissipate the energy in a radiative manner, dissociate giving radicals ion, populate the triplet of one reaction partner, etc.

(b) *Effect on energy transfer processes* Association of molecules in their ground states can also affect classical energy-transfer processes. Perhaps the most trivial of examples is where the quenching of an excited state of a molecule D by A is due to the formation of a non-exciting complex AD. Provided A does not quench the excited state of D, the efficiency of the quenching process is governed by the efficiency of ground-state complex formation:

$$A + D \rightleftharpoons AD.$$

It can be easily shown[3] that if D fluoresces, the equilibrium constant for complex formation (K) is given by the following expression:

$$\frac{\Phi_F - \Phi_q}{\Phi_q} = K[A] \tag{1}$$

where Φ_F = quantum yield of fluorescence of D in the absence of A,

Φ_q = quantum yield of fluorescence of D in the presence of A at a concentration $[A]$.

More often than not the situation is much more complex since A usually also acts as a quencher of D*. Under these circumstances it is possible to get both *static* and *dynamic quenching*:†

$$A^*_{S_1} + D_{S_0} \underset{k_{-D}}{\overset{k_D}{\rightleftharpoons}} [AD]^*_{S_1} \overset{k_E}{\to} A_{S_0} + D_{S_0}$$

dynamic quenching

$$A + D \overset{K}{\rightleftharpoons} [AD]_{S_0}$$ (2)

static quenching

Under these circumstances, a modified Stern–Volmer equation[4a, b] has to be used in order to evaluate the rate constants for quenching of the excited state. Weller[4a] has obtained the following expression for use with steady-state measurements

$$\frac{\Phi}{\Phi_0} = \frac{1 - \gamma\alpha}{1 + \gamma k_D \tau_0 [D]}$$ (3)

in which:

Φ = quantum yield of fluorescence of A in presence of D at a concentration $[D]$,

Φ_0 = quantum yield of fluorescence in absence of quencher D,

$\gamma = k_E \tau_E$, where τ_E = lifetime of $[AD]^*_{S_1}$,

α = fraction of directly excited AD_{S_0} complexes,

$= 1 + \dfrac{\varepsilon_A}{\varepsilon_{AD}} \dfrac{1}{K_{[D]}}$ where ε_A and ε_{AD} are the molar absorption coefficients at the exciting wavelengths for A and AD respectively.

Systems examined, which exhibit this behaviour include quenching of anthracene fluorescence by carbon tetrabromide[5] and the quenching of the fluorescence of N-methylacridinium and tropylium cations by anions.[6]

From equation (3) it can be seen that if association between A and D in the ground state occurs to a significant extent the quenching kinetics are going to be substantially affected. Another effect of ground-state complex formation is that it makes the efficiency of quenching of the excited state wavelength dependent. This wavelength effect arises because the fraction of AD and A molecules excited is dependent on the concentration and the values of ε_A and ε_{AD} for these species. The ε values are, of course, wavelength dependent. A thoroughly studied example which illustrates this effect is the quenching of anthracene fluorescence by carbon tetrachloride.[7]

---

† Superscript S indicates a singlet state; likewise subscript T (*vide infra*) indicates a triplet state.

2. Aggregation

(a)*Effect on production of reactions of excited states* The dissolution of ionic compounds in non-polar solvents and of non-polar compounds in polar solvents can lead to aggregation. Aggregation affects the energy levels of excited states and this is manifested by the absorption and emission spectra of the compounds being perturbed.[8] Aggregation can also affect other photophysical properties of the molecules, e.g. aggregation of some dyes leads to enhanced intersystem crossing.[9]

Another effect of aggregation is that the likelihood of extremely short lived excited-states undergoing bimolecular reactions is increased. It has been found, for example, that thymine derivatives photo-dimerize and that it is the triplet state of the pyrimidine which is responsible for reaction.[10a, b] In aqueous solution reaction also occurs from the short-lived excited singlet state.[10b, 11a, b] Osmometric measurements indicate that the pyrimidines aggregate in solution.[11b, 12] This aggregation takes the form of stacking of pyrimidine molecules. Under these circumstances the excited-singlet state can react since it is situated adjacent to its reaction partner.

Whilst the aggregation of pyrimidines is thought to be due to dipole–dipole interaction, many molecules aggregate through hydrogen-bond formation. This is particularly likely to occur in solutions of compounds containing O—H, N—H, S—H and similar groups. This effect can become extremely important when such solutions are frozen. The formation of hydrogen-bonded dimers has been elegantly used in a photochemical process for oxidation of unactivated methylene groups in steroids and long-chain alcohols.[13] The steroid acid (1) complexes with the benzophenone derivative (2) in carbon tetrachloride solution. Irradiation leads to the triplet carbonyl-group selectively attacking C-16 of the steroid. By suitable treatment of the reaction products the ketone (3) was obtained in good yield.

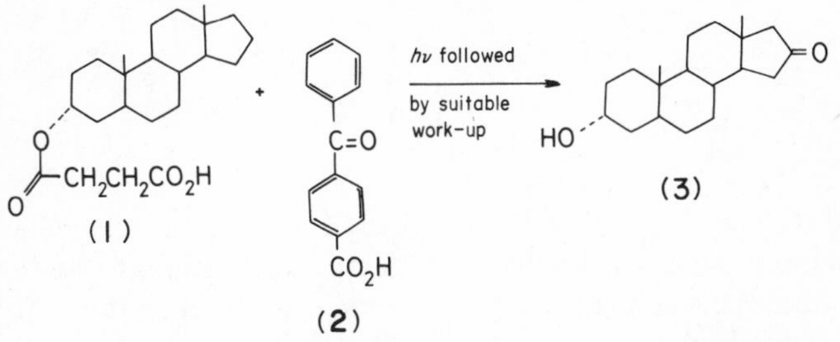

(b) *Effect on energy-transfer processes* Normally, bimolecular energy-transfer processes are affected by the efficiency with which the reaction partners diffuse together. Aggregation through hydrogen-bonding will therefore upset the kinetics of such processes. However, hydrogen bonding can also upset energy transfer in other ways. An early example of such a process being affected by hydrogen bonding was the observation that acridine quenched the fluorescence of carbazole.[14] If an alcoholic solution of the two components is frozen the acridine hardly affects the carbazole fluorescence. On the other hand quenching is extremely efficient when a hydrocarbon solvent is used. The investigators suggested that the two molecules form a hydrogen-bonded complex in the non-polar solvent. This complex has an excited singlet state of lower energy than either of the components and consequently the energy becomes localised in this state and it is from this state that the energy is dissipated non-radiatively.

Another similar example is that of the quenching of benzophenone phosphorescence by carbazole.[15] This occurs extremely efficiently in non-polar matrices in which hydrogen bonding between the two components can occur. When solvents are used which do not allow this type of hydrogen bonding, quenching does not occur.

Primary and secondary aromatic amines aggregate in rigid non-polar matrices and this leads to a bathochromic shift of the fluorescence bands and to a marked quenching of the phosphorescence of the amines.[16] It has yet to be demonstrated that the photochemistry of the amines is affected by the aggregation process but it can be seen that determination of the energies of excited singlet and triplet states will be affected. Another recently found example is that of thioxanth-9-one whose phosphorescence spectra and triplet lifetime are greatly affected by aggregation.[17]

Aggregation, caused by hydrogen-bond formation has also been shown to affect some acid-base reactions of excited molecules. An example is that of the 7-azaindole (4) which in hydrocarbon solutions at room temperature forms hydrogen-bonded pairs.[18] On excitation, these pairs undergo a double proton-transfer reaction.

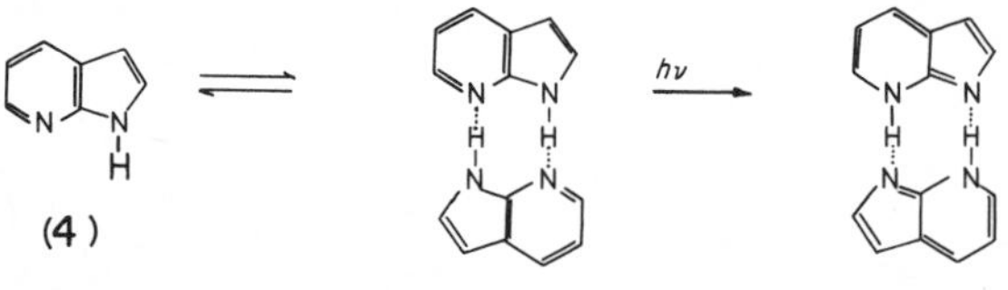

(4)

One of the classical and well-used methods for determining which excited state of a molecule is responsible for a chemical reaction is to study the effect of added quenchers upon the reaction. If very high concentrations of quenchers are used, there is always the possibility that the nearest-neighbour molecule to the excited molecule is a quencher molecule, i.e. employment of high concentrations of quenchers can lead to an "artificial" aggregation situation. Wagner and Kochevar[19] have made a thorough study of nearest-neighbour quenching in relation to triplet–triplet energy transfer. They utilized a modified form of equation (3) to evaluate their data.

The use of photo-sensitizers in higher concentrations is also not without its problems. Excitation of a particular sensitizer molecule, having other sensitizer molecules as nearest neighbours can lead to energy hopping. This may, as suggested by Birks,[20] involve an excited complex

$$^1M_a^* + M_b \rightleftharpoons {}^1(M_aM_b)^* \rightarrow M_a + {}^1M_b^*$$

3. Structural factors

Many compounds have now been synthesized in which two chromophores may interact through space because of the way in which they are connected by the other atoms in the molecule. A particular example is that of the cyclo-phanes.[21] In the case of (5) having m = n = 2, the rings are so close that their

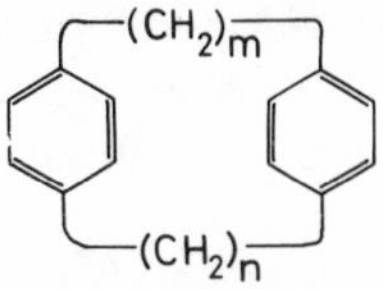

(5)

electron clouds interpenetrate. It is not therefore surprising that the electronic absorption spectrum is somewhat different from that of p-xylene. There is still some interaction, as evidenced by the bathochromic shift of the long wave-length absorption band in the compound m = n = 3. There is apparently no interaction though in the compound having m = n = 4. However, with this latter compound excitation populates the first excited singlet state of one benzonoid residues which then interacts with the other benzenoid residue to give an excimer[22] (Section B).

Several examples of intramolecular donor–acceptor complexes having a cyclophane-type structure have been prepared, e.g. (6)[23] and (7).[24] In the case of (7) it was found that the most intense CT absorption band was observed when the 1,4-dimethoxybenzene ring system overlapped with the imide system. The actual distance between the two rings does not seem to be particularly significant.

B. Complex Formation by Excited States

1. Complex formation due to dipole–dipole interactions

Many ground-state bimolecular reactions occur by the reaction partners initially coming together through a dipole–dipole interaction, e.g. Diels–Alder reaction. This can equally well occur with reactions involving excited states.

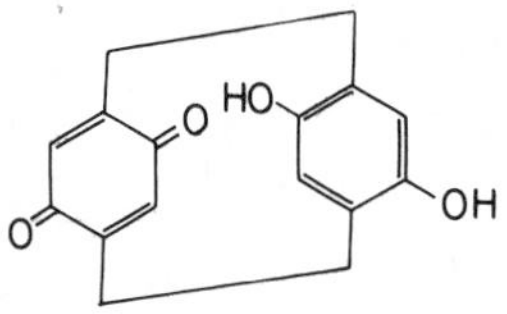

(6)

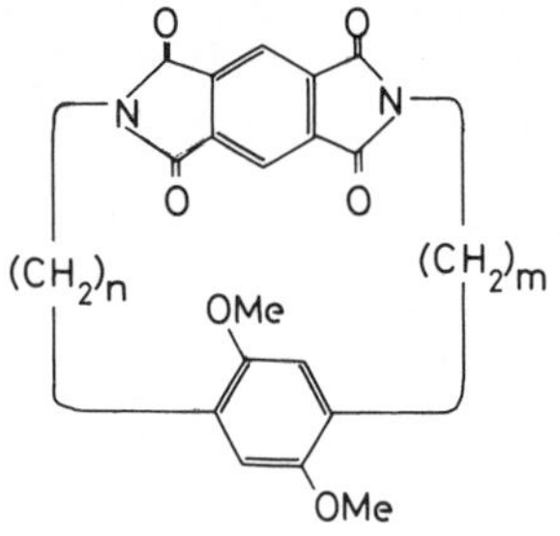

(7)

As yet there is little information with regard to the stability of complexes formed by such a process. Dipole–dipole interactions may be expected to have an effect upon the stereochemical course of the reaction. This point has been demonstrated by Challand and de Mayo[25] in a reaction involving the photo-addition of the unsaturated ketone (9) to the olefin (8). They showed that the ratio of the stereoisomeric products (10) and (11) was solvent dependent.

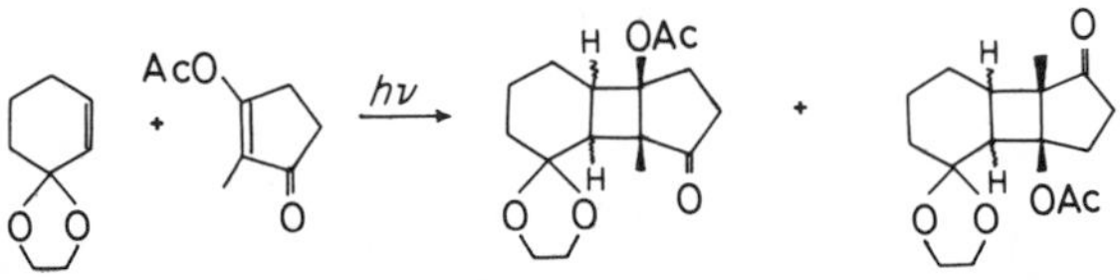

(8)   (9)   (10)   (11)

In non-polar solvents an exceptionally high yield of (10) was obtained. When the polarity of the solvent was increased the yield of this product decreased and that of (11) increased. The change in the ratio of the yields of the products was shown to correlate with change in polarity of the solvent. This solvent effect can be rationalized by postulating that in non-polar solvents the regioselectivity of the reaction is determined by the dipole–dipole interaction of the reaction partners. Increasing the polarity of the solvent decreases this interaction and hence the energetics of formation of the two isomers become similar. It should be emphasized that the dipole–dipole interaction is not the only factor affecting the stereochemistry of addition of enones to olefins. Usually, prediction of the stereochemical outcome of these reactions is extremely difficult because of the number of factors which have to be taken into account.[26]

2. Complex formation due to charge-transfer interactions

(a) *Description of excited complex formation as an adiabatic process*   Several examples of complex formation by excited states have been known for several years, e.g. association of the excited singlet-state of pyrene with a ground-state pyrene molecule,[27] protonation of the excited states of aromatic hydrocarbons and amines.[4a] Förster has described these reactions as adiabatic processes since reaction takes place on the energy surface of the excited state and produces products in an excited state.[28] In the more common type of photochemical reaction products are formed in their ground state and this type of reaction is described as a diabatic process.

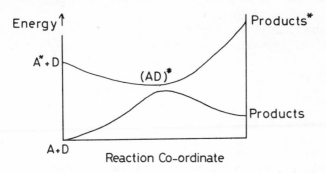

FIG. 1. Potential energy surfaces for interaction of A* with D and of A with D.

From the description of an adiabatic process it will be seen that if excited products are formed by excitation of a product of a photochemical reaction there is the finite possibility that some of the reaction partners may be regenerated in an excited state—i.e. reaction occurs from right to left along the energy surface depicted in Fig. 1. This idea has in fact been realized.[29] Excitation of the photodimer of 9-methylanthracene produced small but finite amounts of the excited singlets of 9-methylanthracene and the 9-methyl-anthracene excimer. This latter species corresponds to (AD)* in Fig. 1, i.e. it is an excited complex. The stability of such complexes is shown in the Figure as being the difference between the energy of (A* + D) and (AD)*. In the following sections we will discuss what factors affect this energy gap and how these excited complexes participate in photochemical reactions.

(b) *Energetics of excited-complex formation*   By far the most common type of excited-complex formation involves some degree of charge transfer. This is not surprising when one considers that excited states should be more prone to undergo redox reactions than their ground-state counterparts.[28] Thus they are more likely to act as electron donors because of the ease of loss of an

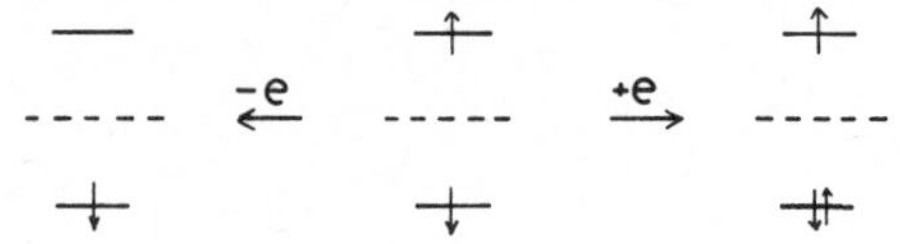

FIG. 2. Pictorial description showing how an excited-singlet state may act as an electron donor or acceptor.

electron from the half-occupied lowest antibonding orbital, and more likely to act as acceptors because of their desire to have the half-occupied highest-bonding orbital completely filled.

If stabilization of the complex is solely due to a charge-transfer interaction then the energetics of formation of the complex will be governed by the ionisation potential of the donor (I_D), the electron affinity of the acceptor (EA_A) and the energy gained by bringing two charges of the opposite polarity together (coulombic energy term).[30]

i.e.
$$E_{(D^- A^+)} = I_D - EA_A - C \tag{4}$$

where $E_{(D^- A^+)}$ is the energy of pure CT state relative to the ground-state energy of A and D, and C is the coulombic energy term. Consideration of this relationship leads to a number of useful conclusions. (i) Many singlet excited complexes exhibit fluorescence and the energy associated with this emission will approximate to the energy of the excited complex. Thus we might anticipate that the frequency of this emission will bear a close relationship to I_D and EA_A. This has been found to be true for a large number of excited complexes formed between the excited singlet states of aromatic hydrocarbons and aromatic amines,[30, 31a] anthracene and alkylbenzenes,[31b] and 1-cyanonaphthalene and olefins.[31c] (ii) The energetics of excited-complex formation will be reflected in the rate constant for complex formation. Thus the more readily formed the complex, so the rate constant will approach the diffusion controlled limit. For a particular excited acceptor we would anticipate that the rate constant for reaction with a donor will increase as the value of I_D or the oxidation potential of the donor decreases. Such a relationship has been found for the quenching of triplet benzophenone[32a] and fluorenone[32b] by a series of amines[32a] and for the quenching of triplet acetone by a series of electron donors such as amines and sulphides.[33] Similarly a good correlation exists between the rate constant for quenching of series of triplet ketones (acceptors) by triethylamine (donor).[34] These examples illustrate the point that if a reaction is suspected of involving an excited CT complex then a linear relationship between the logarithm of the rate constant of the reaction and the appropriate redox potential of the reactants should be sought. (iii) Since the energy of a complex is dependent upon the coulombic term we may anticipate that the energy of a complex will be affected by the polarity of the solvent. Thus, the greater the solvating power of the solvent the lower the energy of the complex should be. Such a relationship has been found for a number of excited complexes, e.g. aromatic hydrocarbon–amine exciplexes,[30, 35a] and excited charge-transfer complexes formed between aromatic hydrocarbons and N-alkyltetrachlorophthalimides.[35b] From the solvent dependence of the fluorescence bands of aromatic hydrocarbon–amine exciplexes it has been found possible to calculate the dipole moments of the excited complexes.

So far the only stabilization factor considered has been that of charge-transfer. However, there is also the possibility that the complexes derive some stabilization through the interaction of the highest-occupied bonding orbital and lowest-unoccupied anti-bonding orbital in the donor and acceptor molecules respectively.[30, 36] In cases where the complex is formed between identical molecules (excimers) this type of interaction will be a maximum and will account for most of the stability of the complex. That there is little stabilization of excimers through a CT process is attested by the fact that the frequency of excimer fluorescence is not affected by solvent polarity.

(c) *Effect of excited-complex formation upon the stereochemical aspects of photochemical reactions*   If the two components of a complex carry some charge, as is the case in an excited electron donor–acceptor complex we may expect that any addition reaction between the donor and acceptor will be affected by this charge. Bryce-Smith and Gilbert have observed[37] that the thermal and photochemical addition of tetrachloro-o-benzoquinone to

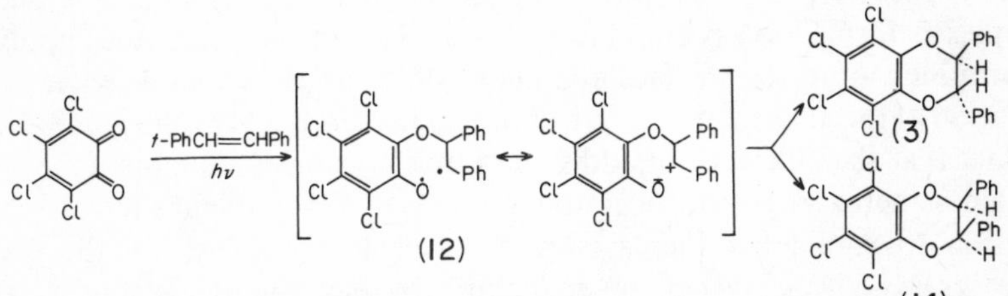

(12)

(3)

(14)

trans-stilbene gives (13 to the almost total exclusion of (14. They rationalized this apparent lack of orbital-symmetry control by postulating (12) as an intermediate.

Turro and co-workers have made an elegant study[38] of the photo-induced addition of carbonyl compounds to olefins and have shown that the electronic requirements of the olefin determine its mode of attack upon the excited

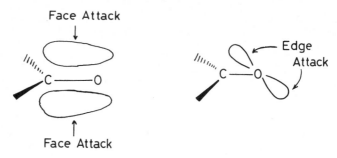

FIG. 3. "Face" (electrophilic) and "Edge" (nucleophilic) attack on $n\pi^*$-state of a carboxyl group.

carbonyl group. They pointed out that the $n\pi^*$-state of the carbonyl group has two reaction sites. Thus, attack may occur in the π-system which is electron rich, i.e. face attack (Fig. 3) and attack may occur at the electron-deficient oxygen atom, i.e. nucleophilic edge attack. In the case of nucleophilic attack it is conceivable that an electron-rich olefin may react by donating an electron to the half-occupied non-bonding orbital, i.e. a CT process occurs, e.g.

$$R_2C{=}O^* + \underset{RO\quad OR}{\diagup\!\!\!=\!\!\!\diagdown} \;\rightarrow R_2\dot{C}{-}\bar{O} + \left(\underset{RO\quad OR}{\diagup\!\!\!=\!\!\!\diagdown}\right)^{\!\!\overset{+}{\cdot}}$$

Attack of an electron-deficient olefin upon the π-system of the triplet ketone may occur via a similar process. Both these examples are extreme cases, i.e. where a complete electron is transferred. What is more probable is that a complex is formed via partial electron-transfer but the same rules apply—electrophilic attack on the face and nucleophilic attack upon the edge. Confirmation of the amphoteric nature of the excited carbonyl group comes from the fact that the efficiency of addition of excited-singlet alkanone to cyano-ethylenes increases as the "electron poorness" of the olefin increases and addition to enol ether increases as the "electron-richness" of the olefin increases.[39] The efficiency of quenching by electrophilic olefins is also related to the energy of its lowest-unoccupied antibonding orbital and that of nucleophilic olefins to that of its highest-occupied orbital. Evidence supporting the thesis that electron deficient olefins attack on the face of the carbonyl group comes from the observation that introduction of axial methyl substituents in positions 3 and 5 of cyclohexanone reduces the efficiency of attack by these olefins.[40] Attack by electron-rich olefins is hindered by substituents which occupy the space near the edge of the carbonyl group.

The regioselectivity of the addition reactions can also be rationalized on the basis of a CT intermediate. Edge attack by electron-rich olefins produces an intermediate in which the olefin has radical-cation character. In the case of the ethoxyethylene radical cation, the charge densities at C_1 and C_2 are about the same and therefore it is not surprising that the addition of ethoxyethylene is not regioselective, i.e. there is no preferential alignment of the enol ether along the edge of the carbonyl group in the intermediate.

Since the geometry of the intermediate precludes direct formation of the product loss of stereospecificity is observed. This may be due to isomerisation of the olefinic species in the intermediate or to collapse of the intermediate into a 1,4-biradical from which loss of stereochemistry occurs. Addition of electron-deficient olefins is both regio- and stereo-specific. Face attack by the olefin produces an intermediate, stabilized by CT interactions which is most favourably set up for bond formation which leads to the observed product. The inefficiency of product formation in these reactions is put down to the intermediates reverting to ground-state ketone and olefin.

(d) *Excited-complex formation and energy wastage*   The examples discussed in Section II.B.2c illustrate the point that excited-complex formation can result in energy wastage. Thus, although olefins may interact with excited-state carbonyl groups extremely efficiently, product formation is inefficient. In the case of these addition reactions there is also the possibility that the energy is lost via a 1,4-biradical.[41a]

$$R_2C{=}O + R_2'C{=}CR_2' \xrightarrow{\ h\nu\ } \left[\begin{array}{c} R_2C{=}O \\ R_2'C{=}CR_2' \end{array} \right]^* \longrightarrow \begin{array}{c} R_2\overset{\cdot}{C}{-}O \\ | \\ R_2'\overset{\cdot}{C}{-}CR_2' \end{array}$$

$$\downarrow$$

$$R_2CO_{S_0} + R_2'C{=}CR_2'$$

Presently available evidence suggests that it is the complex which is responsible.[41b, c]

Excited complexes are good energy dissipators. This has led to their being postulated as intermediates in reactions where energy wastage occurs, e.g. the dimerisation of olefins (indene)[42a] and unsaturated lactones (coumarins),[42b] self-quenching of triplet sensitizers,[42c] quenching of singlet oxygen by amines,[42d] and sulphides.[42e] Usually direct spectroscopic evidence for the intermediates is not available in these reactions and one has to rely solely upon kinetic evidence.

Whilst energy wastage may be readily ascribed to dissipation via an excited complex, there is also the very real possibility that it is due to a reversible photochemical reaction. This was found to be the case for the photochemical reaction of benzophenone with aniline.[43] Whilst product formation is extremely inefficient, radical formation is very efficient.

$$Ph_2CO_{T_1} + PhNH_2 \rightarrow Ph_2\dot{C}OH + Ph\dot{N}H$$

$$Ph_2\dot{C}OH + Ph\dot{N}H \rightarrow Ph_2CO + PhNH_2$$

Thus for all systems in which energy dissipation occurs, it is necessary to inquire carefully as to whether it is occurring *via* a complex or by a chemical process.

In a normal bimolecular energy-transfer process, the donor and acceptor have to come together by diffusion and the energy transfer takes place whilst they are in the vicinity of one another. When the energy transfer process is less than 20 J mol^{-1} (5 kcal mol^{-1}) exothermic, the rate constant starts dropping below the diffusion controlled limit. If excited-complex formation takes place between the donor and acceptor molecules, these relatively inefficient processes are aided by the fact that there is more time available for the transfer energy to occur. The phenomenon of "phantom triplets" in some triplet–triplet energy-transfer processes may be due to excited-complex formation.[44]

(e) *Excited termolecular complexes and reactions* via *these species*   Recently it has been found that some excited complexes are capable of undergoing bimolecular reactions to give termolecular complexes.

$$A^* + D \rightarrow (AD)^* \xrightarrow{A} (A_2D)^* \rightarrow A{-}A + D$$

Saltiel and Townsend have demonstrated,[45] in a most convincing way, that *trans,trans*-hexa-2,4-diene, which is a quencher for the excited singlet-state of anthracene, actually promotes the photodimerization of anthracene. They rationalized this observation by saying that the diene–hydrocarbon exciplex undergoes a bimolecular reaction with a ground-state anthracene molecule to give the anthracene dimer plus hexa-2,4-diene. The interesting, and as yet unexplained, observation, that piperylene catalyses the photodimerization of 9-phenylanthracene (a hydrocarbon which does not dimerize on direct irradiation)[46] may also be due to dimerization occurring *via* a diene–hydrocarbon complex.

Another example, which may involve a termolecular complex is that of the photodimerization of enol ether (15) which only occurs in the presence of benzenoid compounds bearing electron-withdrawing substituents.[47] Dimerization does not occur on direct irradiation of the enol ether. Since excited complexes of electron donors and such compounds as dimethyl naphthalate have been previously characterized[48] it is reasonable to expect that such species be intermediates in these reactions.

That termolecular excited complexes can in fact exist has been demonstrated in the naphthalene–1,4-dicyanobenzene system.[49] Fluorescence from both the bimolecular and termolecular complexes was observed. From the fact that

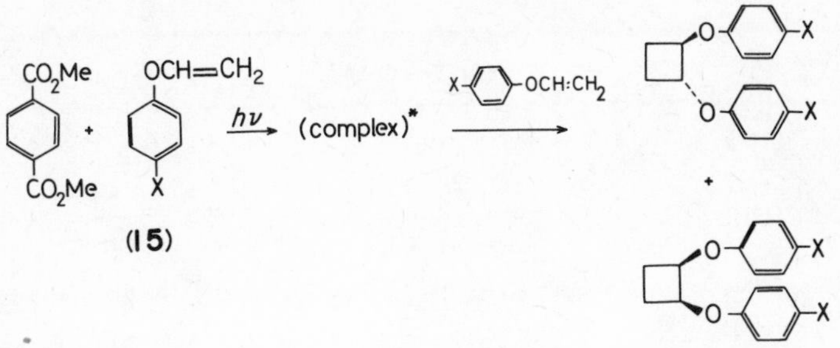

(15)

the termolecular complex has a high dipole-moment (shown by the solvent dependence of its fluorescence spectrum) it was concluded that it has the unsymmetrical structure $\overset{+\;-}{DDA}$ rather than the symmetrical structure $\overset{-\;+}{DAD}$. Termolecular excited CT complexes have been observed by means of their absorption spectra in some aromatic hydrocarbon–tetrachlorophthalic anhydride and pyromellitic anhydride complex systems.[50]

As will be seen in Section III, exciplex formation is often in competition with radical–ion formation.

$$D^* + A \longrightarrow \begin{cases} (DA)^* & \text{Exciplex} \\ \\ D_s^+ + A_s^- & \text{Solvent-separated} \\ & \text{radical-ions} \end{cases}$$

Recently a few examples of reactions have been reported in which the radical cations produced by the above process attack a ground-state donor molecule.

$$D^* + A \to D^{\cdot+} + A^{\cdot-}$$
$$D^{\cdot+} + D \to D^{\cdot}{-}D^+$$
$$D^{\cdot}{-}D^+ + A^{\cdot-} \to D{=}D + A$$

One of the earliest examples of this type of reaction appears to be dimerization of N-vinylcarbazole (16) which is sensitized by many electron acceptors.[51] More recently the dimerization of indenes[52] and 1,1-diphenylethylene (17)[53] electron acceptors in polar solvents has been described.

III. EVIDENCE FOR THE FORMATION AND MODES OF DECAY OF EXCITED COMPLEXES

A. Excited Charge-transfer Complexes

1. Evidence for formation of excited charge-transfer complexes
The principal evidence is: (i) observation of fluorescence from excited singlet

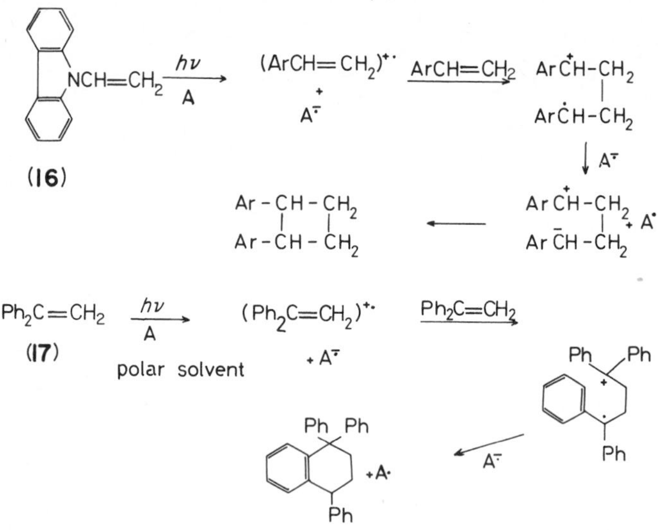

$$\text{(16)} \quad \underset{\text{NCH}=\text{CH}_2}{\bigodot} \xrightarrow[\text{A}]{h\nu} (\text{ArCH}=\text{CH}_2)^{+\cdot} \quad \underset{\text{A}^{\bar{}}}{} \quad \underset{\text{A}^{\bar{}}}{\overset{+}{\text{ArCH}=\text{CH}_2}} \quad \underset{\text{Ar}\overset{\cdot}{\text{CH}}-\text{CH}_2}{\overset{+}{\text{Ar}\overset{\cdot}{\text{CH}}-\text{CH}_2}}$$

$$\begin{array}{c} \text{Ar}-\text{CH}-\text{CH}_2 \\ | \qquad | \\ \text{Ar}-\text{CH}-\text{CH}_2 \end{array} \longleftarrow \begin{array}{c} \text{Ar}\overset{+}{\text{CH}}-\text{CH}_2 \\ | \qquad |\;{}^{+}\,\text{A}^{\cdot} \\ \text{Ar}\,\overline{\text{CH}}-\text{CH}_2 \end{array}$$

$$\text{(17)} \quad \text{Ph}_2\text{C}{=}\text{CH}_2 \xrightarrow[\underset{\text{polar solvent}}{\text{A}}]{h\nu} (\text{Ph}_2\text{C}{=}\text{CH}_2)^{+\cdot} \quad {}^{+}\text{A}^{\bar{}} \quad \underline{\text{Ph}_2\text{C}{=}\text{CH}_2}$$

complexes[54a, b, c] and phosphorescence[55a, b] and delayed fluorescence[56a, b] from triplet complexes; (ii) observation of the absorption spectra of singlet[57a, b] and triplet[58a, b] complexes; (iii) esr signals from triplet complexes.[59]

As might be expected the wavelength for maximum fluorescence of the excited complexes is extremely solvent dependent.[54b, c] From these data dipole moments of excited complexes have been calculated, e.g. complexes between aromatic hydrocarbons and tetrachlorophthalic anhydride have dipole moments† of the order of 10 D.[54c] In many cases the observed fluorescence occurs from the excited singlet which is produced by thermal activation of the triplet state of the complex. The singlet–singlet absorption spectra of excited complexes have been observed by means of the technique of laser flash photolysis.[57a, b] In many cases the spectra either resemble the radical anion of the acceptor or the radical cation of the donor, thus confirming the CT character of the excited state. A particular case in point is that of the excited complex between 1,2,4,5-tetracyanobenzene and toluene whose absorption spectrum resembles that of the tetracyanobenzene radical ion.[57a]

Phosphorescence from triplet complexes can only be observed when the energy of this state lies below that of the triplet state of either the donor or acceptor.[60] Usually the phosphorescence spectra resemble the phosphorescence spectra of either the donor or acceptor.[55b] Similarly, the triplet–

---

† The values of dipole moment are quoted in debye units (D). These units are related to the *SI* unit, the coulomb metre (C m) by:

$$D = 3{\cdot}335\,640 \times 10^{-30}\,\text{C m}$$

triplet absorption spectra of these complexes resemble the triplet–triplet absorption spectra of either the donor or acceptor.[58b]

A detailed study of the 1,2,4,5-tetracyanobenzene–toluene complex has shown that the ground state and excited-tripet state have a similar geometry (18) whereas that of the excited-singlet state is depicted by (19).[61]

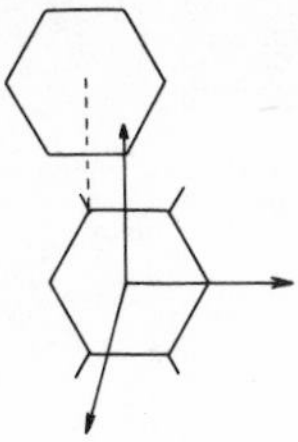

(18) 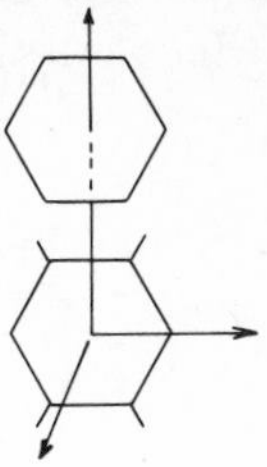   (19)

Thus formation of the excited-singlet state of the complex requires re-orientation of the donor and acceptor molecules as well as re-orientation of the solvent molecules. The latter process has been observed by time-resolved fluorescence spectroscopy.

2. Modes of decay of excited charge-transfer complexes
Excitation of CT complexes produces a complex in which the donor and acceptor molecules are orientated as in the ground state. This state has been termed the Franck–Condon excited state. This excited state may relax by re-orientation of the donor and acceptor molecules to give an "equilibrium excited-state".[61] Solvent re-orientation has also to accompany this relaxation process since the excited state has a different electronic configuration and hence dipole moment, to the ground state. Decay of either the Franck–Condon or equilibrium excited-state may involve intersystem crossing to give a triplet state,[62] dissociation into radical ions[63] (favoured by polar solvents) or internal conversion to the ground state by a radiationless pathway. The equilibrium excited-state may undergo chemical reaction, as previously noted, or fluoresce. Population of triplet complex is particularly facilitated by the presence of heavy atoms in either the donor or the acceptor.[64] The lifetime of the triplet state of the complexes formed between tetrahalogeno-phthalic anhydrides and polycyclic aromatic hydrocarbons decreases as the halogen is changed from chlorine to bromine. Triplet complexes also have a number of routes by which they may decay. Polar solvent aids their dissociation into radical ions.[65] Other routes include thermal activation to give the excited-singlet complex, and phosphorescence to give the ground-state complex. In those cases where the energy of the triplet complex lies above that

of either the triplet donor or acceptor, population of the energetically lowest state occurs.[60] Formation of donor and acceptor triplets by electron transfer between radical anions and cations produced by dissociation of the excited-singlet state of the complex has been observed.[66] The lifetimes of the triplet donor and acceptors are usually found to be less when they are complexed than when they are uncomplexed.[56a] Chart I depicts the various intermediates which may be formed from excited charge-transfer complexes

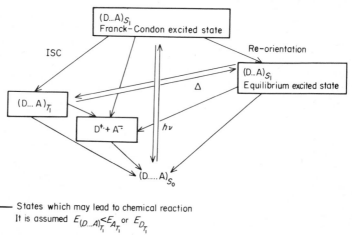

— States which may lead to chemical reaction

It is assumed $E_{(D...A)_{T_1}} < E_{A_{T_1}}$ or $E_{D_{T_1}}$

Chart 1

and which may lead to chemical reactions. The nature of the solvent will of course play a very important part in determining which states are preferentially formed.

B. Exciplexes and Excimers

1. Evidence for the formation of exciplexes and excimers

The principal evidence for the formation of these complexes comes from (a) observation of fluorescence from singlet exciplexes and excimers, (b) from the absorption spectra of singlet exciplexes.

The fluorescence spectra of a wide variety of exciplexes have been recorded, e.g. aromatic hydrocarbon–amines,[67a, b, c, d, e, 68a–d] aromatic hydrocarbon–olefin[31c, 45] and aromatic hydrocarbon–halogen-ion[69] complexes. In the case of the hydrocarbon–amine complexes it has proved possible to see both the birth and decay of the exciplex fluorescence by means of time-resolved single-photon counting.[70] The time required to develop the maximum intensity of fluorescence is dependent upon many rate constants, the viscosity of the solvent and the concentration of the reaction partners.[71] This latter

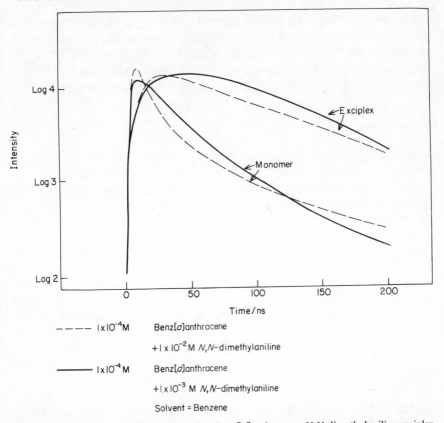

FIG. 4. Rise and decay of the fluorescence of the benz[a]anthracene–N,N-dimethylaniline exciplex and the fluorescence of benz[a]anthracene.

point is illustrated in Fig. 4 in which one can see that the time required for maximum build-up of fluorescence of the benz[a]anthracene–N,N-dimethyl-aniline exciplex increases as the concentration of the amine is lowered. The wavelength for maximum fluorescence intensity of the hydrocarbon–amine exciplexes is very dependent on solvent polarity. Increased solvent polarity causes a bathochromic shift.[67a, c, 72] From a study of these shifts the dipole moments of. many exciplexes have been determined.[67c, 72] Usually they have high values, e.g. naphthalene–triethylamine exciplex has a dipole moment of 13·6 D.[72] The quantum yield of exciplex formation is also decreased by increasing the polarity of the solvent. In highly polar solvents, e.g. acetonitrile, fluorescence from intermolecular exciplexes cannot be observed.[67d] This has been shown to be due to radical-ion formation competing with exciplex formation (see Section III.B.2). It can be observed from intramolecular exciplexes.[68a–d]

The singlet–singlet absorption spectra of a number of aromatic hydro-carbon–amine exciplexes have been obtained,[73] e.g. N,N,N',N'-tetramethyl-phenylenediamine–biphenyl and N,N-diethylaniline–pyrene exciplexes. Absorption bands due to the radical ions of the donor and acceptor molecules are present as well as other bands due to reverse CT transitions of the type

$$^1(D^+A^-)^* \xrightarrow{h_v} {}^1(D^+{}^*A^-)$$

$$^1(D^+A^-)^* \xrightarrow{h_v} {}^1(DA^-{}^*)$$

where D^{+*} and A^{-*} indicates excitation energy localized in D^+ and A^- respectively. Thus the absorption spectra of these complexes complement the data on dipole moments and are a further demonstration of the charge-transfer character of the amine–hydrocarbon exciplexes.

Fluorescence from exciplexes formed between 1-cyanonaphthalene and some mono-olefins and diens have been observed.[31c] Once again the emission is sensitive to the solvent and quite high dipole-moments (*ca.* 10 D) have been calculated for these complexes. Whilst in these complexes a considerable degree of stabilisation is due to a charge-transfer contribution, this may not be the case for exciplexes derived from unsubstituted polycyclic aromatic hydrocarbons and conjugated dienes. Unfortunately fluorescence from this type of exciplex has not been observed. However, from the insensitivity of the rate constant for quenching of aromatic-hydrocarbon fluorescence by dienes with change in solvent polarity, it has been concluded that the major stabilisation factor is the exciton resonance term.[74]

Many excimers have been characterized by means of their fluorescence spectra.[75a, b, c, d, e] The spectra are not sensitive to changes in solvent polarity since stabilization of these complexes is mainly due to the exciton resonance term. The absorption spectra of a few excimers have been recorded.[76]

There is very little spectroscopic evidence for triplet exciplexes. Calculations indicate that triplet aromatic hydrocarbon–amine exciplexes and excimers will be very much less stable than their singlet counterparts and it is believed to be this instability which has hindered their detection. Phosphorescence, attributed to triplet excimer formation, has been observed for benzene, and chlorobenzenes.[77]

There is also very little direct evidence for the formation of triplet excimers by reaction of an excited-triplet molecule with a ground-state molecule.

$$^3M^* + M_{S_0} \rightarrow {}^3(MM)^*$$

As has been noted the "self-quenching" of many triplet sensitizers has been attributed to this process; the evidence is mainly kinetic.[42a–c]

Triplet exciplexes, derived by reaction of an excited-triplet state with a ground-state molecule have been postulated as intermediates in many photochemical reactions.

$$^3M^* + A \rightarrow {}^3(M^+A^-)^*$$

Usually the evidence for such complexes is kinetic.[32–34, 78] However, spectroscopic evidence is now available for an exciplex formation between benzophenone and a tertiary amine—tri-p-tolyamine.[79] The exciplex (possibly a triplet) was generated by addition of the amine radical cation to the ketone radical anion

$$Ar_3\overset{+\cdot}{N} + Ph_2\overset{-}{\overset{\cdot}{C}}O \rightarrow {}^3(Ar_3\overset{+\cdot}{N}Ph_2\overset{-}{\overset{\cdot}{C}}O)^*$$

$$Ar = p\text{-tolyl}$$

and it was characterized by its emission spectrum.

As previously noted, excimers and exciplexes have been postulated as intermediates in many photochemical reactions although spectroscopic evidence for the complexes is not available. Usually the authors have come to the conclusion that such complexes are involved since the rate constants for reaction are directly linked with the redox potentials of the reaction partners. Another method of study has recently been proposed.[80] Quenching due to exciplex formation is often a reversible process—a point overlooked by many authors. A consequence of this reversibility is that quite frequently there is a discrepancy between the rate constants for the quenching process determined by steady-state measurements and those determined from lifetime measurements. Since the quenching rate-constant can be obtained from the fluorescence lifetime of the species being quenched, the observation of an emission band of complex is not a necessity. It remains to be seen how applicable this method is.

2. Modes of decay of exciplexes and excimers

Formation of fluorescent exciplexes has been proposed as occurring *via* a non-fluorescent, non-relaxed complex. This complex relaxes to the equilibrium complex (fluorescent) by reorientation of the donor and acceptor molecules and the solvent molecules. Much effort has recently been put in to discovering which of the two complexes is responsible for chemical reactions for, and production of, reactive intermediates.

One of the earliest observations on the aromatic hydrocarbon–exciplex system was that increasing the polarity of the solvents lead to a diminution of fluorescence from the exciplex although quenching of the hydrocarbon fluorescence was still efficient.[67a, b, 81a, b] Microsecond flash spectroscopy revealed that in polar solvents the quenching process led to the production of radical ions and the hydrocarbon triplet-state.[67a, d, 81b]

$$ArH_{S_1} + Am \xrightarrow[\text{solvent}]{\text{polar}} ArH^{\cdot -} + Am^{\cdot +} + ArH_{T_1}$$

$$ArH = \text{aromatic hydrocarbon,} \qquad Am = \text{amine}$$

That radical-ion formation occurs from the non-relaxed complex has been demonstrated by laser flash-photolysis.[82, 83a] These experiments showed that the radical ions were produced during the time of the laser pulse. It has also been suggested that radical ions can be produced by dissociation of the equilibrium exciplex.[67e] The basis for this suggestion is that the lifetime of some intermolecular exciplexes has been found to decrease as the polarity of the solvent is increased. However, the validity of this assumption has been challenged on the basis of the observation that the lifetime of some intra-molecular amine–hydrocarbon exciplexes actually increases as the polarity of the solvent is increased.[68c] The origin of the aromatic hydrocarbon triplets when non-polar solvents are employed has been shown to be the equilibrium exciplex.[73] Thus the decay of the exciplex is matched by the growth of the triplet. The origin of the triplets in polar solvents is a polemical problem. Two groups of workers[73, 83a, b] have reported on their laser flash-photolysis investigations in which they found that the triplets were produced during the light pulse from the laser. The conclusion reached was that the triplets are being produced from the non-relaxed complex. Triplets are also produced in these systems by electron transfer between radical ions—a relatively slow process.

$$ArH^{\cdot -} + Am^{\cdot +} \rightarrow ArH_{T_1} + Am_{S_0}$$

Another group of workers,[84] again using laser flash-photolysis have shown that triplet formation matches the decay of the radical ions and hence propose that the electron-transfer process is solely responsible for triplet production.

In other singlet exciplex-forming systems there is much less spectroscopic evidence for the formation of radical ions and triplets. An exception is the quenching of aromatic hydrocarbon fluorescence by halogen anions; this has been shown to lead to triplet production and a non-fluorescent exciplex proposed as an intermediate. Spectroscopic evidence for the formation of radical ions in the reaction of triplet ketones with aromatic amines has been obtained.[86] Thus flash photolysis of an acetonitrile solution of benzo-phenone in the presence of tri-p-tolylamine produced the amine radical-cation and ketone radical-anion.

To summarize we may say that in exciplex-forming systems reaction may be expected to occur from the exciplex, radical ions, or triplet species. Chart 2 depicts the various processes involved.

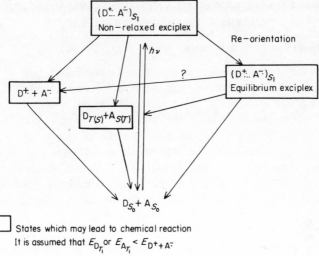

Chart 2

In the case of excimer-forming systems we may expect reaction to occur from the excimer and this has been demonstrated in a number of systems. A particularly nice example is that of (20) which exhibits intramolecular excimer formation and forms (21) on irradiation.[75e] Excited-singlet excimers

$$\xrightarrow{h\nu(\lambda > 280\,\text{nm})}$$

(20)   (21)

may also play a part in intersystem crossing by producing triplets via the triplet excimers

$$\text{ArH}_{S_1} + \text{ArH}_{S_0} \rightarrow (\text{ArHArH})^*_{S_1} \rightarrow (\text{ArHArH})_{T_1}$$
$$\downarrow$$
$$\text{ArH}_{S_0} + \text{ArH}_{T_1}$$

Very little work has been done on this aspect of excimer chemistry. Medinger and Wilkinson[87] have shown that pyrene excimer formation actually leads to a decrease in quantum yield for triplet pyrene formation. A similar result has been observed for benz[a]anthracene.[88] On the other hand, benzene excimer formation leads to a very much increased triplet yield.[89]

IV. PHOTOCHEMICAL REACTIONS INVOLVING EXCITED CHARGE-TRANSFER COMPLEXES

A. Reactions Involving π-Donors

1. Aromatic hydrocarbon–oxygen and related complexes

That aromatic hydrocarbons, e.g. benzene[90] and toluene[91] and even alkanes, e.g. cyclohexane[90, 92] interact with oxygen is demonstrated by the fact that these compounds in the presence of oxygen exhibit electronic absorption bands which are not characteristic of either the hydrocarbon or oxygen. The stabilization energy of these complexes is extremely low ($< kT$) and they have therefore been termed contact charge-transfer complexes.[90, 93] Excitation of the aromatic hydrocarbon–oxygen systems *via* the contact CT absorption band gives rise to excited complexes. Birks[94] and Hoytink[95] have thoroughly examined the variety of complexes which can be produced.

Oxygen quenches the excited-singlet state of aromatic hydrocarbons.[94, 96a–h] This quenching may occur by both static and diffusional processes.[94, 96a–f] There is excellent evidence to substantiate the point that the quenching leads to enhanced intersystem crossing.[96c, d, e] In some cases the quantum yield of triplet formation for some fluorescent hydrocarbons is increased to unity.[96d] Solvent polarity effects,[96d] and the observation that the efficiency of quenching increases as the ionisation potential of the hydrocarbon decreases[96b] lends credence to the suggestion that the quenching process involves CT interactions. Not surprisingly, oxygen has been found to quench excited-singlet excimers of aromatic hydrocarbons.[96g, h] Termolecular complex complex formation between the excimer and oxygen was suggested in order to account for the efficiency of quenching.[96b]

Oxygen quenching of aromatic hydrocarbon triplet-states has been the subject of thorough investigations. The quenching process leads to the formation of the chemically highly reactive species, singlet oxygen.[97a, b] It has been suggested that the carcinogenicity of many hydrocarbons is due to their ability to photosensitize singlet-oxygen production which then attacks DNA.[98] An interesting observation concerning the quenching process is that for exothermic systems, the rate constant for the quenching varies inversely with the triplet energy of the hydrocarbon.[99] This result was analyzed in terms of the quenching involving a collision complex between the triplet- and ground-state oxygen.

Irradiation of many polycyclic aromatic hydrocarbons, e.g. anthracene,[100, 101] dibenzoperylene,[102] in the presence of oxygen leads to the formation of endoperoxides, e.g. (22), (23).[101] Much of the earlier work has been reviewed by Bowen.[100] There is good kinetic data to support the view that these oxidations involve the excited-triplet hydrocarbon sensitizing

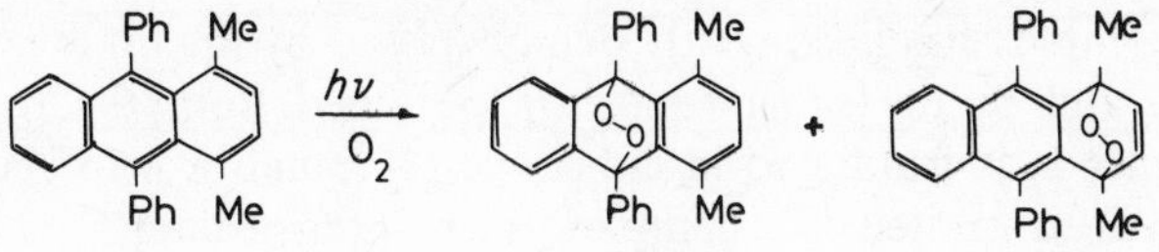

(22)   (23)

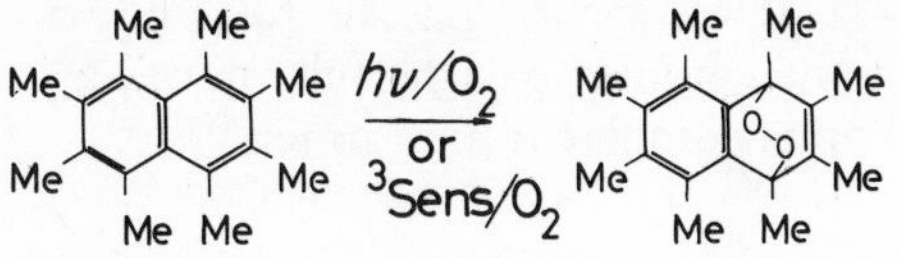

(24)

singlet-oxygen formation and that this then reacts with a molecule of the hydrocarbon in its ground state.[96e, 103] The efficiency of photo-oxidation is solvent dependent.[104] In many cases it is found that triplet-sensitized oxidation of aromatic hydrocarbons, e.g. (24),[105] gives the same products as on direct irradiation. This is further evidence for the photo-oxidation reactions involving singlet oxygen.

Many alkylated benzenes and polycyclic aromatic hydrocarbons undergo oxidation at the benzylic positions on irradiation in oxygenated solutions. It would appear that there is a fine balance as to whether oxidation takes place in the ring or in the side chain. Thus (24) is oxidized in the ring whereas 2-methyl- and 2,3-dimethylnaphthalene undergo oxidation at the methyl groups.[106] Other compounds[91, 107, 108] which undergo side-chain oxidation include toluene,[91] (25)[107] and (26).[107] Oxidation of benzylic C—H bonds

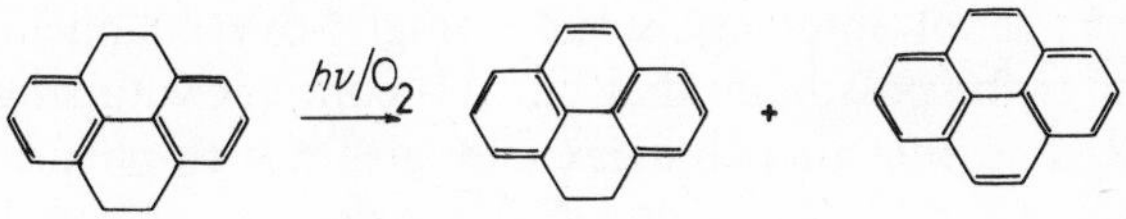

(25)

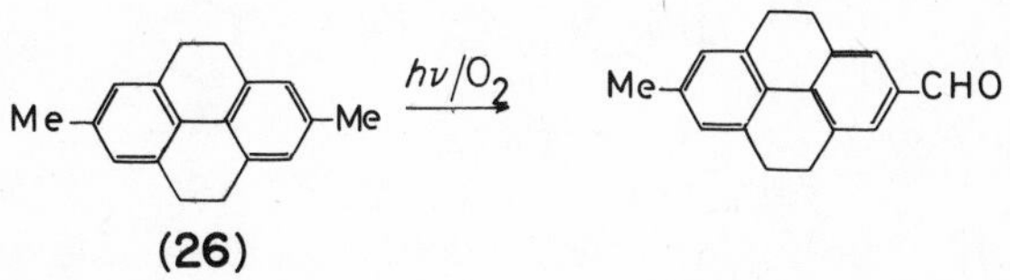

(26)

of alkyl aromatics can occur via a non-photochemical process[108] and this leads one to question whether the light catalyzed reactions are not purely photo-initiated autoxidation reactions. Kinetic measurements have demonstrated that the reaction is not a chain reaction.[91, 109] The intermediacy of a CT complex has been neatly demonstrated by the finding that the same products, and in the same ratio, are obtained from the toluene–oxygen system when it is irradiated in either the CT band or a toluene absorption band.[91] If an excited CT complex is an intermediate in these reactions, as seems likely, we may well ask how does this species gives rise to the products? One possible mechanism which the author suggests is that the excited complex, e.g. (27), decomposes via an intramolecular proton-transfer process. Whether the complex has the full charge, as depicted, or whether it is a singlet or triplet complex is hard to tell.

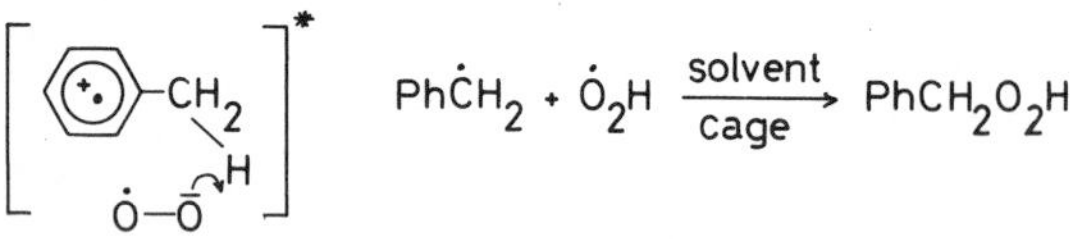

$$Ph\dot{C}H_2 + \dot{O}_2H \xrightarrow[\text{cage}]{\text{solvent}} PhCH_2O_2H$$

(27)

Some dye sensitized photo-oxidation reactions of alkyl aromatic compounds have been found to give products via oxidation of benzylic C—H bonds.[110] In determining the mechanism of these reactions one runs into the problem of ascertaining whether the primary chemical reaction is between the excited dye and the hydrocarbon or between the singlet oxygen and the hydrocarbon. It is conceivable that many aromatic hydrocarbons will have sufficiently low ionization potentials to form excited complexes with singlet oxygen. Such complexes may collapse to give endoperoxides, attack benzylic C—H bonds or give rise to physical quenching. Similar problems arise when one looks into the mechanism of the dye-sensitized photo-oxidation of phenols.[111] Usually it is difficult to ascertain the relative importance of excited-dye–phenol interactions and singlet-oxygen–phenol interactions. Again there is the possibility that phenols complex with singlet oxygen, e.g. to give (28), and that this complex may give rise to phenoxyl radicals.[97b]

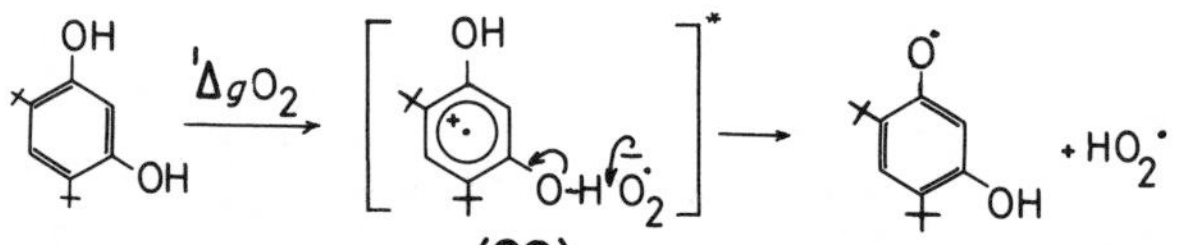

(28)

In many cases the same products are obtained from dye-sensitized oxidation of phenols as from direct irradiation experiments.[111] Whether, under the latter conditions, the triplet phenol is an intermediate or whether a phenol–oxygen complex is involved is not known.

Another mode of reaction of singlet oxygen with olefins and which also occurs on direct irradiation of the olefins, e.g. (29)[112] or (30)[113] with

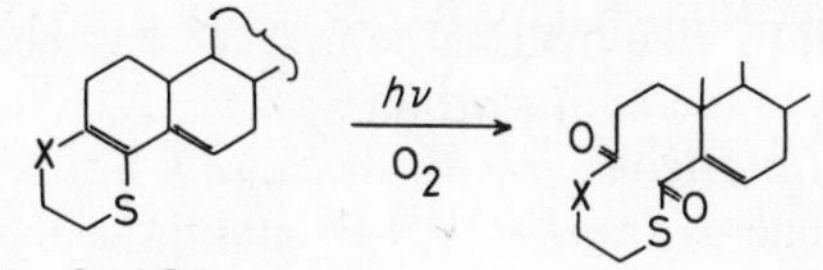

X = 0 or S

(29)

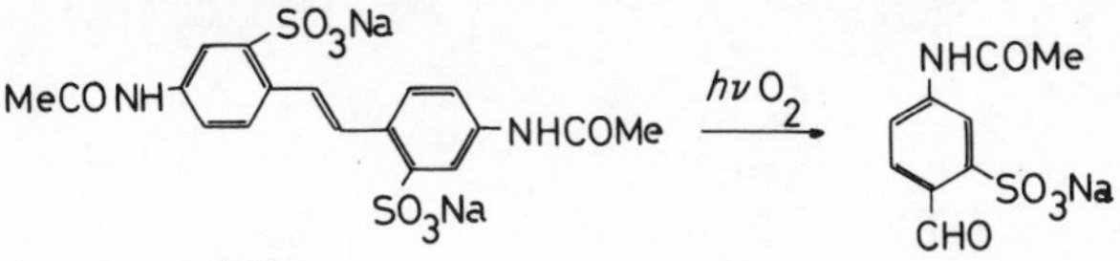

(30)

oxygen, is that of cleavage of the olefinic bond to give two carbonyl groups. The direct irradiation reactions have been postulated as occurring via CT complexes.

As mentioned at the beginning of this section, alkanes also exhibit contact CT complex formation with oxygen. Excitation of these and related systems via the CT absorption band leads to oxidation,[114, 115] e.g. cyclohexane gives cyclohexylhydroperoxide. This reaction may occur in a similar manner to that for the alkylated aromatic hydrocarbons, i.e. an intermediate such as (31) may undergo a proton-transfer reaction.

$$\left(\begin{array}{cc} S_+ & S_- \\ RH \cdots O_2 \end{array} \right)^* \xrightarrow[\text{transfer}]{\text{proton}} R^{\cdot} + \dot{O}_2 H$$

(31)

2. Complexes of aromatic hydrocarbons with anhydrides and imides

There is a wealth of literature reporting intermolecular complex formation between aromatic hydrocarbons and anhydrides and imides.[1, 54c] A few cases of intramolecular complex formation have also been reported.[24, 54b] Formation of excited-singlet complexes in these systems is attested by the fact that irradiation into the CT absorption band gives rise to fluorescence.[54b] The fluorescence emission usually bears an object-to-mirror image relationship to the CT absorption band. Excitation of aromatic hydrocarbon–anhydride complexes in polar solvents leads to radical-ion formation[116] via either the excited-triplet[65] or singlet complex[117] (Franck–Condon complex). In non-polar solvents radical-ion formation via the excited-triplet state of the complex occurs.[117] Phosphorescence from some aromatic hydrocarbon–pyromellitic dianhydride complexes has been observed and found to resemble that of the hydrocarbon.[116]

Compound (32) is of particular interest because it has been used to study intramolecular triplet energy transfer.[118] Excitation of this compound in a *t*-butanol-*i*-pentane matrix (3:7) produces phosphorescence identical with

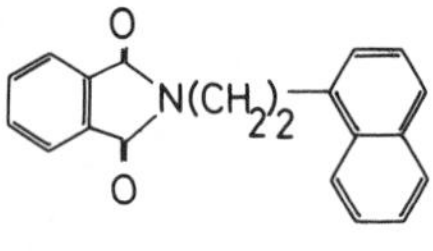

(32) 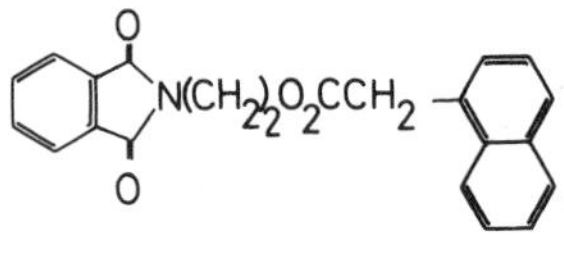   (33)

that from 1-methylnaphthalene. This was taken as evidence for efficient energy transfer. However (32) and the closely related compound (33)[119] exhibit CT fluorescence at both room and low temperature. Thus, it would appear that in rigid medium, a significant degree of complexing (intramolecular?) occurs. This supposition is also supported by the observation that a considerable enhancement of the CT absorption band occurs on freezing solutions of (32) or (33). The CT complex is therefore probably involved in the energy-transfer process.

Aromatic hydrocarbons form complexes with maleic anhydride and related compounds. Irradiation of benzene solutions containing maleic anhydride gives (34, X = O).[121a, b, c, d] Substituted benzenes, containing electron-donating groups, e.g. methyl or methoxyl, react in a similar manner, whereas

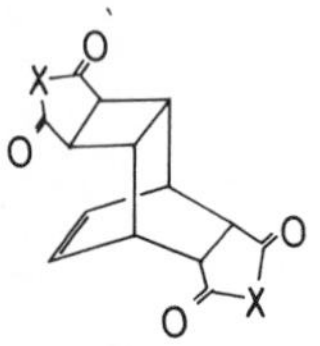

(34) 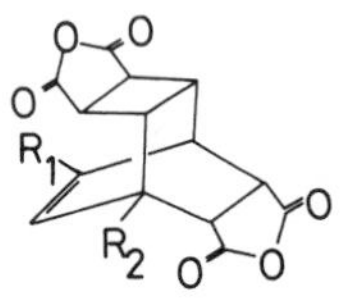   (35)   a, R_1=Me, R_2=H
                                b, R_1=H, R_2=Me

reaction does not occur when electron-withdrawing groups, e.g. nitro or cyano are present.[122] Reaction with toluene gives two products (35a and b) the relative yields of which are determined by the temperature at which the irradiation is carried out. Thus at 20°C the relative yield of 35a to 35b is 1:2 and at 100°C it is 3:2. This increased yield of 35a with increase in temperature is attributed to there being an activation energy barrier along the potential energy surface for formation of this product. Other methylated benzenes behave in a similar manner to toluene. Quinol forms an orange 1:1 complex with maleic anhydride which on irradiation ($\lambda > 290$ mn) in the molten state led to adduct formation.[121b] This and the other photo-addition reactions described can be sensitized by benzophenone.

It has been proposed that such addition reactions occur via a CT complex which ultimately leads to the σ-bonded complex (36).[121a, 123] This complex reacts with a further mole of the anhydride to give the adduct (34, X = O).

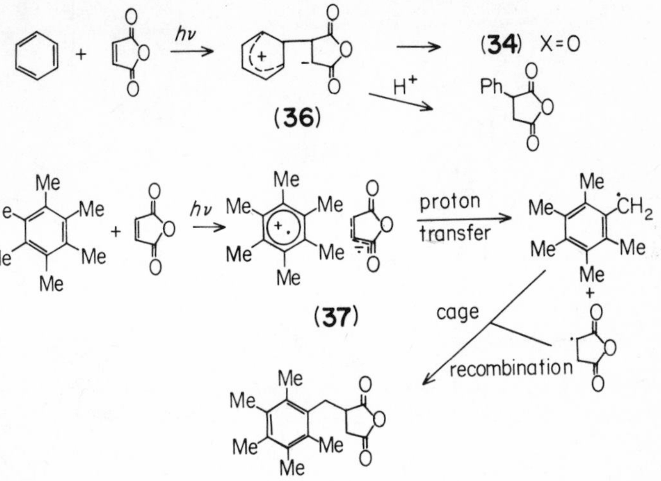

However, if a strong acid, e.g. trifluoroacetic acid, is present in the reaction mixture, protonation of (36) appears to occur and the isolated product is phenylsuccinic anhydride.

It is interesting to find that hexamethylbenzene reacts with maleic anhydride in a slightly different way. Thus, in the absence of acid, irradiation of this complex gives the anhydride (38) and it is reasonable to postulate the radical-ion pair (37) as an intermediate.[125] In this case, collapse of (37) to a σ-bonded complex akin to (36) does not appear to occur.

The photoreaction of the strong acceptors dichloro- and dibromo-maleic anhydride with benzene gives the aryl succinic anhydrides (41) and (43) which react further to give the products (42) and (44). The authors[126] suggest a biradical intermediate (39) although it is possible to rationalize the reactions as occurring via a σ-bonded complex such as (40).

Thiomaleic anhydride undergoes photo-addition to benzene to give the adduct (34, X = S).[128]

Although polycyclic aromatic hydrocarbons form complexes with anhydrides such as maleic anhydride, photo-additions do not always occur on irradiation of the complex, e.g. naphthalene does not add to maleic

I

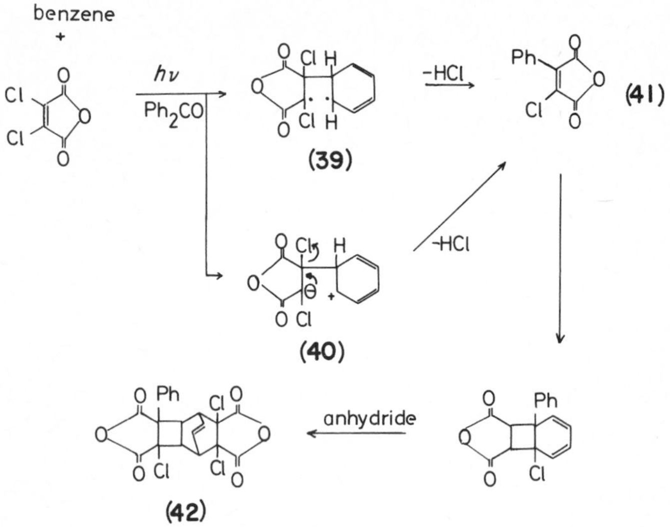

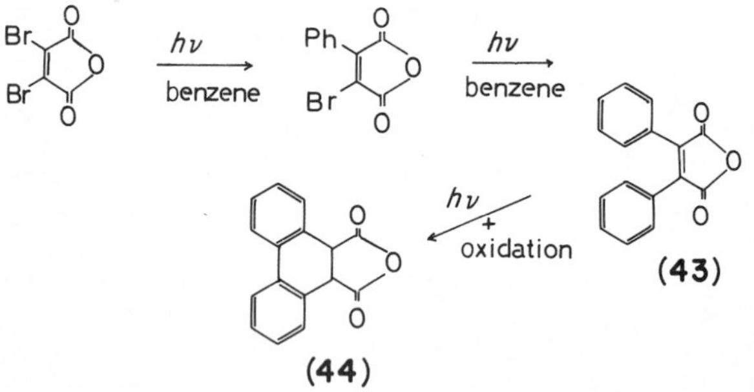

(39) **(40)** **(41)** **(42)** **(43)** **(44)**

anhydride.[129] Irradiation of acenaphthylene in the presence of maleic anhydride leads to polymers but if benzophenone is present in the reaction mixture the adduct (45) is obtained. That this product is formed via a triplet

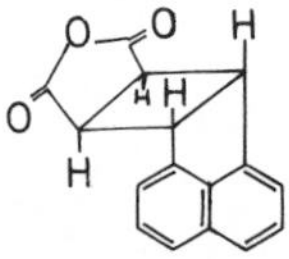

(45)

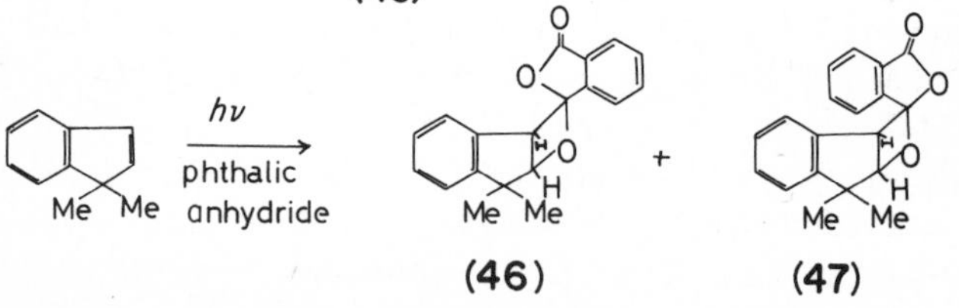

(46) **(47)**

species is demonstrated by the fact that (45) can be obtained from the unsensitized reactions if a solvent containing a heavy atom is used (see Section VIII.C). Indene and closely related compounds form complexes with anhydrides, e.g. maleic, phthalic and 1,8-naphthalic anhydride.[132] Irradiation of these complexes in non-polar solvents leads to oxetane formation, e.g. (46) and (47). With the aromatic anhydrides formation of an isomer having the aromatic rings relatively close, as in (47), readily occurs. Presumably in the ground state of the complex, overlap of the aromatic rings occurs. When polar solvents are employed, no oxetane formation takes place and instead dimerization of the indene occurs.[52] This has been rationalized as involving radical-ion formation (Section II.2e). The possibility that electron transfer between the radical ions occurs to generate triplet indene which acts as the precursor for the dimers was not considered. The isomer distribution of the dimers favours the radical cation rather than the triplet mechanism.

Much less work has been done on the photo-addition reaction of hydrocarbon–imide complexes.[121d, 133] Bryce-Smith has reviewed the work on the addition of maleimide to benzene,[121d, 133] and addition products such as (34, X = NH) have been isolated. These addition reactions are not affected by the presence of acid and product formation is therefore rationalized as occurring via addition of the excited-singlet state of the imide to benzene and as not involving dipolar intermediates. Benzene and alkylbenzenes do not appear to complex with maleimide in the ground state but electron transfer may well take place in the excited complex.

3. Complexes of aromatic hydrocarbons and olefins with cyano-substituted olefins, aromatic cyano-compounds and quinones

The quenching of aromatic hydrocarbon fluorescence by the powerful electron acceptors 1,4-dicyanobenzene,[134a] 1,2,4,5-tetracyanobenzene and tetracyanoethylene[134b, c] has been shown to lead to radical-ion formation when polar solvents are employed. In low polarity solvents fluorescence from hydrocarbon–dicyanobenzene complexes has been detected.[134c] Chemical reaction has been observed on irradiation of the toluene–1,2,4,5-tetracyanobenzene complex and the intermediacy of the radical ions (48) postulated.[135] Similar results were obtained when the toluene–7,7,8,8-tetracyanoquinodimethane complex was irradiated.[136] Compound (49) was isolated from this reaction.

There is good evidence for the formation of complexes between aromatic hydrocarbons and quinones.[120] Irradiation of p-benzoquinone in benzene solution does not lead to product formation unless a good proton donor, e.g. trifluoroacetic acid is present.[123] Under these conditions 4-phenoxyphenol is obtained in high yield and this appears to be good evidence for reaction occurring via the zwitterionic species (50). Phenanthraquinone reacts

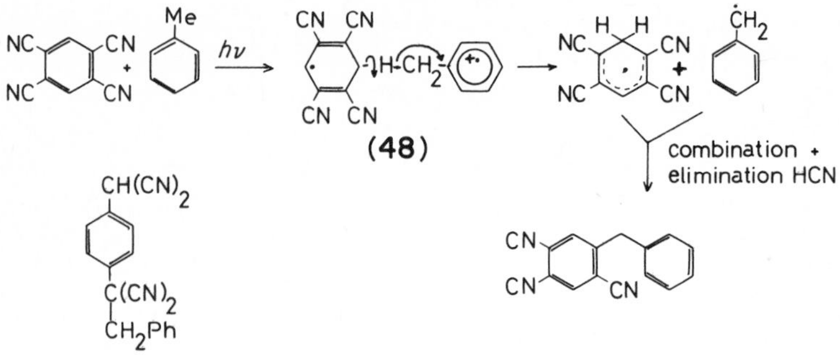

(48)

(49)

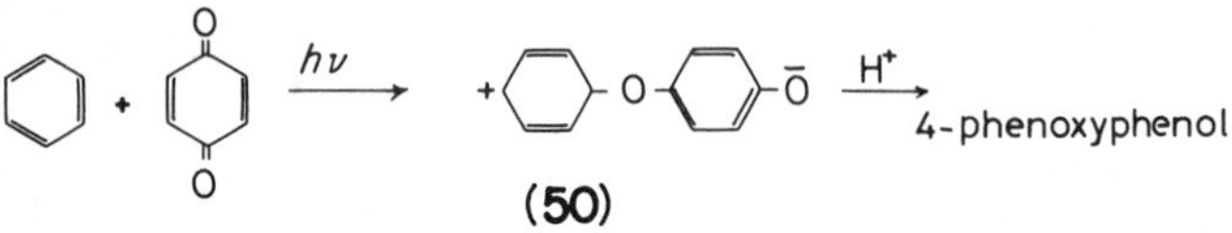

(50)

with benzene[137a, b] to give a number of products, e.g. phenanthraquin-hydrones, biphenyl and 9-hydroxy-10-phenoxyphenanthrene.[137a] The formation of the first two named products suggests that the excited quinone can abstract hydrogen from benzene and indeed radical intermediates have been detected by the technique of flash photolysis.[137c] As will be seen in Section VII.I, other excited carbonyl compounds react in a similar manner. Reference has already been made to the photo-addition of o-quinones to stilbenes.[37]

The rather limited, but excellent work that has been done on the reactions of the complexes of olefins with 1,2-dicyanoethylene and other acceptors[138] has been previously reviewed by Foster.[1]

B. Reactions Involving n-Donors

1. Complexes of ethers, sulphides and amines with oxygen

The formation of contact charge-transfer complexes between oxygen and ethers and amines has been the subject of a detailed study.[139] On the other hand, the question as to what excited complexes may be formed in these systems appears to have been neglected. Recently, the quenching of singlet oxygen ($^1\Delta_g O_2$) by sulphides[140] and by amines[141a, b] has been shown to increase in efficiency as the ionization potential of these compounds decreases. Because of this trend it was proposed that the quenching involves complex formation. The complex may decay by undergoing intersystem crossing via a spin–orbit interaction.[140] We may well ask whether the singlet oxygen–

amine complex for instance, can be produced by excitation of the ground-state amine–oxygen complex. Absorption of light by this complex will produce an excited-triplet complex which should have the same configuration as the singlet oxygen–amine complex, i.e. as in (51). Intersystem crossing

$$R_3N \overset{\delta+}{\cdots} \overset{O}{\underset{O}{\big|}} \delta-$$

(51)

between the two complexes should be very efficient. The finding that many of the chemical reactions which ensue from direct excitation of amine–oxygen complexes are similar to those proposed as occurring via attack of singlet oxygen on amines lends some support to the supposition that a common intermediate is involved. There is also the possibility that excitation of uncomplexed amine in the presence of oxygen will lead to singlet-oxygen production via an energy-transfer process involving the triplet amine. The singlet oxygen may then attack the amine to form products. Such a mechanism has been proposed for the photo-oxidation of aniline.[142] Chart 3 summarizes the reactions just described.

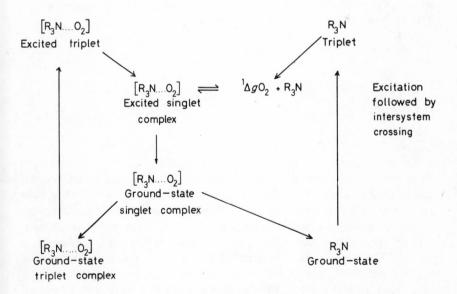

Chart 3.

Another reaction which has been observed is that of electron transfer from amines to oxygen to give the amine radical-cation and presumably the superoxide anion. This reaction is facilitated by the use of polar solvents. Formation of the radical-cation can in principle occur in two ways: (a) electron transfer occurs in an excited amine–oxygen complex or (b) electron ejection from the amine occurs and the oxygen acts as an electron scavenger. The formation of radical ions is only likely to occur when the combined energies of the ions is lower than the energy of the excited complex. If one chooses a system containing a radical cation and the superoxide anion, the combined energies of which are greater than that of singlet oxygen, the radical ions should react to give singlet oxygen. This has in fact been observed in the ferrocenium–superoxide ion system.[143]

Very few studies have been made on the photoreactions of ethers with oxygen. One detailed paper which is available[144] reports that the product of photo-oxidation of diethyl ether is ethyl acetate and that this product is formed from α-hydroperoxy-diethyl ether.

The interaction of singlet oxygen with sulphides in the gas phase has been studied[140] and the rate constants for quenching of the excited oxygen determined. Chemical reaction occurs albeit relatively inefficiently. In solution, singlet oxygen reacts with sulphides and disulphides to give sulphoxides.[145] Sulphoxide formation was shown to occur, by means of crossover experiments, by reaction of an oxygen–sulphide complex, e.g. (52) with another molecule of the sulphide. If the sulphide is present in only a very low concentration, the complex collapses to give a sulphone.

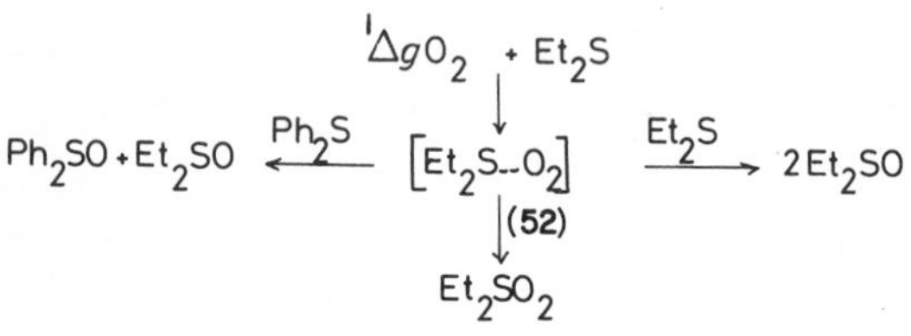

$$^1\Delta g O_2 \;+\; Et_2S$$

$$Ph_2SO + Et_2SO \;\xleftarrow{\;Ph_2S\;}\; [Et_2S{-}O_2] \;\xrightarrow{\;Et_2S\;}\; 2\,Et_2SO$$

$$\Big\downarrow (52)$$

$$Et_2SO_2$$

The story concerning the interaction of amines with oxygen is still far from clear. One indisputable point is that amines act as *physical* quenchers for singlet oxygen. Numerous rate constants have been measured[141a,b] in both the gas and solution phase and all support the finding that the quenching efficiency is tertiary > secondary > primary, i.e. the order which one would expect on the basis of the known ionization potentials of these compounds. The question as to whether chemical reaction occurs as a result of amine–singlet-oxygen interaction is a controversial topic. Singlet oxygen is often generated in solution by means of a sensitizer. Thus in these experiments there is present in solution the sensitizer, the amine and oxygen.

Therefore the possibility exists of reaction between the excited sensitizer and the amine (Mech I) or between singlet oxygen and the amine (Mech II).

$$\left.\begin{array}{l} \text{sens}_{T_1} + \text{amine} \xrightarrow{k_r} \text{sens}^{\cdot}\text{H} + \text{Am}^{\cdot} \\ \text{Am}^{\cdot} + \text{O}_2 \longrightarrow \text{products} \end{array}\right\} \quad \text{Mech I}$$

$$^1\Delta_g\text{O}_2 + \text{amine} \xrightarrow{k_{\text{O}_2}} \text{products} \qquad \text{Mech II}$$

The relative efficiency of these two processes will depend upon the magnitude of k_r and k_{O_2} and the concentration of the amine, oxygen and the sensitizer.[97b] The products of photo-oxidation of amines can be equally well formulated by either mechanism. It is likely that in the amine–singlet-oxygen complex,

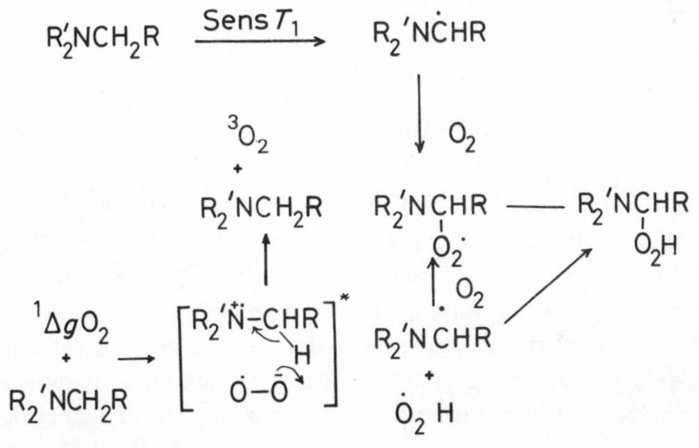

FIG. 5. Diagram showing possible routes by which an amine can be oxidised in a photosensitised reaction.

full charges are not developed.[141b] Reaction of amines with singlet oxygen has also been formulated as occurring via the σ-bonded complex (53).[146] Products isolated from these reactions include α-hydroperoxides. Thus the benzophenone sensitized oxidation of cyclohexylamine leads to an isolable

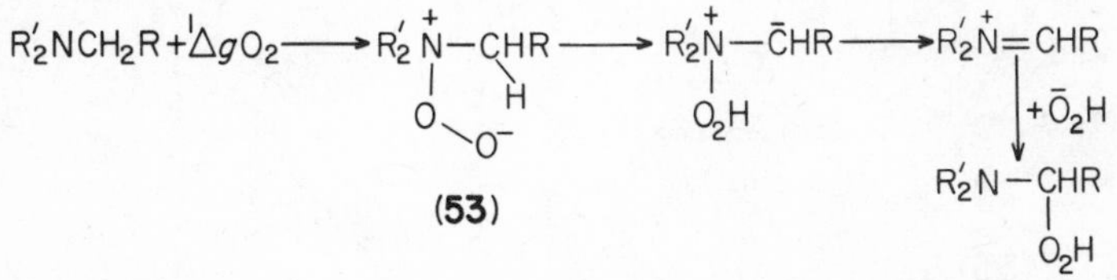

(53)

hydroperoxide.[147] Usually photo-oxidation of amines leads to dealkylation,[148a] (e.g. demethylation[148f, g]), oxidation of N-methyl groups to
formyl groups[148a, b] (e.g. N,N-dimethylaniline gives N-methylformani-
lide[148d]) and Schiff-base formation.[148a–e] Pyrrolines are oxidized to
pyrroles.[149]

What does seem clear about the interaction of singlet oxygen with amines
is that the quenching process is very much more efficient than chemical
reaction. Thus compound (54a) is very stable towards singlet oxygen whereas
the closely related compound (54b) is quite reactive, i.e. the tertiary amino-
group in (54a) is effectively stabilizing the compound towards photo-oxidation.[150] For the rose bengal sensitized oxidation of triethylamine,[151]

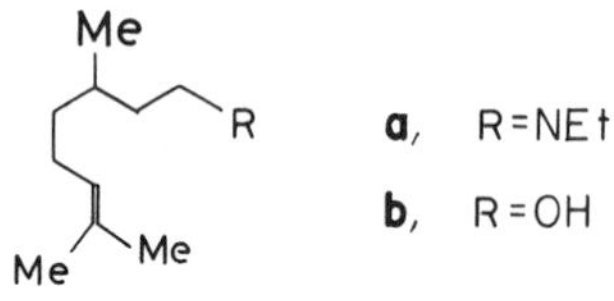

$a,$ R=NEt
$b,$ R=OH

(54)

which is claimed to involve singlet oxygen, it was calculated that there are
nine quenching collisions for every collision that results in reaction. In what
is perhaps the most thorough kinetic investigation of oxidation of amines[152]
it was found that the amount of aniline consumed in the photosensitized
oxidation was proportional to the singlet-oxygen concentration. Unfortunately the relative efficiency of the quenching reaction to chemical reaction
was not assessed. Other evidence that single oxygen can chemically react
with amines, albeit with unknown efficiency, comes from the observation
that singlet oxygen, generated in the gas phase reacts with codeine in either
viscous solution or on a solid support. Reaction causes N-demethylation of
the alkaloid.[148g]

As we have noted before, direct irradiation of oxygenated amine solutions
often leads to the same products as from sensitized oxidation reactions. Thus
cyclohexylamine gives similar products from the two modes of oxidation.[148, 153] The benzoisoquinoline alkaloid (55) and some related compounds

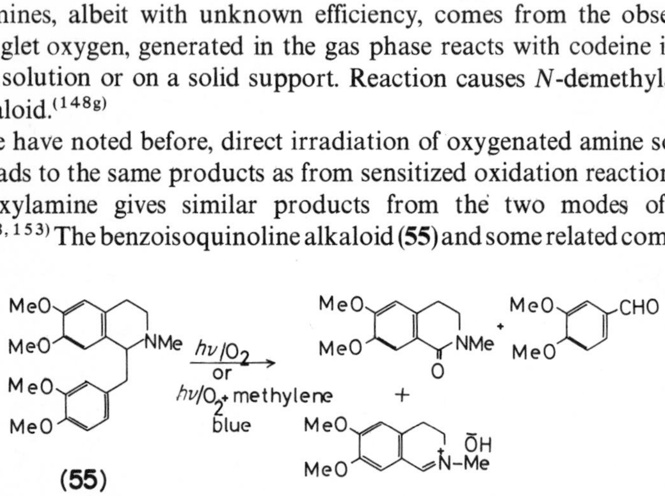

(55)

give similar products although in slightly different ratios on direct and sensitized oxygenation.

The electron-transfer reaction which occurs on direct irradiation of amines in oxygenated solution has been observed with aromatic amines. Flash photolysis of oxygenated acetonitrile solutions of tri-p-tolylamine produces the radical cation.[79] This reaction does not take place in non-polar solvents. Thus for reaction to occur stabilization of the radical cation via a solvation process is required. Similar results were obtained with N,N-dimethylaniline and N,N-dimethyl-p-anisidine.[155] In this work it was concluded that electron transfer took place either via the excited-triplet amine–oxygen complex or via reaction of the excited-singlet amine with oxygen. It appears that reaction of the triplet amine with oxygen did not lead to electron transfer. Irradiation of oxygenated ethanolic phenothiazine solution leads to formation of the amine radical.[156] This was thought to occur via the amine radical-cation which was deprotonated to give the neutral radical.

2. Complexes of amines with halogen-substituted compounds

Evidence for complex formation between aliphatic[157] and aromatic amines[158] with compounds such as chloroform and carbon tetrachloride comes from absorption spectra measurements. Excitation into the CT band produces excited complexes which apparently readily undergo electron-transfer reactions. Fluorescence from excited complexes

$$R_3N + CX_4 \xrightarrow{h\nu} R_3\dot{N}^+ + \dot{C}X_3 + X^-$$

formed between aromatic amines and aryl halides has been observed.[160] A number of studies have been made of the quenching of amine fluorescence by halogeno-compounds. The correlation between quenching efficiency and the electron affinity of the acceptor[160, 161, 162] is indicative of the quenching occurring by a charge-transfer process.

Some of the photoreactions of amines in the presence of halocarbons have already been reviewed by Kosower.[163a] The reaction of n-butylamine with carbon tetrachloride has been the subject of a very thorough study.[163b] In the absence of oxygen quantum yields of between 1 and 100 for the reaction were obtained, i.e. this reaction is an example of a photo-initiated free-radical reaction. Products of the reaction include the hydrochloride of the amine, chloroform, hexachloroethane, and butylidenebutylamine. The reaction may be rationalized as follows:

$$CH_3(CH_2)_3NH_2 + CCl_4 \xrightarrow{h\nu} (CH_3(CH_2)_3\overset{+\cdot}{N}H_2 + \dot{C}Cl_3 + Cl^-$$

$$CH_3(CH_2)_3\overset{+\cdot}{N}H_2 + Cl^- \longrightarrow CH_3(CH_2)_2CH{=}NH + HCl$$

$$CH_3(CH_2)_2CH{=}NH + CH_3(CH_2)_3NH_2 \rightarrow CH_3(CH_2)_2CH{=}N(CH_2)_3CH_3$$

I*

$$CH_3(CH_2)_3NH_2 + \dot{C}Cl_3 \rightarrow CHCl_3 + \begin{cases} CH_3(CH_2)_2\dot{C}HNH_2 \\ or\ CH_3(CH_2)_3\dot{N}H \end{cases}$$

$$\begin{cases} CH_3(CH_2)_3\dot{C}HNH_2 \\ or \\ CH_3(CH_2)_3\dot{N}H \end{cases} + CCl_4 \rightarrow CH_3(CH_2)_2CH{=}NH + HCl + \dot{C}Cl_3$$

Dimethylamine reacts with carbon tetrachloride in a light catalyzed reaction to give N,N,N',N'-tetramethylmethanediamine.[164] Aliphatic amines also react with aromatic halogeno-compounds, e.g. irradiation of a mixture of 4-chlorobiphenyl and triethylamine gives a high yield of biphenyl.[162] The order of reactivity of the amines is tertiary > secondary > primary. Aromatic amines react with alkyl halides[165] and aromatic halogeno-compounds.[160, 165, 166, 167, 168] n-Decyl bromide reacts with N,N-diethylaniline to give n-decane, N-ethylaniline and o- and p-n-decyl-N,N-diethylanilines. These reactions can be envisaged as involving free n-decyl radicals (56, R = n-decyl).[165] Reaction of N-alkylanilines with aryl halides gives

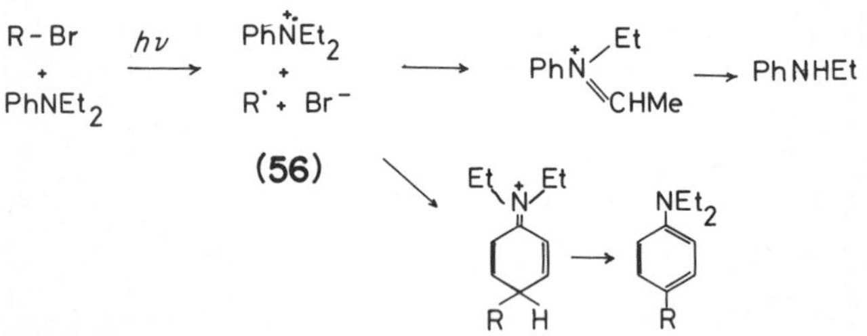

(56)

products which can be rationalized as occurring via free aryl radicals, e.g. reaction between chlorobenzene and N,N-dimethylaniline gives benzene, biphenyl, N-methylaniline and o- and p-dimethylaminobiphenyl.[160] A very interesting observation is that irradiation of aqueous acetonitrile solutions of N,N-dimethylaniline containing chloroform gives 4-dimethylamino-benzaldehyde, i.e. a photo Reimer–Tiemann reaction has occurred.[167] The lack of abstractable hydrogens in arylamines enables radical-cation formation to be observed when these amines react with carbon tetra-chloride[166] and chloroform.[168] The dehalogenation of bromopyrimidines which occurs on irradiation of these compounds in the presence of diethyl-amine was shown to be a chain reaction.[169] Indole reacts with ethyl chloro-acetate to give ethyl indoleacetates.[170] Most positions of the indole ring were found to be susceptible to radical attack.

3. Complexes between amines and quinones and nitro-compounds

The thermal and photochemical reactions of amine–quinone complexes have been reviewed and it was noted that light frequently increased the rate of thermal reactions.[163a] Several acids of the type $ArXCH_2CO_2H$ (where X = O, S or NH) undergo decarboxylation on irradiation in the presence of quinones and aryl ketones.[171, 172] These reactions are discussed in further detail in Section VII.I.

Absorption spectra measurements have demonstrated that a wide variety of amines complex with nitro-compounds.[1] Intramolecular complex formation between amino and nitro groups has been observed and these compounds usually exhibit CT fluorescence.[54a, 173] Photoreduction of aromatic nitro-compounds by aliphatic amines, e.g. diethylamine and triethylamine, gives numerous products including the corresponding arylamine and hydroxylamine.[174] The authors of this work discounted an electron-transfer mechanism on the grounds that reaction via the triplet nitro-compound would be endothermic to the extent of $130 \, J \, mol^{-1}$ ($33 \, kcal \, mol^{-1}$). The lack of any absorption due to complex formation between the aliphatic amines and nitro-compounds was also quoted as mitigating against an electron-transfer process. Aromatic amines, on the other hand, form stable complexes with aromatic nitro-compounds. Irradiation of nitro-compounds in the presence of N-alkylanilines caused reduction of the nitro to an amino group.[175] When N,N-dimethylaniline was used as a reductant dealkylation and oxidation of the methyl to a formyl group was observed. This result was particularly interesting in the light of an earlier report concerning the use of aromatic nitro-compounds to sensitize the decarboxylation of N-aryl-aminoacetic acids, e.g. (57).[176] In these reactions decarboxylation and dealkylation were observed. Presumably the same type of radical is involved in the decarboxylation and reduction reactions. Whilst the products of the two

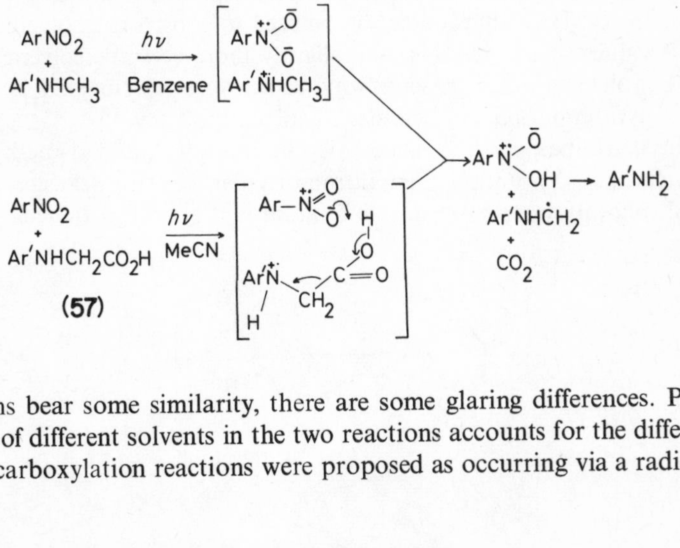

reactions bear some similarity, there are some glaring differences. Perhaps the use of different solvents in the two reactions accounts for the differences. The decarboxylation reactions were proposed as occurring via a radical-ion pair. If the excited nitro-compound had undergone a hydrogen-abstraction reaction, products derived by abstraction from the N—H bond should have been observed. Furthermore, nitro-compounds such as 1-nitronaphthalene which are very poor hydrogen abstractors in their triplet states sensitize the decarboxylation efficiently.

The intermolecular reactions of aromatic nitro-compounds with N-arylaminoacetic acids are relevant to a consideration of the mechanism of decarboxylation of N-(o-nitrophenyl)aminoacetic acids.[177a, b] These acids decarboxylate on direct irradiation and a mechansim involving intra-molecular attack of the nitro-group on the amino-group has been pro-posed.[177b] The intramolecular hydrogen abstraction reactions of o-nitro-alkylbenzenes are suppressed when tertiary amines are added to the reaction mixture and reduction of the nitro-group now occurs.[178] Bearing this result in mind and the fact that nitro-groups can intermolecularly sensitise the de-carboxylation of N-(aryl)aminoacetic acids, it would seem a worthwhile exercise to have another look at the photo-induced decarboxylations of N(o-nitrophenyl)-aminoacetic acids and S-(o-nitrophenyl)glycollic acids[179] in order to determine whether the reactions are either unimolecular or bi-molecular.

C. Reactions of Complexes in Which Anions Act as Donors

1. Quenching of excited pyridinium and related cations by anions

Kosower has made a detailed study of the absorption spectra of a number of pyridinium iodides.[163a] The frequencies at which the CT absorptions of these complexes occur are extremely sensitive to solvent polarity (Chapter 3). Kosower has shown that CT absorption measurements utilizing 1-ethyl-4-carbomethoxypyridinium iodide can be used to construct a solvent polarity scale (Z-values scale). Another, and slightly more versatile solvent polarity scale (E_T values) utilizes the sensitivity of the intramolecular CT absorption band of pyridinium phenol betaines to solvent polarity.[180]

1-Alkylpyridines,[181, 182] 1-methylcollidinium[182] and 1-methylquino-linium iodides[182] undergo an intramolecular electron-transfer reaction on flash photolysis, to give an iodine atom and a neutral radical, e.g. (58).

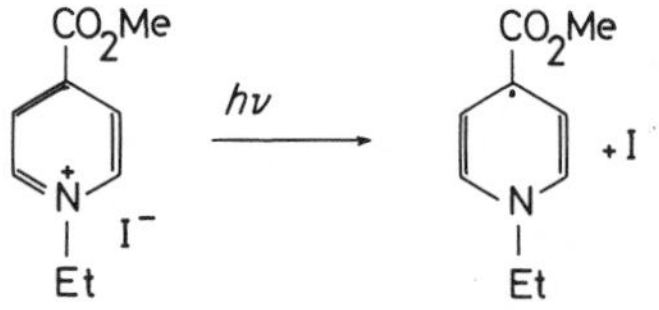

(58)

The course of a number of photochemical reactions of salts, e.g. anilinium and sulphonium salts, have been explained by postulating the occurrence of such an intramolecular reaction as a primary photochemical process. However this process will only occur

e.g. $$R_4\overset{+}{N} + Hal^- \xrightarrow{h_\nu} R_4\overset{+\bullet}{N} + Hal^\bullet$$

effectively if the salt is associated to a reasonable degree in the solvent employed. If the salt is fully dissociated, the anion and excited cation will have to diffuse together for quenching to occur. Thus the degree of quenching by the counter ion is governed by the Stern–Volmer equation.

$$\frac{\Phi_0}{\Phi_q} = 1 + k_q \tau [Q] \tag{5}$$

where Φ_0 = quantum yield of reaction in the absence of quencher; Φ_q = quantum yield of the reaction in the presence of a quencher at concentration $[Q]$; k_q = rate constant for quenching; τ = lifetime of excited state being quenched. For a 1×10^{-4} mol l^{-1} solution of a completely dissociated salt, the quenching of the anion amounts to 1 per cent if the lifetime of the excited state is assumed to be 10 ns and the quenching rate-constant given a value of 1×10^{10} l mol^{-1} s^{-1}. Rate-constant measurements for the quenching of N-methylacridinium cation fluorescence by a variety of ions in a number of solvents have been measured.[6] When extensive dissociation of the acridinium salt occurs, quenching occurs by a dynamic process. Similar measurements have been made for the quenching of lucigenin fluorescence by anions.[183a] The quenching was attributed to electron transfer from the anion to the cation. However this mechanism was rejected by Whitten and co-workers who proposed that quenching involves nucleophilic attack of the anion upon the excited cation.[183b]

The electron-transfer reactions of quaternary salts of 4,4′-bipyridyls have been the subject of much interest because of the use of these compounds as herbicides.[184] Ledwith has shown[184, 185] that irradiation of these salts in alcoholic solution leads to oxidation of the alcohol and formation of the radical cation of the heterocycle. Reaction via electron transfer from the alcohol to the excited-singlet bipyridyl was proposed and the available kinetic evidence supports this view.[186] Flash photolysis of 1,1′-dimethyl-4,4′-bipyridinium dichloride in degassed aqueous solution (pH < 5) produced the radical cation of the base and a chlorine atom.[187] In this case it appears that electron transfer from the counter-ion to the excited heterocycle has occurred. It would be interesting to know to what extent this reaction participates in the reactions in alcoholic solution. Bipyridyl salts of acids such as formic, oxalic, and benzilic acids exhibit CT absorption bands.[188] Irradiation of these salts leads to the liberation of carbon dioxide. Electron transfer from the carboxylate anion to the excited heterocycle was proposed as being the primary photochemical reaction.

2. Photoreactions of anilinium, sulphonium, and phosphonium salts

Rationalization of results in this area are particularly difficult because of the lack of knowledge as to the degree of association of the salts in the various solvents which have been employed. It would appear that in many cases, direct excitation of the cation and irradiation of the associated salts, leads to the same products.

Direct irradiation of anilinium salts in methanol gives benzene plus the appropriate free amine, e.g. trimethylanilinium salts gave benzene and trimethylamine,[189] The highest yields were obtained from salts having iodide as the counter-ion. Reaction was therefore proposed as occurring via electron transfer from the anion to the excited anilinium cation. Aromatic amines e.g. N,N-dimethylaniline have been shown to sensitize the photo-decomposition of anilinium salts in ethanol solution.[190] In this case it was

$$PhNMe_3^+ \; I^- \overset{h_v}{\rightarrow} I^{\cdot} + Ph\overset{+\cdot}{N}Me_3$$

$$Ph\overset{+\cdot}{N}Me_3 \rightarrow Ph^{\cdot} + Me_3N$$

proposed that electron transfer to the anilinium cation occurred from the excited amine, i.e. the anion played no part at all. Triplet-sensitized decomposition of anilinium salts has also been accomplished.[191] The excited-triplet anilinium cation was proposed as undergoing homolysis giving a phenyl radical. Homolysis of excited anilinium cations has been observed by the technique of flash photolysis.[192]

Triphenylsulphonium iodide exhibits a CT absorption and irradiation of the salt at the wavelength of this band produces benzene, iodobenzene, and diphenyl sulphide.[193] Reaction was proposed as occurring via the excited CT singlet state of the salt. This excited species can either collapse to diphenyl sulphide plus iodobenzene or undergo intramolecular electron-transfer from the iodide to the sulphur cation. The sulphur radical so produced can then break up giving a phenyl radical plus diphenyl sulphide.

Products from irradiation of phosphonium salts have also been studied.[194, 195] The solvents used were a benzene–ethanol mixture and acetonitrile. It is therefore probable that the observed reactions emanate from excited ion-pairs. Electron transfer from the counter-ion to the phosphonium cation was proposed. Decomposition of the salt (59) was found to be the most efficient when iodide was the counter-ion.

3. Miscellaneous photo-reactions involving complex formation

Charge-transfer absorption bands have been observed for complexes formed

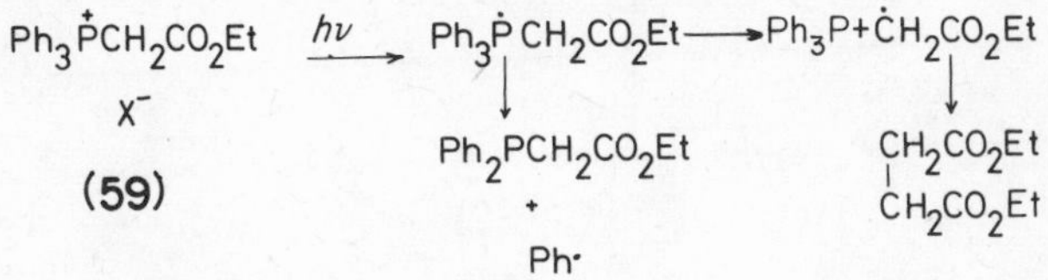

$$Ph_3\overset{+}{P}CH_2CO_2Et \xrightarrow{h\nu} Ph_3\overset{\cdot}{P}CH_2CO_2Et \longrightarrow Ph_3P + \overset{\cdot}{C}H_2CO_2Et$$

$$X^-$$

$$\mathbf{(59)} \qquad\qquad Ph_2PCH_2CO_2Et \qquad\qquad CH_2CO_2Et$$

$$+ \qquad\qquad\qquad |$$

$$Ph\cdot \qquad\qquad\qquad CH_2CO_2Et$$

between a number of heterocyclic compounds, e.g. furan, thiophen and pyrrole, with acceptors such as tetracyanoethylene, chloranil and maleic anhydride.[196] Photo-addition of furan[197] and thiophen[198] to 2,3-dimethylmaleic anhydride has been reported. Triplet sensitizers were employed in these reactions. Due to a lack of information one cannot tell whether the reaction occurs via an excited-triplet complex or not.

Aromatic amines form complexes with imides, e.g. phthalimide[199] and maleimides.[200] Fluorescence from the excited-singlet state of (**60**, n = 2, 3 or 4) has been observed. Irradiation of N-phenylmaleimide in neat

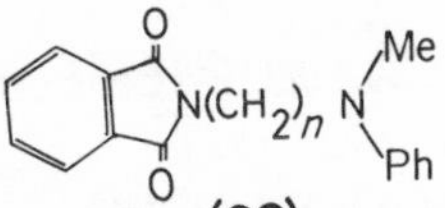

(60)

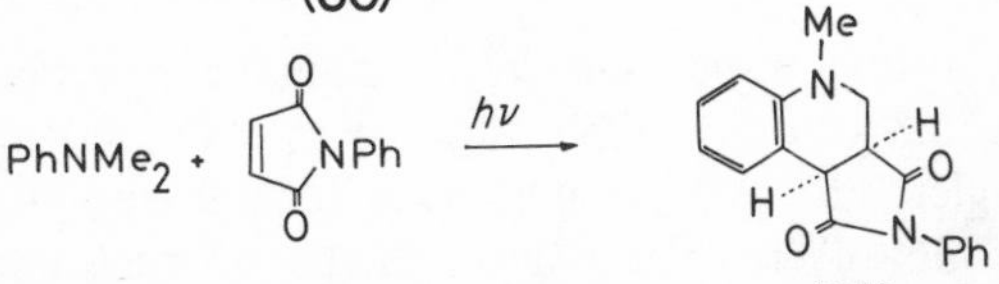

$$PhNMe_2 + \quad [\text{N-phenylmaleimide}] \quad \xrightarrow{h\nu} \quad [\text{product}]$$

(61)

N,N-dimethylaniline gave (**61**).[201] Presumably, under these conditions reaction occurs via the amine–imide complex.

Aromatic nitro-compounds form complexes with aromatic hydrocarbons[1] and as yet very little work has been done on the photo-reactions of these systems. A most interesting observation is that 1-methoxynaphthalene undergoes ring cleavage to give (**63**) on irradiation in the presence of aromatic nitro-compounds.[202] Compound (**62**) was suggested as being an intermediate and this seems reasonable since the photo-addition of nitro-compounds to olefins has been previsously observed.[203]

Irradiation of 2-methyl-6-nitroindoles in aerated solution causes oxidation of the methyl to a formyl group.[204] It is difficult to ascertain whether reaction occurs via intermolecular attack of an excited nitro-group upon the methyl group or whether an indole–oxygen complex is involved. Rather interestingly, 5-nitro-2,3-dimethylindole, under similar conditions undergoes cleavage to give (**64**). However, the question remains: is (**64**) formed via

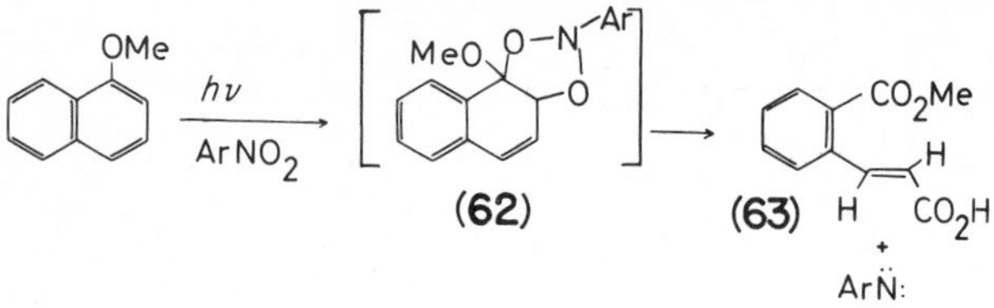

(62)   (63)

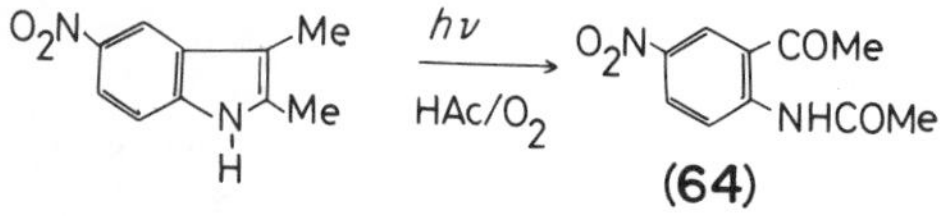

+

ArN:

(64)

addition of the nitro-group to the indole or is the cleavage reaction brought about by singlet oxygen?

Irradiation of the complex formed between N-alkylcarbazoles and tetranitromethane gives N-alkyl-4-nitrocarbazole and nitroform.[205]

V. REACTIONS INVOLVING EXCIMER FORMATION

A. The Photodimerization of Aromatic Hydrocarbons

Many derivatives of simple polycyclic aromatic hydrocarbons, e.g. 9-cyanoanthracene[206] (65, R = CN) undergo photodimerization from their first excited-singlet states. In some cases, e.g. 9-methylanthracene, the hydrocarbons exhibit excimer emission and therefore the question arises as

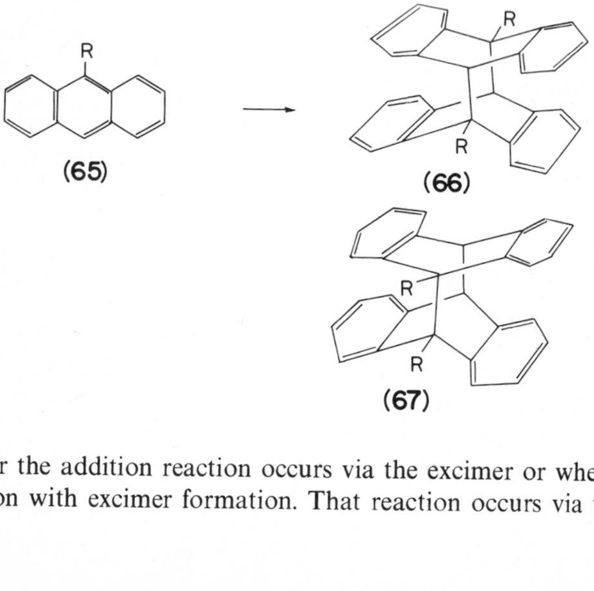

(65)   (66)

(67)

to whether the addition reaction occurs via the excimer or whether it is in competition with excimer formation. That reaction occurs via the excimer has been demonstrated in the following ways. (*a*) By careful kinetic analysis of the reactions.[75b] (*b*) Irradiation of anthracenes in oxygenated solution gives photo-adducts and photo-peroxidation. An increased concentration of the hydrocarbon, which favours excimer formation, increases the amount of adduct formation compared with peroxide formation.[207] (*c*) Irradiation of crystalline 9-cyanoanthracene produces excimer emission. This decreases in intensity as irradiation is continued and photo-adduct formation is observed.[208]

It is also observed that the quantum yields of photo-adduct formation for aromatic hydrocarbons which do not exhibit excimer emission, e.g. anthracene, are higher than those for hydrocarbons which form excimers, i.e. excimer fluorescence is leading to wastage of electronic energy.[75d] This observation is compatible with either of the proposed mechanisms.

For photo-adduct formation to occur an excited-singlet hydrocarbon and a ground-state molecule have to diffuse together. They will eventually approach close enough for excimer formation and then close enough for adduct formation, i.e. excimer formation is on the reaction path for adduct formation. If adduct formation is particularly facile it will act as an extremely good internal quenching mechanism for excimer emission. Anthracene does not exhibit excimer fluorescence in solution because of the efficient dimerization reaction. Introduction of alkyl substituents into the 9-position of anthracene slows up the addition reaction and consequently excimer emission can be observed. 9,10-Dialkylanthracenes, e.g. 9,10-dimethylanthracene exhibit excimer formation but do not photodimerize, i.e. the substituents cause sufficient steric congestion to prevent close enough approach of the two hydrocarbon molecules. If the substituents in the 9 and 10- positions are particularly bulky, e.g. in 9,10-diphenylanthracene, excimer formation is also suppressed. That substituents in aryl rings can affect the stability of excimers and hence the amount of observable excimer at room temperature has been demonstrated by the finding that the excimer of phenylcyclohexane is 0·1 eV less stable than the benzene excimer.[209a] Another example is that of 9,10-dimethylanthracene whose excimer binding energy has been shown to be lower than that of 9-methylantracene.[209b]

Although anthracene excimer fluorescence cannot be detected in fluid solution at room temperature it can be observed by a special technique using rigid matrixes. Chandross[210] photolysed the anthracene photodimer (**66**, R = H) in a rigid matrix at low temperatures, at relatively short wavelengths (<300 nm). In this way, pairs of anthracene molecules were produced in an environment in which the constituents of the pairs could interact. These sandwich pairs exhibit excimer emission.

Many mono-[206, 211, 212, 213, 214] and di-substituted[211] anthracenes dimerize on irradiation. In the case of monosubstituted compounds two products (66) and (67) may be formed. When R = CHO, CO_2Me or CH_2OH isolated products having structure (67) have been obtained[214] whereas when R = Br or CN, products having structure (66) have been isolated.[213] As yet no reason has been put forward for this divergent behaviour. Disubstituted anthracenes, e.g. 9-methoxyl-10-methylanthracene can also form photodimers.[215] Crossed photo-addition products can be produced on irradiation of solutions containing two differently substituted anthracenes.[215, 216] As has been mentioned, the photodimers of anthracene can be cleaved by irradiation.[210] Menter and Förster[29] have very elegantly shown that both excited-singlet anthracenes and excimers can be formed in these reactions.

Several naphthalene derivatives exhibit excimer emission, e.g. 1,6-dimethylnaphthalene.[217] 2-Methoxynaphthalene exhibits excimer emission[218a, b] and photodimerizes to (68) and (69), the former being produced in the greater yield.[218b]

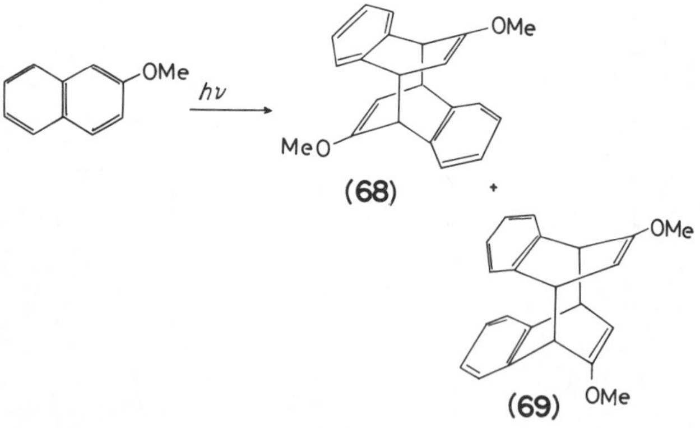

Other 2-alkoxynaphthalenes dimerize in an analogous manner as long as the alkyl group is not branched.[219] When this is the case addition and excimer formation are not observed. Presumably steric congestion does not allow the reaction partners to approach close enough.

Acenaphthylene (70) is a non-fluorescent hydrocarbon which on irradiation gives the dimers (71) and (72).[220a, b] The anti-compound (71) was shown to be produced predominantly from the triplet hydrocarbon. This point was substantiated by the demonstration that the yield of (71) is increased by the use of heavy atom containing solvents and by triplet quenching experiments. The syn-isomer (72) was proposed as being produced via a singlet excimer although no emission from such a species has been detected.[220a] A factor in favour of this suggestion is the low efficiency of dimerization since this points to an energy wastage step which could easily result from excimer formation. It

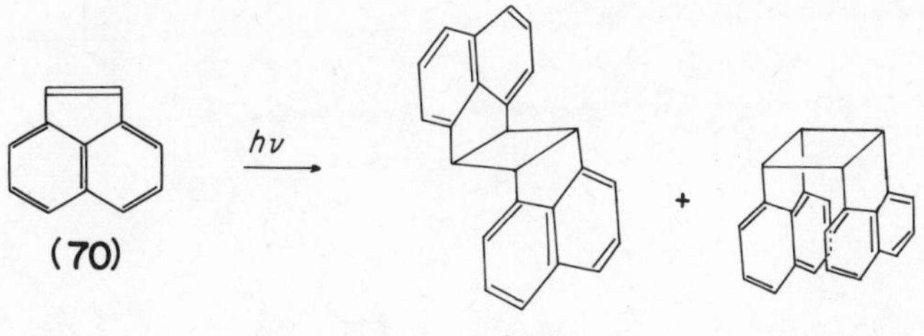

(70) → (71) + (72)

will be noticed that in (72) the aromatic rings are favourably orientated for intramolecular excimer formation. Fluorescence attributable to this species has been detected.[221]

Phenanthrene is a peculiar hydrocarbon in that it neither photodimerizes nor exhibits excimer formation. Even when two phenanthrene molecules constitute a sandwich-pair, no excimer emission can be observed on excitation.[222] 9-Cyanophenanthrene and the anhydride of 9-phenanthroic acid do photodimerize in the expected manner. However, these compounds do not produce detectable excimer emission.

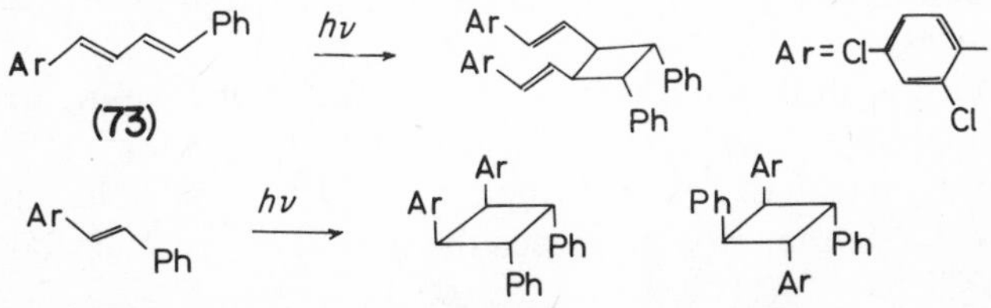

(73) → (75) + (76)   Ar = Cl-⟨⟩-Cl

(74)

The 1,4-diaryldiene (73)[223] and the substituted stilbene (74)[224] photodimerize and exhibit excimer emission in the crystalline state. The orientation of the molecules of (73) and (74) is such that favourable interaction between the constituent molecules occurs. The stilbene (74) dimerizers in solution to give (75) and (76), whereas (75) only is produced when its crystals are irradiated. This is an example of the molecular arrangement in crystals determining the course of a photochemical reaction, i.e. it is an example of topochemical control.

B. Photomerization of Coumarin and Pyrimidines

The mechanistic aspects of the dimerization of coumarin have been the subject of a detailed study by Morrison and coworkers.[225, 226] Irradiation of solutions of coumarin produces the dimers (78a–c) the yield of (78c) being very low. Triplet sensitization of the reaction, or employment of very dilute solutions of coumarin produce (78b and c). This and other evidence establishes that (78a) is derived from the excited-singlet state of (77). Formation of (78a) is favoured by high concentrations of (77) and by *lowering* the temperature of the reaction. This latter point was taken as being indicative of (78a) being formed via an excimer. Coumarin does not exhibit excimer fluorescence in solution although when contained in rigid matrices, emission attributed to the excimer has been observed. The dimerization of coumarin is also very inefficient and the singlet excimer is an obvious culprit for energy wastage.

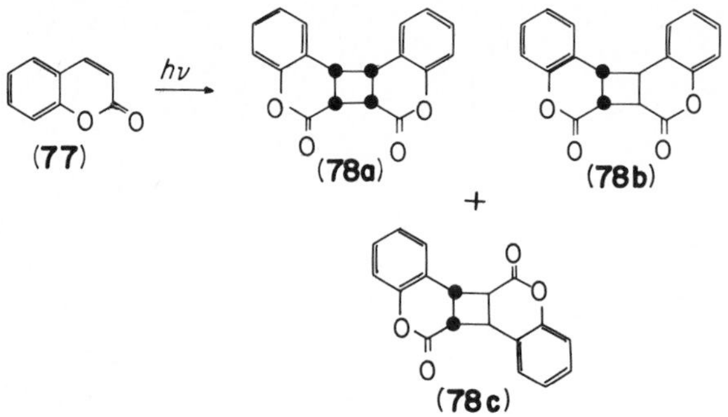

Attention has already been drawn to the fact that thymine and related pyrimidines dimerize on irradiation and that aggregation of these bases favours reaction via the excited-singlet state.[10a, b, 11a, b] The products of dimerization of thymine (79) have been the subject of a careful study because

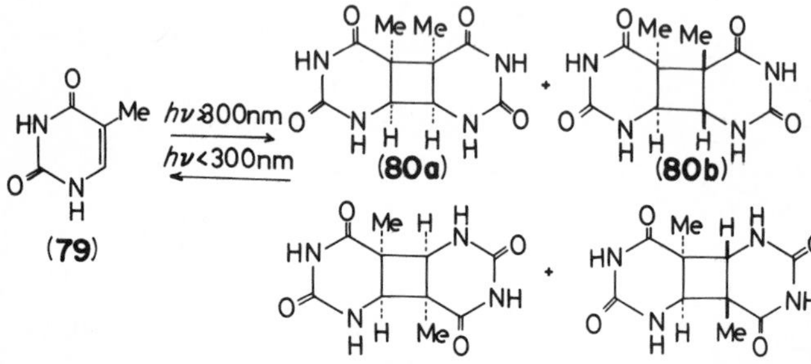

of the importance of this reaction in relation to photo-damage of DNA.[227] In native DNA dimerization occurs between adjacent thymines on the same strand and leads to the *cis-syn*-isomer (80a). Irradiation of thymine dimers at short wavelengths or in the presence of sensitizers, e.g. anthraquinone-2-sulphonic acid, leads to their break up and regeneration of thymine.[228, 229]

There has been much discussion as to whether the dimerization of thymine, involving its excited-singlet state, actually involves an excimer. Pairs of thymine molecules, correctly orientated in a matrix do not exhibit excimer emission but do dimerize extremely efficiently.[230] Thus by analogy to the previously discussed dimerization of anthracene, one may argue that the dimerization of thymine in these circumstances is acting as an extremely effective internal quenching mechanism for the excimer. Presumably the introduction of bulky groups into thymine would slow down the dimerization and aid detection of the excimer. In the case of 1,3-dimethylthymine, Morrison has shown[11b] that dimerization via the excited-singlet state shows a negative temperature effect and that not all the excited singlets undergo reaction. Consequently the conclusion was reached that a singlet excimer was involved.

VI. REACTIONS INVOLVING INTRAMOLECULAR EXCIMER FORMATION

A. Introduction

The subject of intramolecular excimers has recently been reviewed by Klöpffer and in particular the physical properties of these complexes has been dealt with very thoroughly.[231] The following is a short summary of some of the main points. (a) Intramolecular excimer emission is concentration independent. (b) Increasing the lifetime of the initially excited species will aid excimer formation by giving it more time to take place. (c) The conformational requirements of the chain atoms linking the two groups will affect the efficiency of excimer formation. (d) The lack of intramolecular excimer fluorescence at room temperature may indicate a low binding energy for the species. Lowering the temperature should enable the fluorescence to be seen. (e) The most favourable conformation is the one in which the two interacting chromophores adopt a sandwich configuration with a separation of approximately 2·5 Å between the rings. (f) In solid solution at low temperature, excimer emission is not normally observed. The conformation of the molecule having the chromophores correctly oritated in fluid solution is probably one of the least favourable and therefore on freezing most of the molecules will be in conformations which preclude interaction of the chromophores. There are examples, e.g. with purines, in which such factors as hydrogen-bonding hold the interacting groups together as the solutions are being frozen. (g) A lowering of the temperature will disfavour the thermally activated back-crossing of the excimer to an excited monomer. (h) Viscosity effects are less for intra- than inter-molecular excimers and their role in intramolecular excimer formation is not clear. (i) The entropy loss for intramolecular excimer formation is less than that for its intermolecular counterpart. This favours the detection of weak excimer formation.

B. Intramolecular Excimers of Aromatic Hydrocarbons

One of the earliest reports of intramolecular excimer formation was concerned with its occurrence in 1,3-diphenylpropane and related compounds.[232] It was shown that diphenylalkanes having an interconnecting chain of one, two, four, five or six methylenes did not exhibit this behaviour. Excimer formation in 1,3-diphenylpropane is favoured by the fact that the methylene chain allows good overlap of the phenyl rings and the methylene chain can adopt a favourable conformation.

This work was followed by an elegant study on naphthalene excimer formation by Chandross and Dempster.[75e, 233] They demonstrated that 1,3-di(α-naphthyl)propane (20) and 1,3-di(β-naphthyl)propane (β,β-isomer) exhibit intramolecular excimer formation and that (20) photodimerizes to give (21). Interestingly, excimer emission from (20) is much weaker than that from the β,β-isomer and there is the possibility that this is due to the addition reaction acting as an internal quenching mechanism. Irradiation of (21) contained in a rigid matrix at low temperature, with short-wavelength light (254 nm) led to the regeneration of (20) having the naphthalene rings correctly oriented for excimer formation.[234] Excitation of this "sandwich-pair" led to the observation of excimer fluorescence. Also phosphorescence, very similar to that of a simple naphthalene, was observed. This situation is particularly favourable for triplet excimer formation and yet this does not take place. One concludes that triplet excimers of aromatic hydrocarbons are very unstable species. Chandross and Dempster also examined 1,2-di(α-naphthyl)ethane for excimer emission but failed to detect it. It was concluded that in these compounds the energy requirements, which reflect the conformational requirements, for excimer formation are too great. Very weak excimer emission from 1-(α-naphthyl)-3-(β-naphthyl)propane was detected. This excimer is very much less stable than that from (20).

Compound (81) does not exhibit intramolecular hetero-excimer formation.[235] Efficient singlet–singlet energy transfer from the naphthalene to the anthracene chromophore occurs. However, if very dilute solutions of

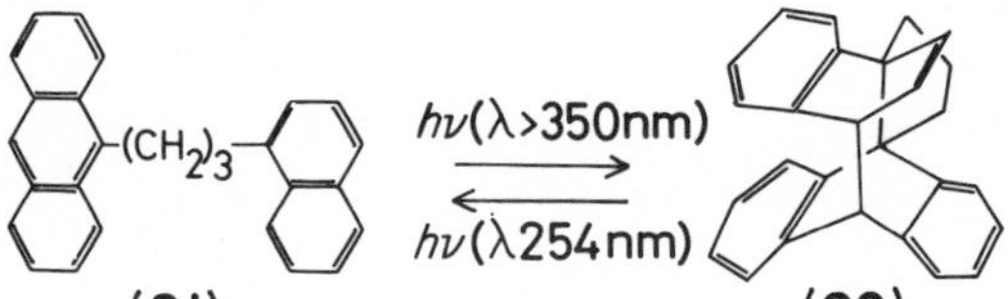

$$ (81) \quad \xrightarrow[h\nu(\lambda\,254nm)]{h\nu(\lambda>350nm)} \quad (82) $$

(81) are irradiated the adduct (82) can be isolated. It would appear that (82) is being formed via a very unstable hetero-excimer. Cleavage of (82) in a rigid matrix at low temperatures produces (81) having the naphthalene and anthracene ring systems correctly orientated for hetero-excimer formation and under these conditions emission from this species can be observed.[236a] Intramolecular hetero-excimer emission has been observed from (83).[236b]

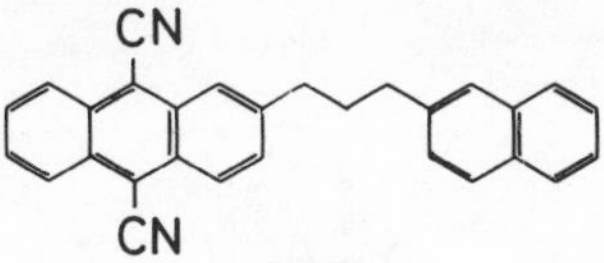

(83)

Interestingly the wavelength of excimer emission is temperature dependent. At low temperatures it is red-shifted and it was proposed that at these temperatures emission was from a ground-state complex between the naphthalene and anthracene systems.

There are a large number of examples of compounds[237a, b, c, d, 238] containing two anthracene units which undergo intramolecular photo-addition reactions across the 9,10-positions, e.g. (84)[237a] (85),[237b] (86).[238] As will be noted, the addition reactions seem to occur irrespective of the type or size of linkage between the two chromophores. Examples (85) and (86) are remarkable in this respect. Intramolecular excimer emission has not been detected in these systems and therefore excimer intermediacy in product formation is a matter of conjecture.

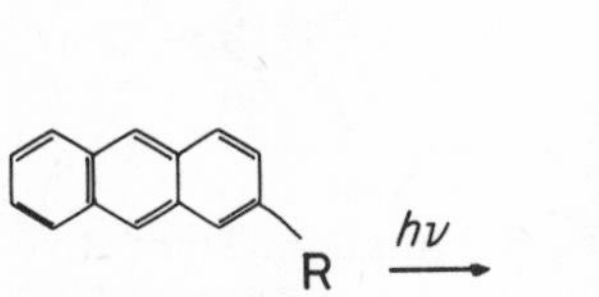

(84) 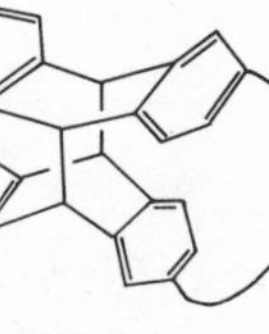   (85)

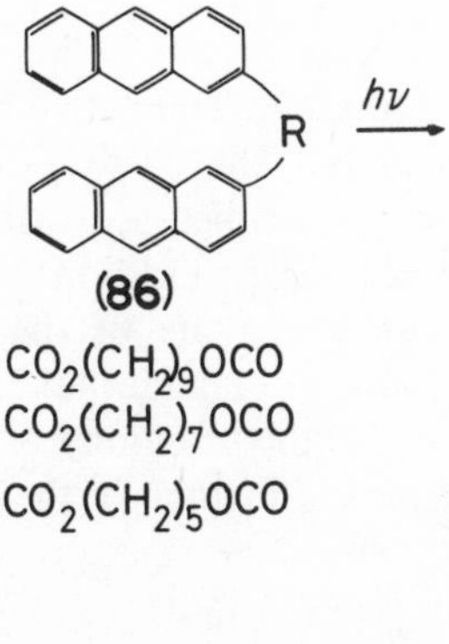

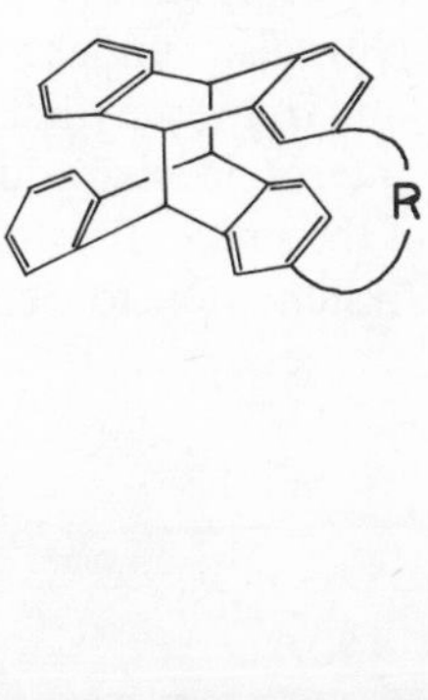

(86)

$R = CO_2(CH_2)_9 OCO$
$= CO_2(CH_2)_7 OCO$
$= CO_2(CH_2)_5 OCO$

C. Intramolecular Cycloaddition Reactions of Unsaturated Lactones and Imides

A number of molecules which readily undergo intermolecular photo-induced cyclo-addition reactions, e.g. coumarins, and maleimides, also undergo intramolecular cyclo-addition reaction when linked by a suitable connecting chain of atoms.[239] The coumarin (87) forms (88a and b), The

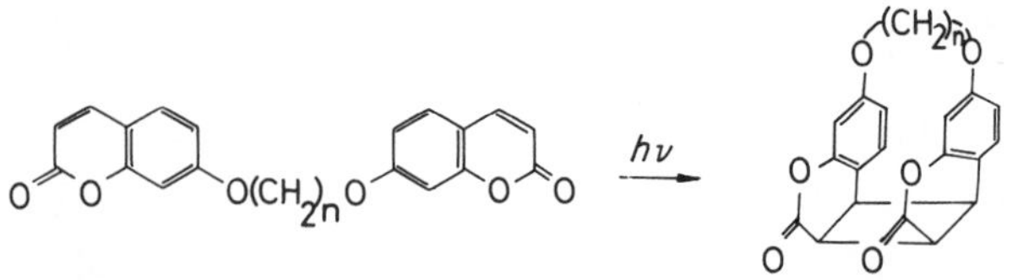

(87)   (88a)

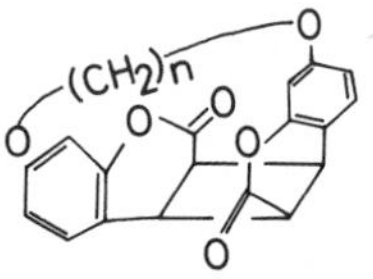

(88b)

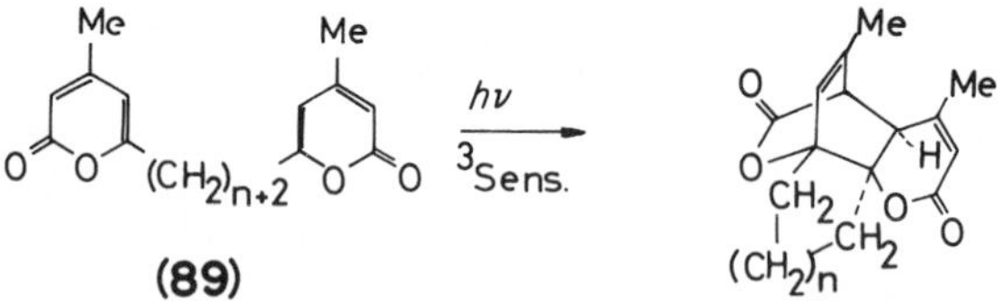

(89)

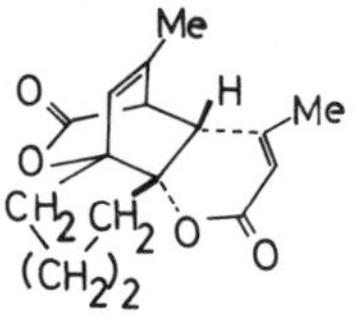

only apparent restriction caused by the linking chain is that it is sufficiently long so that it can bridge the rings in (88a) and (88b) easily. Cyclization of the closely related compound (89) occurs on triplet sensitization.[240] The maleimides (90) also undergo intramolecular addition reactions. This reaction occurs from the first excited-singlet state of the imide.[241] In contrast, the intermolecular addition reactions of imides take place from the triplet state of the imide. Presumably in (90) the proximity of the maleimide rings enables interaction to occur during the lifetime of the excited-singlet

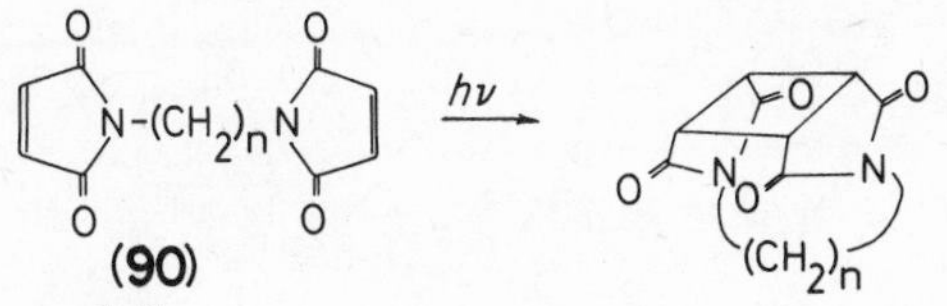

(90)

n = 3–6

state. Although **(90)** did not exhibit an excimer emission, it was postulated that excimers are intermediates in the reaction.

D. Intramolecular Interactions of Pyrimidines and Purines

Some very elegant work has been carried out on the interaction of various pyrimidine and purine constituents of di- and poly-nucleotides in ethylene glycol–water glasses at low temperatures.[242] Even in the frozen glass the bases are held together through such forces as hydrogen bonding and this enables excimer formation to be observed. Interactions between these important heterocyclic bases has also been studied by synthesizing a variety of compounds having two bases linked by a polymethylene chain.[243] Interactions between the bases was found to be maximized when a trimethylene chain was employed. From the ultraviolet absorption spectra of these compounds, hypochromism values were calculated which indicated that the efficiency of base–base interaction was purine–purine > purine–pyrimidine > pyrimidine–pyrimidine. Fluorescence from ethylene glycol–water glasses at 77 K containing the bases was studied and excimer emission observed from the purine–purine and purine–pyrimidine base pairs. In the case of the compounds having two thymine units linked by a trimethylene chain there was no detectable excimer fluorescence. In all probability this lack of excimer emission indicates that the thymine units are not correctly orientated in the glass for excimer emission since thymine emission has been observed in dinucleotides.[242]

Compounds having two thymine units linked by a trimethylene chain, e.g. **(91,** n = 3)[244a, b] and two uracil units[244a, 245] linked in a similar way undergo intramolecular cyclo-addition reactions on direct irradiation. The triplet sensitized photo-addition reactions of **(91)** n = 2, 3, 4, 6 were found to occur with the following order of efficiency, n = 3 > n = 4 > n = 2 > n = 6.[244a] This order just reflects the ease of intramolecular interaction of the thymine units.

E. Some Comments on the Relationship Between Excimer Formation and Cycloaddition Reactions

It will have been noted that in the preceeding section many photo-addition

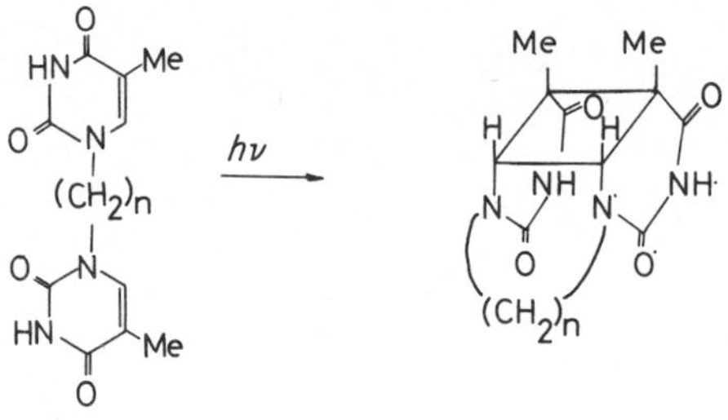

(91)

reactions have been postulated as occurring via excimers although no emission from such species could be detected. Whilst recognizing that the orientation of the molecules comprising the excimer is one which should be on the reaction profile for product formation, we may well ask if the excimer has any real existence, i.e. is it a transition state or an intermediate? If it is an intermediate, how can we prove it? Since excimers are able to undergo efficient radiationless decay to the ground state it is possible that detection of an energy wastage process, that is not due to either the initially excited species or to a biradical intermediate, may constitute evidence for the intermediacy of such a species.

We have also noted that many of the cyclo-addition reactions can be triplet sensitized and that triplet excimers are unstable (as judged by their low binding energy). These addition reactions often give products in which it appears that the reaction partners have come together in such an orientation that a triplet excimer may well have been involved. In such reactions energy wastage due to triplet excimer formation may occur. Potential triplet-excimer intermediates or transition states may affect cycloaddition reaction, e.g. dimerization of acenaphthylene which gives rise to two stereoisomeric products. Reaction via the triplet state will not favour the reaction pathway involving maximum overlap of the reaction partners since this would involve a triplet excimer. Consequently, addition to give the *cis*-photodimer (**77**) will be disfavoured compared with reaction to give the *anti*-isomer (**71**).

VII. PHOTOREACTIONS INVOLVING EXCIPLEX INTERMEDIATES FOR WHICH SPECTROSCOPIC EVIDENCE IS AVAILABLE

A. Photoreactions of Aromatic Hydrocarbons with Amines, Ethers and Phosphines

1. Intermolecular reactions
Many of these reactions have been the subject of a recent review by Lablache-

Combier[246] and the products of the reactions are described in detail.

We have already noted the following relevant points. (a) In non-polar solvents tertiary amines exhibit exciplex formation with aromatic hydrocarbons. (b) Polar solvents lead to radical-ion formation in preference to fluorescent exciplex formation. Another relevant observation is that primary and secondary aromatic amines quench the fluorescence of aromatic hydrocarbons but do not give detectable radical-ion or exciplex formation.

One of the earliest demonstrations of the effect of solvent on the course of reactions of amines with hydrocarbons related to the reduction of anthracene (93)[247, 248] and acenaphthylene[247] by N,N-dimethylaniline and by triethylamine. Reduction only occurs when highly polar solvents, e.g. aceto-

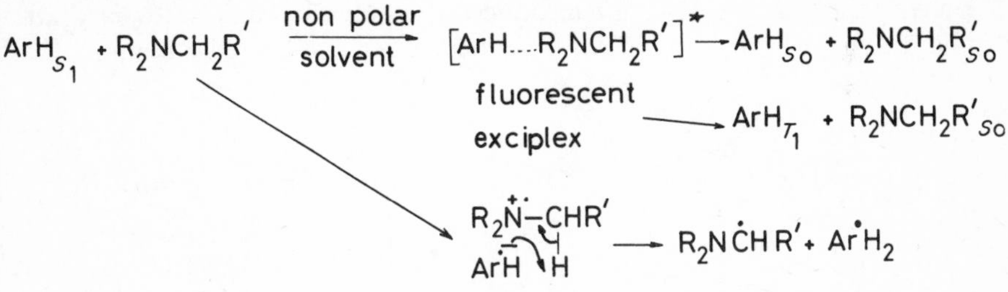

(92)

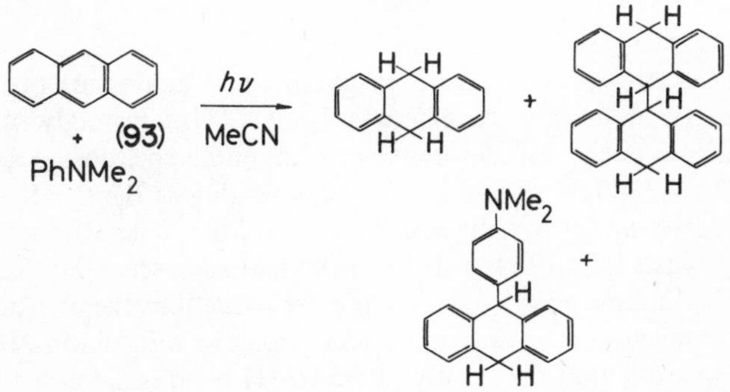

nitrile are employed. In non-polar solvents inefficient photodimerization of the hydrocarbons occurs to the exclusion of reduction.[247, 248, 250] More recently the reduction of naphthalene in polar solvents has been reported.[249] The solvent effects may be readily rationalized by saying that in non-polar solvents exciplex formation occurs which leads to energy wastage whereas use of polar solvents leads to radical-ion formation and ultimately to reduction. For reduction to occur, the radical ion-pair, e.g. (92) has to undergo a proton-transfer reaction to give neutral radicals which may either disproportionate, combine or dimerize. That radical ions are involved in these reactions has been elegantly demonstrated by Pac and Sakurai.[248] They found that addition of D_2O to the hydrocarbon–amine mixtures led to the incorporation of deuterium in the products which was consistent with the intervention of radical-ion intermediates. The quantum yields for photoreduction of a number of aromatic hydrocarbons by aromatic amines has been shown to be higher than for aliphatic tertiary amines.[251] Quite a variety of aromatic hydrocarbons[252] and styrenes[252] have now been reported as being reduced by tertiary amines.

Some of the earliest work on the reduction of hydrocarbons, e.g. benzene, utilized pyrrole as reductant.[253] In this case it was not necessary to use a high polarity solvent. This has been shown to be due to the fact that proton transfer from the N—H bond to the hydrocarbon radical-ion takes place and neutral radicals, e.g. (94), are produced.[254] Primary and secondary aliphatic amines have been shown to reduce benzene in a similar manner.[255]

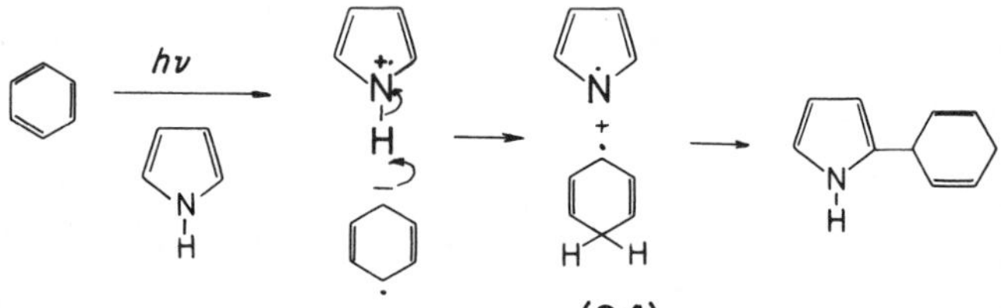

(94)

Naphthalene is also photoreduced by pyrrole and analogous products to those obtained from benzene have been isolated.[256] Interestingly, N-methylpyrrole, which quenches the fluorescence of naphthalene, does not act as a reductant. Presumably, this is due to the low acidity of the C—H bonds of the methyl group in the pyrrole radical-cation. Polycyclic aromatic hydrocarbons are also photoreduced by primary and secondary aromatic amines[250, 257] in non-polar solvents. In these reductions the question arises as to whether reaction is occurring via an exciplex or radical ions. If it is the former, it is possible that the acidity of the N—H bond is increased by some of the excitation energy residing on the amine.

That the formation of reaction products from ion pairs such as (92) is determined by the acidity of suitably placed protons is demonstrated by the reaction of aromatic hydrocarbons with N-arylglycines (95).[258] In these reactions it is found that decarboxylation predominates over reactions involving attack on methylene or N-methyl groups (95, R = Me) or N—H groups (95, R = H), i.e. the acidity of the carboxylic acid proton has completely altered the course of the reaction. Another feature of these reactions is that they take place in non-polar as well as polar solvents. It is possible that the carboxyl group acts as a solvent molecule and aids dissociation of the exciplex to give radical ions which then react in the normal manner. Alternatively the carboxylic acid proton may be sufficiently acidic to force the exciplex to undergo chemical reaction.

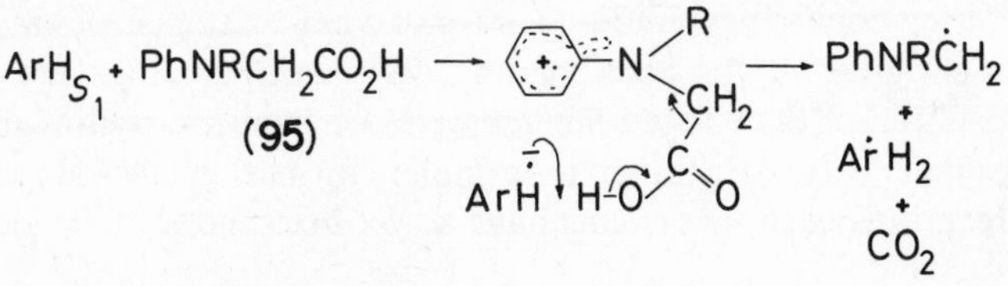

The photoreduction of benzene by tertiary amines is in particular facilitated by the presence of a good proton-donor,[254] e.g. methanol, in the reaction mixture. It was suggested that the methanol served to protonate the aromatic hydrocarbon radical-ion and that the methoxyl anion then abstracted a proton from the amine radical-cation. That hydroxyl groups can participate in the reduction reaction was shown by the fact that the amino-alcohol (96, R = Me) undergoes fragmentation on reaction with excited singlet perylene and anthracene.[259] On the other hand when the compound

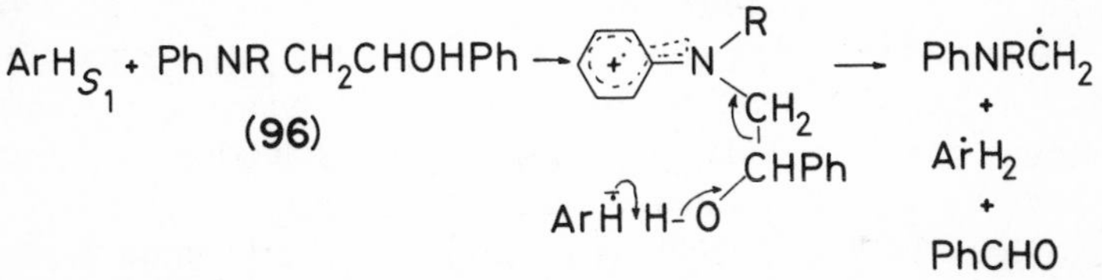

(96, R = H) was utilized no fragmentation was observed. Since the hydrocarbons were consumed in these reactions it was concluded that reaction at the N—H bond had occurred. Thus in these amino-alcohols, the N—H bond is a better proton donor than the O—H group. In the case of the amino-acids, N—H bonds cannot compete with the carboxyl group as a proton donor.

Little work has been done on the photoreactions of aromatic hydrocarbons with ethers. The one outstanding exception is the reported reaction of diethyl ether with benzene.[260] Reaction occurs when a benzene solution of ether containing trifluoroacetic acid is irradiated. It was postulated that reaction occurred via the excited benzeneonium ion. Electron transfer from the ether to this species is suggested as giving (97) which deprotonated to give

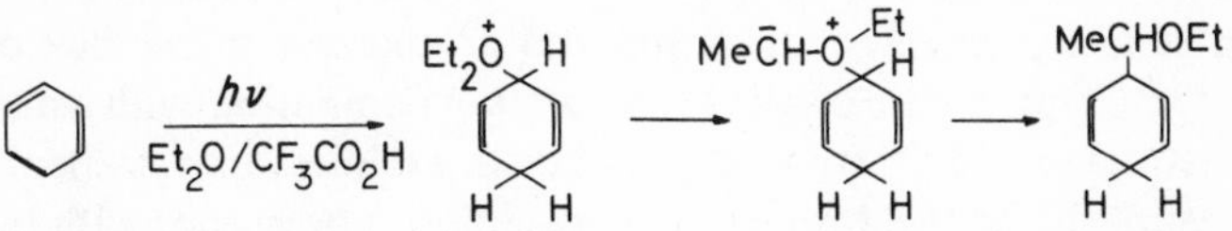

(98). The latter could then undergo an oxonium analogue of the Stevens rearrangement give the observed product, (99).

The quenching of the excited-singlet states of aromatic hydrocarbons by phosphines has been attributed to exciplex formation.[261] However, no chemical reactions of these systems have as yet been reported.

2. Intramolecular exciplex formation and reactions

The formation of intramolecular amine–hydrocarbon exciplexes has been studied for two principal reasons: (a) to examine the conformational require- ments of exciplex formation[262]; (b) to examine its effect on the photoreactivity of the hydrocarbon.[263] Chandross and Thomas,[262] and Brimage and Davidson[263] have reported on exciplex forming properties of (100)–(103). It was found that with n = 1, quenching of the fluorescence of the hydro- carbon occurred but exciplex fluorescence could not be detected. Chandross and Thomas showed that the exciplexes derived from (100) and (101) having n = 3 were more stable than either of the related compounds having n = 2 or n = 4. This is readily understood when one realizes that with n = 3, the lone pair on the nitrogen atom can interact with the π-system of the aromatic ring without causing conformational problems in the linking methylene chain. The importance of the conformational requirements is borne out by the observation that (104, R = Me) exhibits intramolecular exciplex formation even though there are a large number of atoms linking the interacting groups.[264] Some rather unusual observations have been made for com- pounds (105), (106), and (107) which contain an arylamino group as a quencher.[265] Compounds having n = 3 exhibited exciplex formation and this must involve interaction of the aromatic ring of the arylamino group with the polycyclic aromatic hydrocarbon π-system. For (105) having n = 1 and n = 2, exciplex formation was not observable in low polarity solvents but it could be observed in higher polarity solvents. For the compound having n = 1, it is impossible for the two interacting groups to adopt a sandwich configuration. Thus one concludes that for arylamino groups as quenchers exciplex formation can take place even if the interacting groups do not lie parallel to each other. The lack of exciplex emission from (105) n = 1 and 2, in non-polar solvents was attributed to the exciplex energy lying above that of the S_1 state of the hydrocarbon.

An interesting observation relating to (102) n = 2 and (104) R = Me, R' = Ph, is that increasing the polarity of the solvent leads to a decrease in quantum yield for exciplex formation and an increase in the lifetime of the exciplex.[264] Clearly these observations are not compatible with a mechanism which postulates dissociation of a relaxed exciplex into radical ions as solely accounting for the diminution of exciplex fluorescence with increasing solvent polarity. The observation is compatible with a mechanism in which

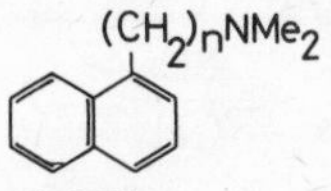

(100) 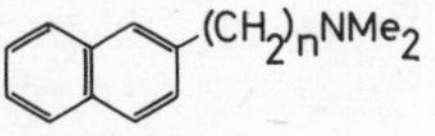   (101) 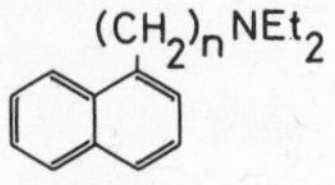   (102)

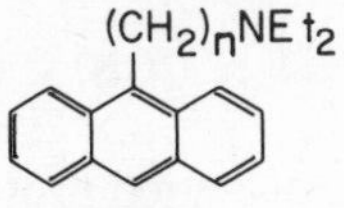

(103) 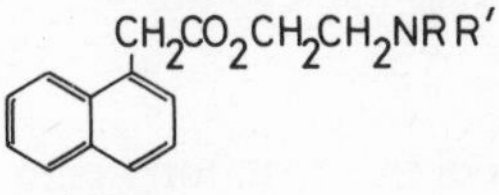   (104)

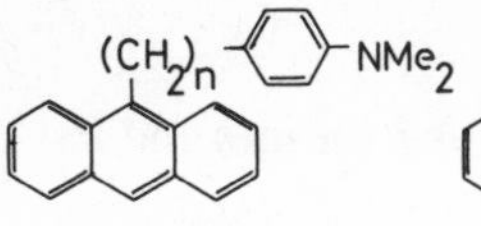

(105) 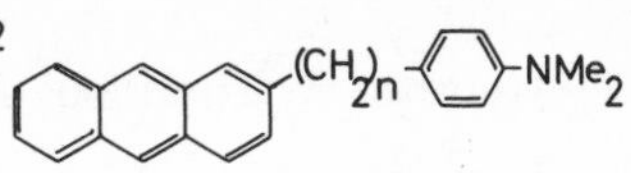   (106)

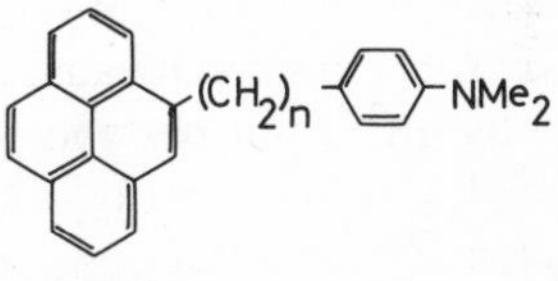

(107)

the polarity of solvent mainly affects the degree of dissociation of the non-relaxed exciplex which is formed on initial interaction of the excited aromatic hydrocarbon with the amine.

The photoreactions of the aromatic hydrocarbons are found to be dramatically affected by intramolecular exciplex formation.[263] In the case of (102), $n = 3$, intramolecular exciplex formation suppressed the intermolecular reaction of the hydrocarbon with amines. Compounds such as naphthalene have been shown to sensitize the photo-oxidation of amines by reacting with the amine to give amine radicals which subsequently react with oxygen to give oxidation products.[251] Compound (102), $n = 3$ was found to be ineffectual as sensitizing this process. In the case of (103), $n = 3$, the amino group suppressed dimerization of the anthracene ring system. It has already been noted that such compounds as 9-methylanthracene dimerize fairly readily but in the case of (103), $n = 3$, quenching of the excited singlet state via the exciplex precludes this reaction. For compounds such as (102), $n = 3$, it was anticipated that in polar solvents intramolecular photocyclization reactions might result via intramolecular radical-ion formation. However, this compound and others proved to be remarkably photostable on irradiation in acetonitrile. Thus if radical-ion or radical formation does take place with these compounds it must be rapidly reversible. 1-Dimethylamino-3-phenyl-propane, however, exhibits intramolecular exciplex formation and also undergoes chemical reaction to give (108) as the major product.[266] Perhaps

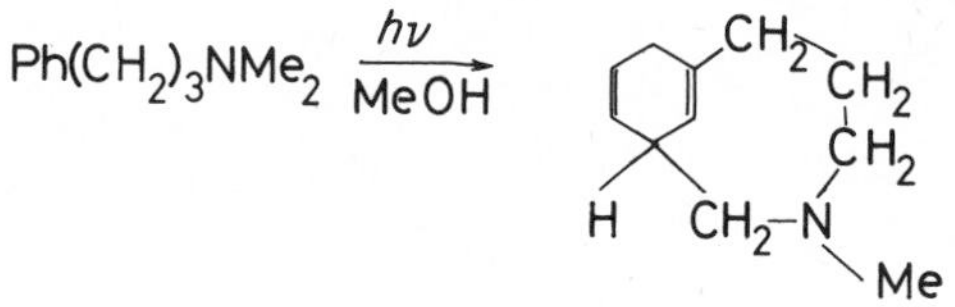

$$Ph(CH_2)_3NMe_2 \xrightarrow[\text{MeOH}]{h\nu}$$

(108)

the success of this reaction lies in the use of a good proton-donor as a solvent. If intramolecular radical-ions are very short-lived because of the rapidity of back electron-transfer to give a neutral molecule a good proton-donor may be able to capture them.

B. Reactions Between Aromatic Hydrocarbons and Purines and Pyrimidines

This topic has received attention because of the possibility that polycyclic aromatic hydrocarbons can cause cancer through their ability to photo-react with the constituent purines and pyrimidines of DNA.

A purine 1,3,7,9-tetramethyluric acid forms exciplexes with pyrene and benz[a]pyrene.[267] In benzene solution the exciplexes exhibit a broad fluorescence band. Benz[a]pyrene has been reported to form products (of unspecified structure) with adenine and guanine.[268]

As yet, exciplex emission from aromatic hydrocarbon–pyrimidine complexes has not been reported. However, products have been isolated from reactions of hydrocarbons such as benz[a]pyrene with thymine[268, 269, 270] and cytosine.[268, 269, 271] The structure of these products has been the subject of some debate. Firm evidence supporting structures (109a) and (109b) for the thymine[270] and cytosine[271] adducts is now available. A thymine-benz[a]pyrene adduct has been isolated from irradiation of benz[a]pyrene in the presence of DNA.[272]

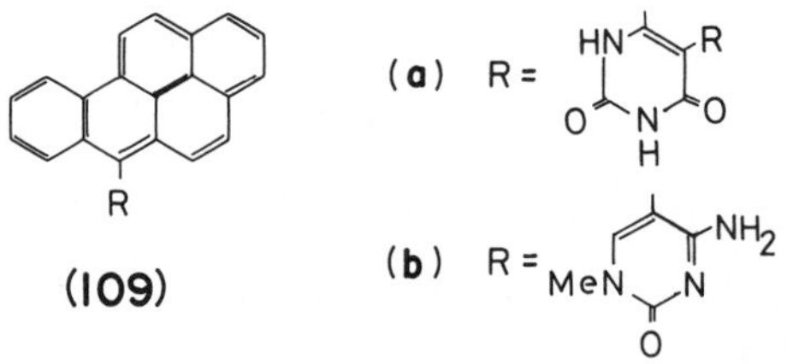

(109)

(a) R =

(b) R =

C. Reactions Between Porphyrins and Aromatic Nitro-compounds. Examples of Triplet Exciplex Formation

The excited singlet and triplet states of pyrochlorophyll[273a, b] and several metalloporphyrins, e.g. zinc etioporphyrin[274, 275, 276] are quenched by aromatic nitro-compounds, e.g. 4-nitrostilbene. Whilst the excited-singlet state complexes have not been detected spectroscopically, the absorption spectra of the triplet metalloporphyrin complexes have been recorded.[275] Surprisingly, the lifetime of the complex increases as the concentration of the nitro-compound is increased until the amount of nitro-compounds affects the dielectric constant of the medium. The increase in lifetime has been suggested as being due to termolecular complex formation

$$^3D^* + A \rightarrow {}^3(DA)^*$$

$$^3(DA)^* + A \rightleftharpoons {}^3(DAA)^*$$

D = porphyrin and A = nitro-compound

Increasing the polarity of the solvent leads to radical ion-pair formation.[276] Rate constants for the quenching of metalloporphyrin triplets by a variety of aromatic nitro-compounds have been measured and as expected there is a linear relationship between the logarithm of the quenching rate constants and the $E_{\frac{1}{2}}$ values for the nitro-compounds.[277] The question of the structure of the complexes has been discussed and it was proposed that there is loose overlap of the molecular orbitals of the components of the complex and that charge-transfer stabilization occurs to only a small degree.[277]

Irradiation of nitrostilbenes in the presence of metalloporphyrins leads to *cis–trans* isomerization of the olefin.[274] This occurs most efficiently when triplet exciplexes are intermediates. A quantitative study of this isomerization reaction showed that little intersystem crossing occurs between the singlet and triplet complexes.

Pyrochlorophyll has been shown to sensitize the reduction of aromatic nitro-compounds by hydrazobenzene.[273b] Because of the low concentrations of reactants employed in these reactions it could be concluded that the triplet state of the pyrochlorophyll was responsible for reaction. Electron transfer from this excited state to the nitro-compound was suggested and the reduction then took the course shown in the following equations:

$$^3PC + ArNO_2 \rightarrow P\overset{+}{\overset{\bullet}{C}}{}^+ + Ar\overset{\bar{}}{N}\overset{\bar{}}{O}_2$$

$$P\overset{+\bullet}{C} + PhNHNHPh \rightarrow PC + Ph\overset{+\bullet}{N}HNHPh$$

$$Ph\overset{+\bullet}{N}HNHPh + Ar\overset{\bar{}}{N}\overset{\bar{}}{O}_2 \rightarrow Products$$

PC = Pyrochlorophyll

D. Intermolecular Photoreactions of Carbonyl Compounds with Amines

1. Products of the reactions

Amines have been shown to be powerful reducing agents for excited carbonyl

K

compounds.[278] Many carbonyl compounds, e.g. xanthone,[278] *p*-aminobenzophenone,[279] and fluorenone,[278, 280] which are not reduced by the good hydrogen-donor propan-2-ol, are reduced by amines. When organic solvents are used for these reactions the ketone is usually reduced to a pinacol.

$$\mathrm{Ar_2CO + R_2NCH_3 \xrightarrow{h\nu} Ar_2\dot{C}OH + R_2N\dot{C}H_2}$$

$$\mathrm{2Ar_2\dot{C}OH \rightarrow \begin{array}{c} Ar_2COH \\ | \\ Ar_2COH \end{array}}$$

If primary aliphatic amines are employed, the amino radical produced in the photoreaction reduces a ground-state molecule of the ketone.[281]

$$\mathrm{Ar_2CO + H_2N\dot{C}H_2 \rightarrow Ar_2\dot{C}OH + CH_2{=}NH}$$

The imine formed in this reaction may then react with a further molecule of the amine:

$$\mathrm{CH_3NH_2 + CH_2{=}NH \rightarrow CH_3N{=}CH_2 + NH_3}$$

Examples have been found in which the amino-alkyl radicals derived from tertiary amines combine with a ketyl radical, e.g. in the photoreduction of fluorenone by triethylamine.[280]

$$\mathrm{Ar_2\dot{C}OH + R_2N\dot{C}HCH_3 \rightarrow Ar_2C(OH)CH(NR_2)CH_3}$$

Radical coupling reactions have been observed in a number of reactions where aromatic amines, e.g. **(110)**, have been used.[282]

$$\mathrm{\underset{\textbf{(110)}}{Ph_2CO + Ph_2NMe} \xrightarrow{h\nu} \begin{array}{c} Ph_2COH \\ | \\ Ph_2COH \end{array} + \begin{array}{c} Ph_2NCH_2C(OH)Ph_2 + \\ Ph_2NCH_2CH_2NPh_2 \end{array}}$$

If aqueous solutions of aliphatic amines are used, the carbonyl compounds are often reduced to hydrols as well as pinacols.[283, 284] The reason for the formation of the former product under these conditions lies in the fact that aqueous solutions of aliphatic amines are sufficiently basic to cause ionization of the intermediate ketyl radicals.

$$\mathrm{Ar_2\dot{C}OH \xrightarrow{Am} Ar_2\dot{C}O^- + HAm^+}$$

$$\mathrm{Ar_2\dot{C}O^- + Ar_2\dot{C}OH \longrightarrow Ar_2CO + Ar_2\bar{C}OH}$$

$$\mathrm{Ar_2\bar{C}OH + HAm^+ \rightarrow Ar_2CHOH + Am}$$

Am = amine

The formation of radical anions in these systems has been detected by esr spectroscopy.[284]

Usually, reaction with amines bearing different alkyl groups occurs at the least substituted group. N-Methyl groups are particularly prone to attack and usually this results in demethylation. For reaction in aqueous solution it was proposed that demethylation involves an electron-transfer reaction.[283]

$$Ar_2CO + \dot{C}H_2NR_2 \rightarrow Ar_2\dot{C}O^- + CH_2{=}\overset{+}{N}R_2$$

Norman has shown, by esr studies, that radicals can undergo such redox reactions.[285] Demethylation is also observed in non-polar solvents[282] and evidence has been presented for imine intermediates.[78] Attention has been drawn to the applicability of this photochemical demethylation procedure in syntheses.[78]

2. A suggested mechanism

In order to account for the very high reactivity of amines, Cohen suggested that the reduction reactions occur via an initial electron-transfer process rather than a hydrogen-atom transfer reaction.[281]

$$R_2CO^* + R_2'NCH_2R' \rightarrow R_2\dot{C}O^- + R_2'\overset{+\cdot}{N}CH_2R$$
$$R_2\dot{C}O^- + R_2'\overset{+\cdot}{N}CH_2R' \rightarrow R_2\dot{C}OH + R_2'\dot{N}CHR'$$

If this mechanism is correct, the rate-determining step will probably be the electron-transfer rather than the proton-transfer process. Thus the efficiency of these reactions should be governed by the reduction potential of the excited ketone and the oxidation potential of the amine. In the light of the aromatic hydrocarbon–amine work we may question what species is or are produced by the electron-transfer process. If highly polar solvents are used for the reaction and the energetics are correct, radical-ion formation may take place. In less polar solvents an intermediate having a degree of charge-transfer and probably a sharing of electronic energy between the reaction partners is produced.

$$R_2CO^* + R_2'NCH_2R' \rightarrow [R_2\overset{\delta-}{C}OR_2'\overset{\delta+}{N}CH_2R']^*$$

If either radical-ion formation or excited-complex formation takes place one may anticipate that energy wastage may occur via the intermediates regenerating the reactants in their ground state.

3. Evidence for the "electron-transfer" mechanism

(a) Detection of radical-ion formation   The feasibility of detecting radical ions in these reactions is determined by (a) the extent to which electron transfer actually occurs in the reaction and (b) the ease with which the proton-transfer and the reverse electron-transfer reaction occur. With regard to point (a) there is little hard experimental evidence. Calculations on the energetics of the triplet acetone–tertiary amine reactions suggest that complete electron-transfer should not occur except when highly polar solvents, e.g. water, are used.[33] However, there is also little experimental evidence to support the supposition that excited carbonyl groups form excited complexes with amines. Exciplex emission has not been observed for such systems except for certain rather special cases, e.g. in the chemiluminescent reaction between triarylamine radical cations and benzophenone radical anion,[79] and from a tetrahydrofuran matrix at low temperature containing benzophenone and a vast excess of tertiary amine.[286] More experimental information is available on point (b). Examination of a number of carbonyl compound–amine reactions by esr spectroscopy[86, 287] has led to the positive identification of protonated ketyl radicals in these systems, e.g.

$$Ph_2CO + PhNMe_2 \xrightarrow{h_\nu} Ph_2\dot{C}OH$$

This result indicates that the proton-transfer reaction is occurring relatively efficiently in these systems.

One approach which was adopted to facilitate detectable radical-ion formation was to utilize triarylamines as the electron donors, i.e. to use amines which lack an abstractable α-hydrogen atom.[86] Examination of the reaction of benzophenone with triphenylamine and tri-p-tolylamine in acetonitrile solution by the technique of flash photolysis revealed the intermediacy of the amine radical-cation and the ketone radical-anion. When non-polar solvents were utilized these species could not be detected thus emphasizing the important role that the solvent plays in these electron-transfer reactions. More recently it has been shown, by the technique of flash photolysis, that fluorenone reacts with N,N-dimethylaniline in acetonitrile to give radical ions.[288] Presumably the weakly basic character of the fluorenone radical-anion slows down the proton-transfer step so as to allow the geminate radical-ions to diffuse apart. Radical-ion formation has also been recently detected in the reaction of 4,4'-dimethylbenzophenone with 1,4-diazabicyclo-[2.2.2]octane (DABCO) by the use of chemically induced dynamic nuclear polarization (CIDNP).[289] This appears to be a particularly powerful method for detecting radical-ion formation and will no doubt find extensive use for this purpose in the future. Chemically induced electron polarization studies of some amine–carbonyl compound reactions have been made and the results interpreted in terms of radical-ion intermediates.[290]

(b) *Energy wastage.* Tertiary amines have been shown to be quenchers of the excited-singlet states of biacetyl,[261, 291, 292] fluorenone,[261, 293, 294, 295, 296] thioxanthone,[297] and some anthraquinone derivatives.[298] In the case of the fluorenone system, very little chemical reaction appears to take place.

This observed physical quenching process cannot be accounted for by a classical energy-transfer process since the energies of the first-excited singlet states of the amines lie above those of the ketone. The electron-transfer mechanism offers a ready explanation for the energy dissipation but it does beg the question as to why chemical reaction does not take place. A possible explanation is that in the geminate radical ion-pair the radicals will be in the singlet state and therefore extremely well set up for participation in the reverse electron-transfer reaction. Under these circumstances it may be impossible for chemical reaction to compete.

The triplet states of a number of aromatic ketones, e.g. benzophenone,[299] fluorenone,[32b, 293] acetone,[33] and biacetyl,[291, 292] are quenched by amines. In many cases this quenching is accompanied by chemical reaction. Thus, triplet fluorenone is photoreduced by a number of para-substituted dimethylanilines.[32b] However, the efficiency of the photoreductions, as judged by the quantum yield for the reactions, does not match the efficiency of the amines in quenching the triplet ketone. This result may be rationalized by postulating that the amine and the ketone react to give a complex which may either result in quenching or in chemical reaction. Many primary and secondary aromatic amines act as quenchers with little chemical reaction taking place at all.[43, 293] However, in these systems one has to be careful to ensure that the lack of observation of chemical products is not due to the reversibility of the photochemical reaction[43] (Section B.2d). A rather interesting amine is 1,4-diazabicyclo[2.2.2]octane (DABCO) because the behaviour of this amine with a number of triplet ketones is totally unpredictable. For some, e.g. 4-amino-benzophenone, it acts as a reductant[279] and for others, e.g. 2-acetylnaphthalene,[300] it acts as a quencher. Presumably an intermediate exciplex is responsible for these observations but the factors determining its mode of decomposition are as yet unknown. Triarylamines act as quenchers for a number of triplet ketones.[86, 291, 292] The increased efficiency of tri-p-tolylamine as a quencher for triplet benzophenone on going from a non-polar to a polar solvent, demonstrates how an increase in the polarity of the solvent can decrease the energetic requirements of the electron-transfer process. A variety of aliphatic amines act as quenchers for triplet ketones and usually this is accompanied by chemical reaction.[299] Data on the rate constants for quenching triplet ketones are shown in Table I.

(c) *Relation between rate constants for quenching excited ketones and redox data*   Cohen and co-workers have determined rate constants for the reaction of benzophenone,[283, 299] 4-benzoylbenzoic acid,[283, 201] and fluorenone[32b] with a variety of amines. These values combined with those in Table II have been used to study the relationship between the efficiency of reaction of the carbonyl compounds with amines and the reduction potential of the

TABLE I. Rate constants (k_q) for quenching triplet ketones by amines.

Ketone	Amine	Solvent	$k_q/\text{mol}^{-1}\,\text{s}^{-1}$	Ref.
Benzophenone	2-Aminobutane	Benzene	2.5×10^8	299
Benzophenone	2-Amino-2,2-dimethylethane	Benzene	7.0×10^7	299
Benzophenone	N-Methyl-2-aminobutane	Benzene	1.4×10^9	299
Benzophenone	Triethylamine	Benzene	2.3×10^9	299
Benzophenone	4-Cyano-N,N-dimethylaniline	Benzene	2.0×10^9	299
Benzophenone	N,N-Dimethylaniline	Benzene	2.7×10^9	299
Benzophenone	4-Methyl-N,N-dimethyl-aniline	Benzene	4.3×10^9	299
Fluorenone	4-Cyano-N,N-dimethylaniline	Benzene	5.4×10^6	32b
Fluorenone	4-Carbethoxy-N,N-dimethyl-aniline	Benzene	4.7×10^7	32b
Fluorenone	4-Bromo-N,N-dimethylaniline	Benzene	3.3×10^8	32b
Fluorenone	4-Chloro-N,N-dimethyl-aniline	Benzene	2.8×10^8	32b
Fluorenone	N,N-Dimethylaniline	Benzene	6.0×10^8	32b
Fluorenone	4-Methyl-N,N-dimethyl-aniline	Benzene	1.6×10^9	32b
Fluorenone	4-Thiomethyl-N,N-dimethyl-aniline	Benzene	6.7×10^9	32b
Fluorenone	4-Ethoxy-N,N-dimethyl-aniline	Benzene	9.6×10^9	32b
Fluorenone	4-Acetamido-N,N-dimethyl-aniline	Benzene	4.1×10^9	32b
Fluorenone	4-Dimethylamino-N,N-dimethylaniline	Benzene	1.6×10^{10}	32b
Acetone	n-Hexylamine	n-Hexane	1.8×10^7	33
Acetone	n-Hexylamine	Acetonitrile	1.3×10^7	33
Acetone	n-Hexylamine	t-Butanol	1.0×10^7	33
Acetone	n-Hexylamine	Water	1.0×10^7	33
Acetone	2-Amino-2-methylpropane	Water	1.9×10^6	33
Acetone	Triethylamine	Water	2.7×10^8	33
Acetone	Triethylamine	Acetonitrile	3.9×10^8	33
Acetone	Triethylamine	t-Butanol	3.3×10^8	33
Acetone	N,N-Dimethylaniline	Water	2.5×10^9	33
Acetone	N,N-Dimethylaniline	n-Hexane	2.0×10^9	33
Biacetyl	i-Propylamine	Benzene	2.8×10^7	292
Biacetyl	Diethylamine	Benzene	2.2×10^7	292
Biacetyl	Di-i-propylamine	Benzene	2.5×10^7	292
Biacetyl	Triethylamine	Benzene	5.0×10^7	292
Biacetyl	Triethylamine	Acetonitrile	2.7×10^8	292
Biacetyl	Tri-n-propylamine	Benzene	8.0×10^7	292
Biacetyl	1,4-Diazabicyclo[2,2,2]-octane	Benzene	5.4×10^7	292
Biacetyl	Aniline	Benzene	5.0×10^8	292
Biacetyl	Aniline	Acetonitrile	1.5×10^9	292
Biacetyl	Diphenylamine	Benzene	1.7×10^9	292

TABLE I.—*continued*

Ketone	Amine	Solvent	$k_q/\text{mol}^{-1}\,\text{s}^{-1}$	Ref.
Biacetyl	Diphenylamine	Acetonitrile	2.8×10^9	292
Biacetyl	Triphenylamine	Benzene	3.1×10^7	292
Biacetyl	Triphenylamine	Acetonitrile	1.6×10^9	292

TABLE II. Rate constants (k_q) for reaction of triplet ketones with amines.

Ketone	Amine	Solvent	$k_q/\text{mol}^{-1}\,\text{s}^{-1}$	Ref.
Benzophenone	2-Aminobutane	Benzene	2.3×10^8	283
Benzophenone	2-Aminobutane	Pyridine–water	1.7×10^7	283
Benzophenone	2-Aminobutane	Acetonitrile	3.8×10^8	283
4-Benzoyl-benzoate	2-Aminobutane	Aq i-propanol	6.3×10^7	283
4-Benzoyl-benzoate	N-Methyl-2-aminobutane	Aq i-propanol	1.6×10^8	283
4-Benzoyl-benzoate	N,N-Dimethyl-2-aminobutane	Aq i-propanol	9.3×10^8	283
4-Benzoyl-benzoate	Triethylamine	Aq i-propanol	6.0×10^8	283
Acetophenone	α-Methylbenzylamine	Benzene	7×10^7	305
4-Aminobenzo-phenone	Triethylamine	Cyclohexane	4.2×10^7	308
Fluorenone	Triethylamine	Cyclohexane	1.7×10^7	280
Fluorenone	N-Methyl-2-aminobutane	Cyclohexane	2.2×10^6	280
Fluorenone	N-Methyl-N-ethylaniline	Cyclohexane	2.8×10^8	78
Fluorenone	N-Methyl-N-isopropylaniline	Cyclohexane	4.8×10^8	78
Fluorenone	N,N-Diethylaniline	Cyclohexane	4.4×10^8	78
Fluorenone	N,N-Di-n-propylaniline	Cyclohexane	2.6×10^8	78
Fluorenone	N,N-Di-i-propylaniline	Cyclohexane	2.3×10^8	78
2-Naphthalde-hyde	Triethylamine	Acetonitrile	8×10^6	300
2-Naphthalde-hyde	Triethylamine	Benzene	5×10^5	300
2-Acetyl naphthalene	Triethylamine	Acetonitrile	6×10^5	300
2-Acetyl-naphthalene	Triethylamine	Benzene	1×10^5	300

carbonyl compound and the oxidation potential of the amine.[32a, 34] An inverse linear relationship should be obtained between the logarithm of rate constant for reaction and the relevant redox potentials if a charge-transfer

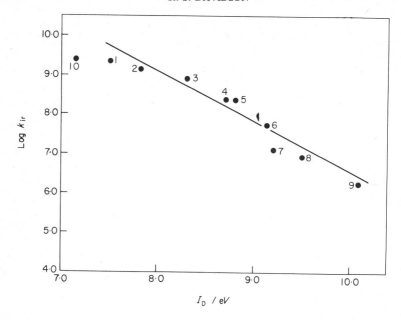

FIG. 6. Relation of rate constant for quenching of benzophenone triplet, k_{ir}, to ionization potential of donor, I_D. Donors: (1) triethylamine; (2) N-methyl-2-butylamine; (3) di-n-butyl sulphide; (4) 2-butylamine; (5) 3-methyl-2-pentene; (6) cis-2-butene; (7) n-propylmercaptan; (8) di-n-propyl ether; (9) propan-2-ol; (10) N,N-dimethylaniline. [Reproduced with permission from J. B. Guttenplan and S. G. Cohen, *J. Amer. Chem. Soc.*, **94**, 4040 (1972)].

process is responsible for the interaction (Section II.B.2b). Figure 6 shows that this is the case. The slope of the line is much lower than one would anticipate if the quenching and/or chemical reaction occurred solely by an electron-transfer process. It was therefore concluded that in these reactions only partial electron-transfer was occurring and that hydrogen-atom abstraction was contributing to the process.[78] The relative contributions of the two processes differ for the aliphatic and aromatic amines. One can perhaps understand the duality of the mechanism if one considers that the reaction involves an excited complex having only partial charge-transfer character.

$$Ar_2CO^* + R_2NCH_3 \underset{k_{-E}}{\overset{k_E}{\rightleftharpoons}} [Ar_2\overset{\delta^-}{CO} \; R_2\overset{\delta^+}{NCH_3}]^*$$

$$[Ar_2\overset{\delta^-}{CO} \; R_2\overset{\delta^+}{NCH_3}]^* \overset{k_r}{\longrightarrow} Ar_2\dot{C}OH + R_2N\dot{C}H_2$$

Such a complex may react (rate constant k_r) by a hydrogen-atom transfer process. The relative magnitudes of k_E and k_r may be similar and consequently both will play an important part in determining the overall efficiency of the quenching process and chemical reaction. Yip and co-workers who have studied the reaction of triplet acetone with amines have come to the conclusion that in these reactions also, complete electron-transfer does not occur.[33] Fluorenone, which has a considerably lower triplet energy than benzophenone does appear to react with amines by an electron-transfer

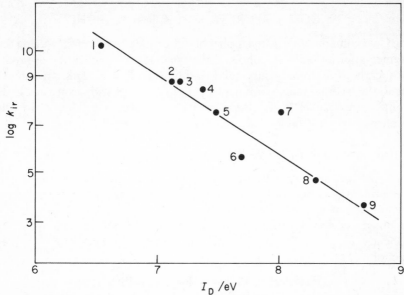

FIG. 7. Log k_{ir} for quenching of fluorenone triplet vs ionization potential of donor. (1) N,N,N',N'-Tetramethyl-p-phenylenediamine; (2) N,N-Dimethylaniline; (3) 1,4-Diaza(2.2.2)bicyclooctane; (4) p-Chloro-N,N-dimethylaniline; (5) Triethylamine; (6) 1-Aza(2,2,2)bicyclooctane; (7) N-Methyl-2-butylamine; (8) Di-n-butyl sulphide; (9) 2-Butylamine. [Reproduced with permission from J. B. Guttenplan and S. G. Cohen, *Tetrahedron Letters*, 2163 (1972)].

process as judged by the slope of the line in Fig. 7. Although there is less electronic energy available in these reactions, the lower reduction potential of fluorenone makes up for the deficiency.

(d) *Chemical evidence for the electron-transfer process.* In the section concerned with reduction of aromatic hydrocarbons by amines (Section VII.A) it was pointed out that for reduction to occur, the initially formed radical-ions had to undergo a proton-transfer reaction. Furthermore, the acidity of the proton transferred plays an important part in determining the course of the reaction. These findings should also hold for the carbonyl–amine reactions and therefore we would anticipate that carbonyl compounds should sensitize the decarboxylation of acids such as (**95**) and amino-alcohols such as (**96**). Quantitative studies showed that the rate constant for decarboxylation of N-(o-chlorophenyl)glycine sensitized by benzophenone is

K*

very similar to that for quenching of triplet benzophenone by N,N-dimethyl-aniline.[172] This finding is indicative of, although not unequivocal evidence for, there being a common mechanism for the reduction and decarboxylation reactions. The decarboxylation reactions also provided some evidence for the occurrence of the following reaction:[171]

$$Ph_2\dot{C}OH + ArNR\dot{C}H_2 \rightarrow Ph_2CO + ArNRCH_3$$

Carbonyl compounds, e.g. benzophenone, are also effective in sensitizing the fragmentation of amino-alcohols of the type (96).[259] Unlike the aromatic hydrocarbons reaction occurs quite readily when $R = H$. This is no doubt due to the fact that reaction of triplet carbonyl groups with aromatic N—H groups is reversible[43] and therefore its only effect is to slow down the fragmentation reaction by wastage of electronic energy.

4. Photoreaction of particular ketones with amines and related compounds

(a) Benzophenone.   The photoreduction of benzophenone by aliphatic and aromatic amines is perhaps the most well studied amine–carbonyl compound reaction and many of the data have been referred to in previous sections.[78] As noted previously, primary and secondary aromatic amines react in-efficiently to give products due to the reverse hydrogen-atom transfer reaction:[43]

$$Ph_2\dot{C}OH + Ph\dot{N}H \rightarrow Ph_2CO + PhNH_2$$

However, products, e.g. (111) have been isolated from reaction with di-phenylamine in t-butanol solution.[302] Benzophenone also reacts with

(III)

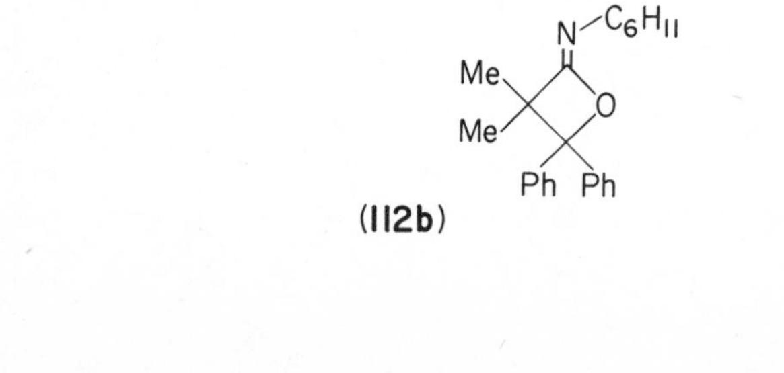

(112a)

(112b)

ketimines to give cyclo-addition compounds, e.g. (**112a** and **b**) and these reactions have been rationalized as occurring via an electron-transfer process.[303]

(b) *o-Hydroxybenzophenone* This ketone is particularly stable with regard to intermolecular reduction reactions. The excited state of the ketone is deactivated by very rapid intramolecular hydrogen-atom transfer from the phenolic hydroxyl group. However, the ketone is reduced by hexamethyl-phosphoric triamide and triethylamine.[304] The reactivity of these reductants was attributed to their ability to bring about the reaction by an electron-transfer rather than a hydrogen-atom transfer process.

(c) *Acetophenone* This ketone is reduced by aliphatic and aromatic amines and quantitative measurements have been made on its reduction by α-methylbenzylamine.[305] Quantum yields of greater than unity were observed and attributed to the occurrence of hydrogen-atom transfer from amino-alkyl radicals to ground-state ketone molecules. Use of an optically active amino-ether, (+)-1,4-bis(dimethylamino)-2,3-dimethoxybutane, as a reductant gave optically active acetophenone pinacol.[306] This appears to be a rare example of a photochemical reaction in which asymmetry has been induced without the use of polarized light.

(d) *Aminobenzophenones and aminoanthraquinones* *p*-Aminobenzophenone is an interesting ketone because the character of its triplet state changes from an $n\pi^*$ to a CT type, as the polarity of the solvent is increased. By use of stilbene as a triplet counter it was shown that increasing the solvent polarity also leads to a decrease in the yield of triplets capable of isomerizing this olefin.[307] One may interpret this finding by saying that either the triplet yield decreases with increase in solvent polarity or by saying that the energy of the CT triplet state is below that of the triplet stilbene. The ketone is photoreduced by a number of amines and the order of reactivity is found to be tertiary > secondary ≫ primary. The reactivity of a particular amine decreases as the polarity of the solvent is increased and this is further evidence for the triplet state of the ketone being the reactive species in the reaction.[308] The lack of reactivity of primary amines may well be due to the ability of these amines to hydrogen bond to the ketone,[16] so facilitating formation of the unreactive CT triplet state. Self-quenching is observed in reactions of *p*-aminobenzophenone[308] and this may be due to the deactivating effect of the amino group present in a ground-state molecule. Supporting evidence for this supposition comes from the fact that *p*-(*N,N*-dimethyl-amino)-benzophenone undergoes photoreduction in inert solvents and it was shown that the *N,N*-dimethylamino-group was acting as a reducing entity.[309]

The structure of a product, isolated from this reaction corroborates this finding. Another observation which demonstrates that the amino group of p-(N,N-dimethylamino)-benzophenone can act as a reducing agent, is that this ketone, in an excited state, reduces benzophenone.[310] p-Aminobenzophenone can act as a sensitizer for the photo-oxidation of amines through its ability to react with amines to give radicals which can subsequently be attacked by oxygen.[148c]

The properties of 1- and 2-piperidino-9,10-anthraquinones are somewhat similar to p-aminobenzophenone. The 2-piperidino derivative is fluorescent and amines quench this emission extremely efficiently.[298]

(e) *Fluorenone* The efficiency of intersystem crossing of this ketone decreases as the solvent polarity is increased.[294, 295, 296, 312] The excited-singlet state of fluorenone is quenched by a variety of amines and the efficiency increases as the ionization potential of the amine decreases. The apparent increase in efficiency of quenching when the solvent polarity is increased is due to an increase in lifetime of the singlet state of the ketone caused by the solvent change.[296] As discussed previously the quenching process in these systems is not accompanied by chemical reaction. The excited-singlet state does react chemically with ketimines to give cyclo-addition products akin to (112).[311]

Triplet fluorenone is reduced by tertiary amines[32b, 278, 280, 284, 313] and less efficiently by secondary and primary amines. The efficiency of photo-reduction decreases as the polarity of the solvent is increased due to the drop off in efficiency of intersystem crossing.[280] Normally the pinacol of fluorenone is produced in the reduction reactions. However, if aqueous or ethanolic solutions of aliphatic amines are used, 9-hydroxyfluorene is also obtained.[284] Since the fluorenone radical-ion has been detected by esr spectroscopy in these systems it is reasonable to propose that this radical is the precursor of the carbinol. Fluorenone is an effective sensitizer for decarboxylation of N-arylglycines[314] and for the fragmentation of the alcohols (96).[259] It also sensitizes the photo-oxidation of a number of amines.[148c]

(f) *2-Acetylnaphthalene and* 1-*Naphthaldehyde* Unlike fluorenone or p-aminobenzophenone, the quantum yield for photoreduction of these carbonyl compounds by triethylamine increases as the solvent polarity is increased.[300] This result is somewhat surprising since the efficiency of intersystem crossing and the nature of the triplet states appears to be hardly affected by the nature of the solvent. This begs the question as to whether the quantum yields reflect either the effect of solvent upon the mode of decay of the exciplex or whether it affects the reactions of the radicals generated in the photochemical reaction. As yet this question has not been resolved. Primary and secondary aliphatic amines are ineffectual reductants for these carbonyl compounds although 2-acetylnaphthalene does photosensitize the photo-oxidation of such amines.[148c] 2-Acetylnaphthalene also photosensitizes the decarboxylation of N-arylglycines.[314]

(g) *Acetone* A number of quantitative measurements have been made on the efficiency of quenching of triplet acetone by amines but as yet the products of the reactions have not been examined.[33, 315]

(h) *Biacetyl* Quantitative studies have been made on the quenching of the fluorescence and phosphorescence of biacetyl by amines.[261, 291, 292] The rate constant for quenching of fluorescence increases as the solvent polarity is increased and this may be visualized as being due to the increase in polarity reducing the energy requirements for the electron-transfer process.[261, 292] Biacetyl, like the other ketones sensitizes the decarboxylation of N-aryl-glycines.[314] Recently biacetyl has been shown to sensitize the fragmentation of such compounds as (113) and an electron-transfer mechanism has been suggested.

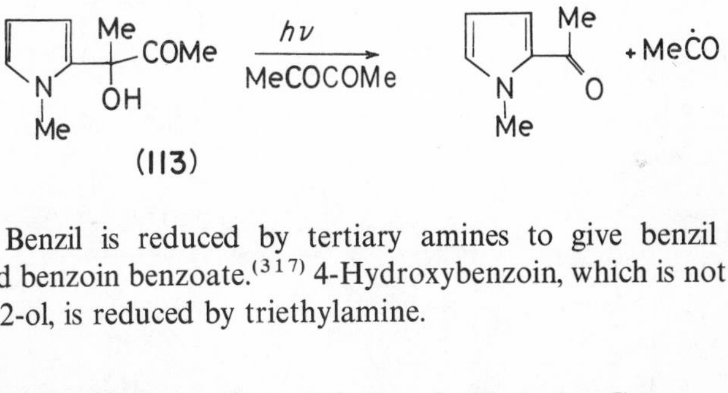

(113)

(i) *Benzil* Benzil is reduced by tertiary amines to give benzil pinacol, benzoin and benzoin benzoate.[317] 4-Hydroxybenzoin, which is not reduced by propan-2-ol, is reduced by triethylamine.

E. Intramolecular Photoreactions of Carbonyl with Amino Groups

1. Product studies

Interest in the photochemistry of acyclic aminoalkyl ketones has centred around the application of the reactions to the synthesis of small ring hetero-cyclic compounds. The earliest report related to a synthesis of azetidinols, e.g. (114)[318] and since then there have been others. The reaction is facilitated by electron-withdrawing substituents being attached to the nitrogen atom[319, 320] e.g. (115). Cyclopropanols, e.g. (117) have been synthesized by irradiation of (116).[321]

The ready reaction of excited carbonyl groups with O—H bonds situated adjacent to a nitrogen atom shows up the reaction of a number of acyl aziridines,[322a, b323] e.g. (117a). In the case of (117b) intermolecular reduction of the carbonyl group by the solvent occurs in preference to intramolecular reaction.[323a] The reason for the unfavourability of the intramolecular hydrogen abstraction reaction is that it requires (117b) to adopt a conformation in which all three substituent groups are situated on the same side of the aziridine ring, i.e. adopt a sterically congested conformation. Reaction of (118) was proposed as occurring via an electron-transfer process since it was

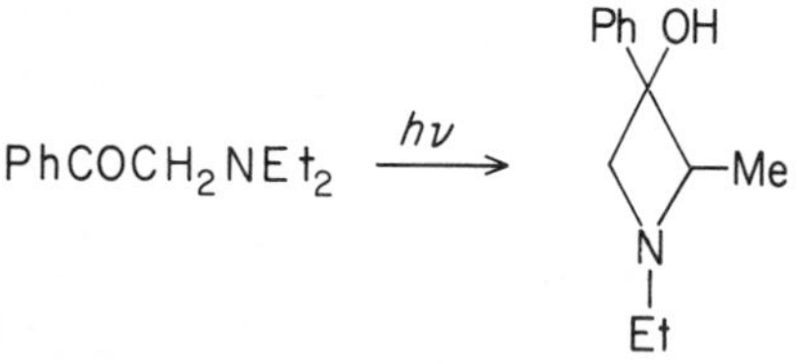

$$PhCOCH_2NEt_2 \xrightarrow{h\nu}$$

(114)

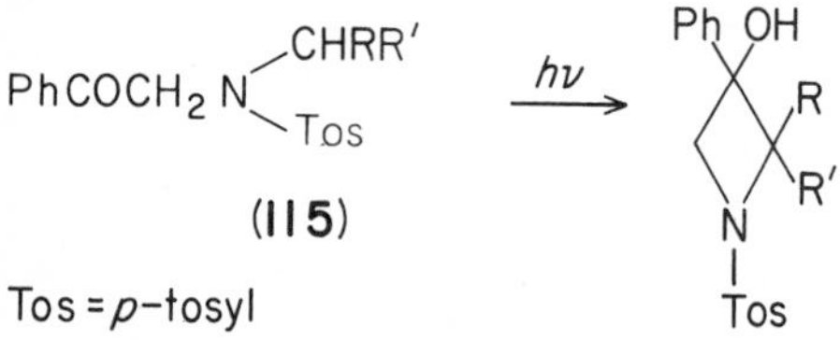

$$PhCOCH_2N\begin{smallmatrix}CHRR'\\Tos\end{smallmatrix} \xrightarrow{h\nu}$$

(115)

Tos = p-tosyl

$$PhCOCHR''\overset{\shortmid}{C}HR'N\!\!\!\diagdown\!\!\!\diagup O \xrightarrow{h\nu}$$

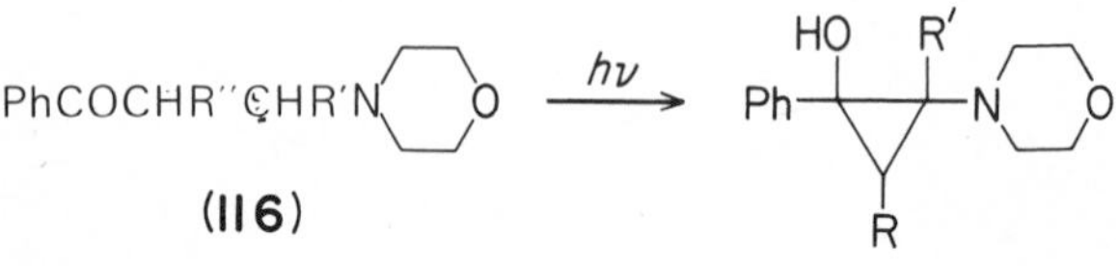

(116)

(117)

shown that the triplet ketone reacts with the amino group with high efficiency and yet observable products are only formed with low quantum efficiency.[324] Intramolecular hydrogen-abstraction reactions have been observed in the photoreactions of acyl azetidines,[323, 325, 326] e.g. (119). In the case of (120) there is competition between the Norrish Type II process, which only involves the acyl group, and reaction of the excited carbonyl group with the ring.[326] It was shown that the Type II reaction occurred from the excited-singlet state whereas attack on the ring occurred from the triplet state of the carbonyl group. Attack on the ring was proposed as occurring by an electron-transfer process. Since such processes are known to be extremely fast it is not surprising that it competes so effectively with the Type II process.

2. Mechanistic studies

The bulk of the mechanistic work has been done on the acylic aminoalkyl ketones.[327, 328, 329, 330] A most significant finding was that in ketones such as (121) the number of methylene groups separating the amino and carbonyl groups has a very great effect upon the efficiency of intersystem crossing. Thus the quantum yield of intersystem crossing in benzene solution for (121)

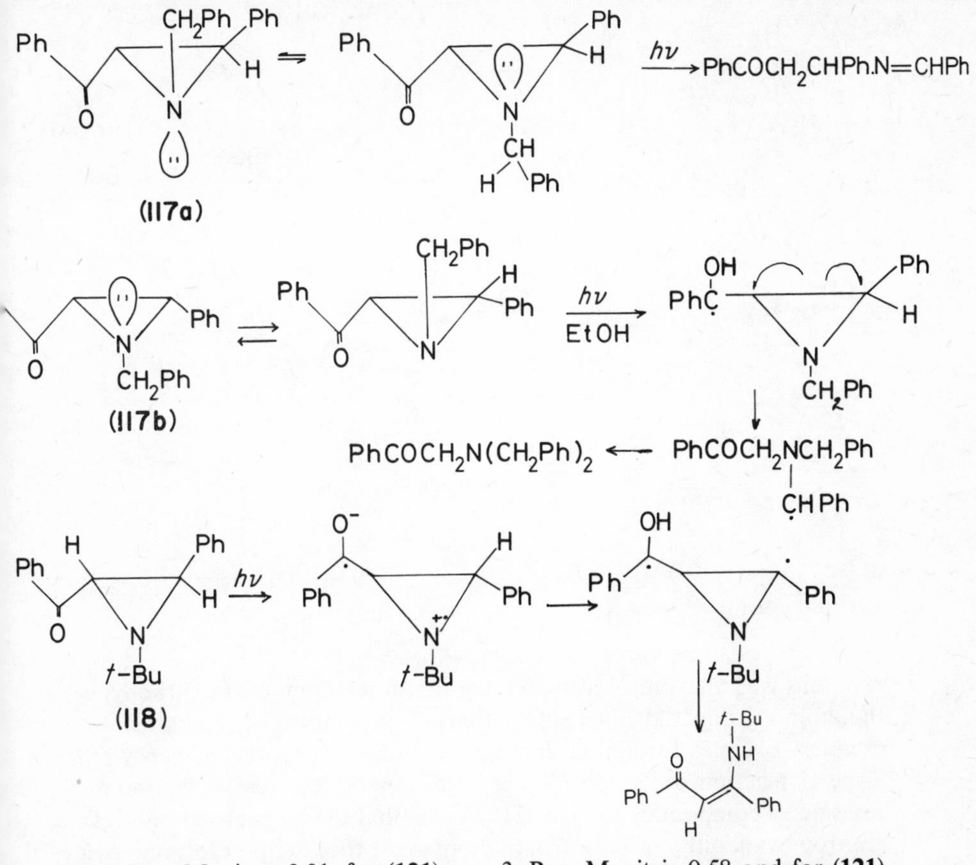

n = 1, R = Me, is <0·01, for (121) n = 3, R = Me, it is 0·58 and for (121) n = 4, amino group–pyrrolidine ring, it is 0·82.[329] The origin of this effect is unknown but there is the possibility that it is a through-bond interaction.

Ketones of the type (121) undergo the Norrish Type II reaction[331] and the detailed knowledge of the mechanistic aspects of this reaction have aided an unravelling of the mechanism of the aminoalkyl ketone reactions.

Detailed kinetic studies on compounds (121) having n = 3 or 4 show that the Type II reaction occurs from the $^3n\pi^*$ state. Only quenching appears to take place from the $^1n\pi^*$-state. Both $^1n\pi^*$ and $^3n\pi^*$-states of the carbonyl group should interact with the nitrogen atom fairly efficiently. It was therefore proposed that the quenching observed from both these states is due to singlet and triplet exciplex formation respectively. However it was not possible to tell whether the Type II reaction occurred from the triplet exciplex or directly from the $^3n\pi^*$-state. The lack of a large solvent effect upon the

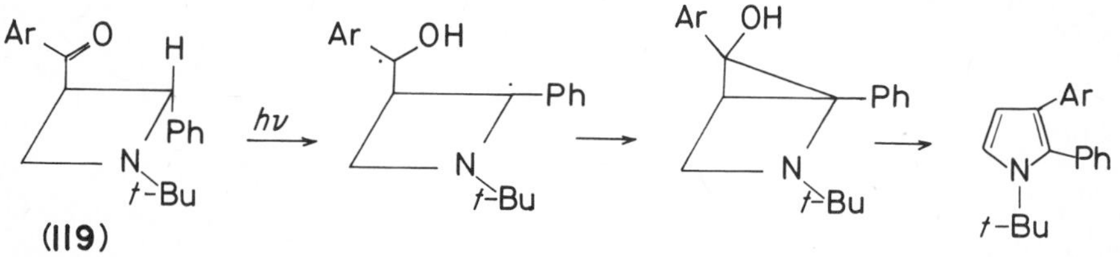

(119)

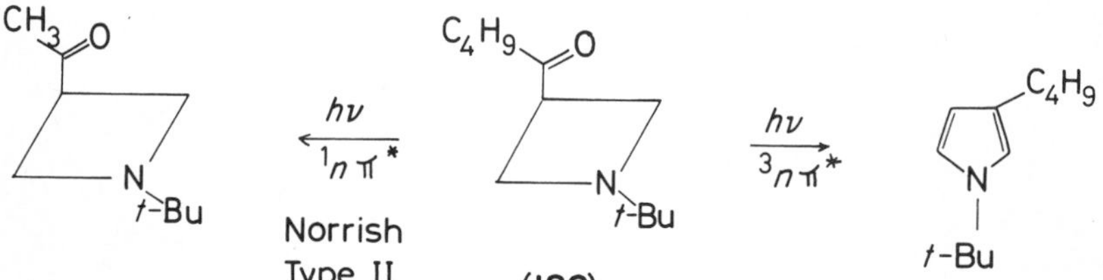

Norrish
Type II   (120)

$$PhCO(CH_2)_nNR_2$$

(121)

$$ArCOCH_2NR_2$$

(122)

reactions was interpreted as indicating that interaction of the carbonyl with the amino group did not lead to total electron-transfer but rather to an exciplex. Exciplex formation in these reactions reduces the efficiency of the Type II reaction quite considerably. Presumably the reason for the ready reaction of compounds such as (115)[321] is that the tosyl group reduces the electron availability of the nitrogen atom and hence exciplex formation does not take place.

α-Aminoalkyl ketones, e.g. (121) n = 1, undergo the Type II reaction but often with low quantum efficiency. Wagner proposes[327] that most of the electronic energy is dissipated in these compounds by radiationless processes which do not include quenching via an exciplex. He considers that the conformational requirements for interaction of the amino with the carbonyl group are energetically too demanding. On the other hand Padwa has observed[330] that compounds such as (122) where Ar = biphenyl or 2-

naphthyl, undergo the Type II reaction and proposes that these occur via an exciplex intermediate. He argues that on the known unreactivity of acyl biphenyls and naphthalenes in hydrogen-abstraction reactions, that the Type II reaction of (122) should not occur. The observed reactivity of these compounds appears to be best explained by an electron-transfer mechanism. The lack of quenching of the reaction of (122) by triplet quenchers suggests that reaction occurs from the $^1n\pi^*$-state. This contrasts with the finding that the $^1n\pi^*$-state of (121) n = 1, fails to undergo chemical reaction.

F. Photoreactions of Aromatic Hydrocarbons and Ketones with Sulphides and Sulphoxides

1. Aromatic hydrocarbons

Diaryl sulphides have been found to quench the fluorescence of polycyclic aromatic hydrocarbons ($k_q \sim 1 - 10 \times 10^9\,\mathrm{l\,mol^{-1}\,s^{-1}}$).[332] Sulphoxides also act as quenchers, e.g. methyl p-chlorophenyl sulphide quenches naphthalene fluorescence ($k_q\,2 \times 10^8\,\mathrm{l\,mol^{-1}\,s^{-1}}$ in acetonitrile as solvent).[333] Since the excited-singlet states of these quenches lie above those of the aromatic hydrocarbons, the quenching cannot be attributed to classical energy-transfer. Exciplex formation was postulated to account for the quenching; the energy of the exciplex being dissipated by the sulphoxide group undergoing inversion. If optically active sulphoxides are used as quenchees they undergo racemization.[333, 334] Interestingly, compound (123) n = 2 undergoes efficient intramolecular photosensitized racemization

(CH$_2$)$_n$SO C$_6$H$_4$CH$_3$(p)   MeCHNHAc

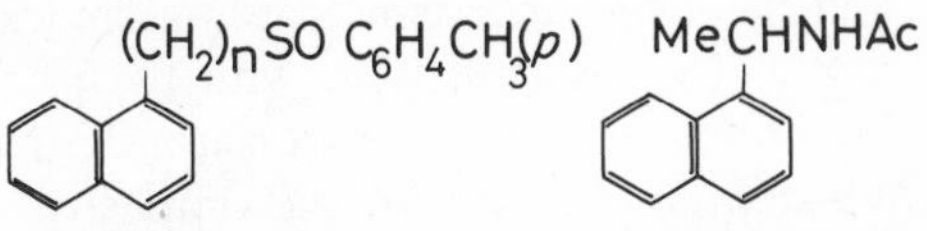

(123)   (124)

presumably via an intramolecular exciplex. Although compound (123) n = 1 undergoes sensitized racemization it also decomposes relatively efficiently. If the singlet state of an optically active aromatic hydrocarbon, e.g. (124) is quenched by a racemic sulphoxide then optical activity in the sulphoxide is induced.[335]

2. Carbonyl compounds

(a) *Intermolecular reaction of ketones with sulphides and thiols*   Sulphides inefficiently quench the fluorescence of biacetyl.[332] They also act as relatively inefficient hydrogen donors for the photoreduction of triplet ketones.[336a, b, c] However, kinetic investigations of these reactions revealed that the triplet carbonyl groups interact with sulphides with quite a high efficiency.[336a, b]

This fact has also been confirmed by studying the efficiency with which sulphides quench the phosphorescence of carbonyl compounds in solution at room temperature.[336b] Thus, once again we meet the state of affairs where the physical process of association of the reaction partners occurs efficiently whereas product formation is inefficient. As stated before, this may be rationalized by postulating an intermediate exciplex which leads to the partitioning of excitation energy between reaction and quenching. This rationalization has been applied to the carbonyl compound–sulphide reactions.

$$Ph_2CO_{T_1} + RCH_2SR \rightarrow exciplex$$

$$exciplex \rightarrow Ph_2CO_{S_0} + RCH_2SR$$

$$exciplex \rightarrow Ph_2\dot{C}OH + R\dot{C}HSR$$

$$Ph_2\dot{C}OH + R\dot{C}HSR \rightarrow products.$$

As might be expected, on the basis of this mechanism, many carbonyl compounds sensitize the decarboxylation of alkyl and arylthioacetic acids.[171, 172]

$$Ph_2CO_{T_1} + RSCH_2CO_2H \rightarrow Ph_2\dot{C}OH + RS\dot{C}H_2 + CO_2$$

$$Ph_2\dot{C}OH + RS\dot{C}H_2 \rightarrow products$$

$$Ph_2\dot{C}OH + RS\dot{C}H_2 \rightarrow Ph_2CO + RSCH_3$$

In these reactions it was very noticeable that the latter reaction occurred relatively efficiently and therefore this begs the question as to the cause of the inefficiency in the photoreduction reactions. Presumably it is a result of both physical (exciplex) and chemical (reverse hydrogen-atom transfer) quenching.

Another observation which is consistent with the proposal that triplet carbonyls react with sulphides via an electron-transfer mechanism is that normally unreactive compounds like 4-acetylbiphenyl are reduced by sulphides.[337] These reactions, which occur relatively efficiently, would not be expected to occur if they involved straightforward hydrogen-atom abstraction process.

Thiols are also relatively good quenchers for triplet ketones such as benzophenone[336b] and acetophenone.[338] Presumably, this quenching results from a combination of exciplex formation and hydrogen abstraction. The values of the rate constants are of particular interest because of the common practice of using thiols as good hydrogen-atom sources for radicals produced in photochemical reactions. Thus they have been used in studies of the Norrish Type II reaction and the decarboxylation reactions just discussed.[171] Incorporation of benzenethiol into these reactions lead to the occurrence of the following reaction

$$RS\dot{C}H_2 + PhSH \rightarrow RSCH_3 + Ph\dot{S}$$

It would appear that the use of thiols as radical scavengers in photochemical reactions of carbonyl compounds does not affect the primary processes provided they, are used in relatively low concentrations.

(b) *Intramolecular interaction of carbonyl with sulphide groups* Keto-sulphides of the type (125) undergo the Norrish Type II reaction.[337] With

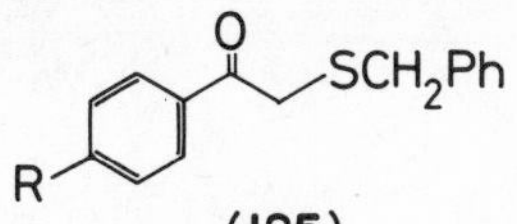

(125)

R = H, quenching studies indicated that the reaction involved a fast reaction of the triplet carbonyl group. However with R = Ph, the reaction was found to be unquenchable by standard triplet quenchers. It was proposed that this latter reaction involved a charge-transfer process. Since the influence of the sulphide group upon the efficiency of intersystem of the carbonyl group has not been determined it cannot be categorically stated as to which excited state is responsible for reaction.

(c) *Reactions of ketones with disulphides* Several aromatic ketones, e.g. benzophenone, have been shown to sensitize the decomposition of di-sulphides, e.g. benzyl[339a] and *t*-butyl disulphide.[339b] With both these compounds the ketones sensitize homolysis of a C—S bond. A wide variety of ketones was found to sensitize the decomposition reaction, e.g. acetophenone, 2-methylanthraquinone and the efficiency of these sensitizers was found not to be related to their triplet energies but to their reduction potentials.[339a] The conclusion was therefore reached that the disulphides were interacting with the excited carbonyl groups by a charge-transfer process.

G. Photoreactions of Unsaturated Carbonyl Compounds and Esters with Amines

Tertiary amines, e.g. triethylamine add to the olefinic double bond of α,β-unsaturated esters and ketones to give derivatives of the type (126).[340a] The reaction appears to be a general one, as judged by the wide variety of reported examples, e.g. (127). An important competing reaction is reduction of the unsaturated carbonyl compound to its saturated counterpart. Similar reactions occur when aromatic amines are utilized, e.g. *N,N*-dimethyl-aniline.[332] Reaction with benzalacetophenone gave (128) to the exclusion of the normal (2 + 2)-cyclo-addition products which are observed on irradia-tion of this ketone. Reduction with 9,10-dihydroacridine gave (129) which is derived by hydrogen abstraction from the carbon rather than the nitrogen atom. All the reactions may be rationalized in terms of an electron-transfer process leading to hydrogen-atom abstraction from the amine followed by

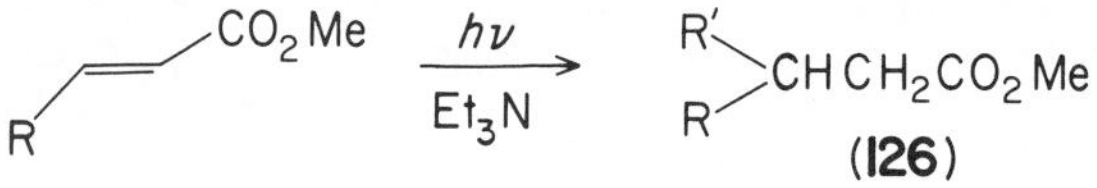

(126)

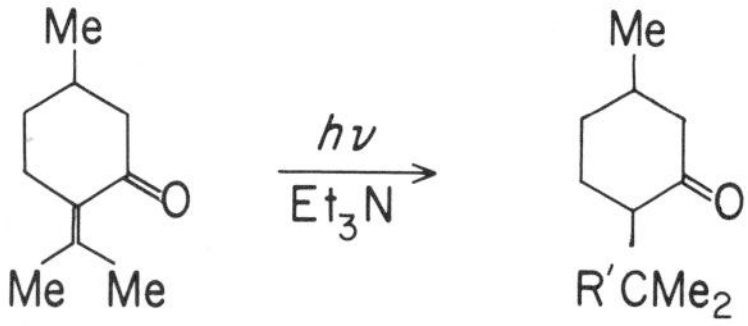

(127)

$R' = Et_2NCHMe$

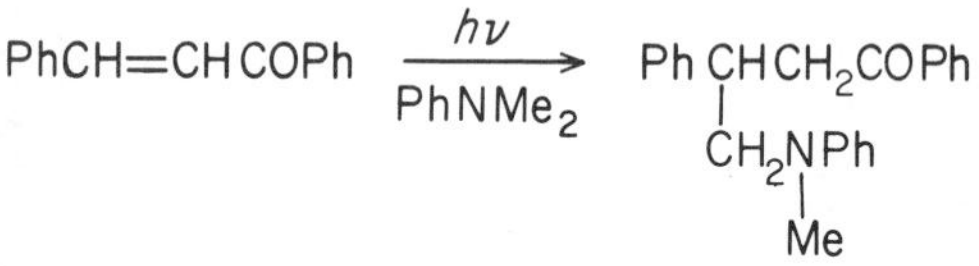

(128)

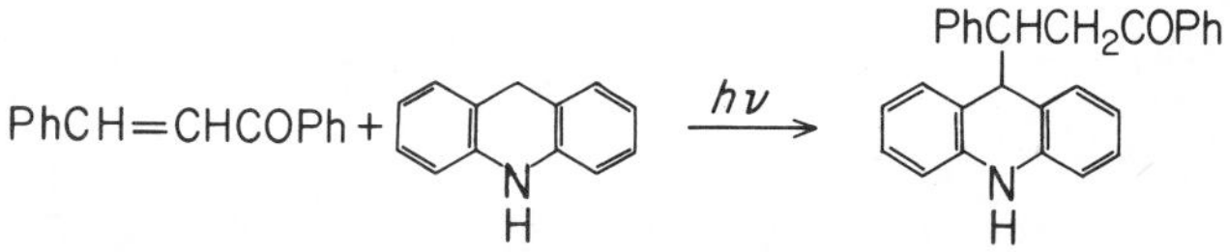

(129)

combination of the radicals so formed. As yet, mechanistic studies on these reactions have not been made.

Aliphatic amines act as reductants for a variety of esters.[340b] The reaction is most effective in the case of benzylic and allylic esters. It would appear that the ester abstract a hydrogen atom from the amine to give a radical such as (130) which then cleaves giving benzoic acid and an alkyl radical. The suggestion was made that the hydrogen-abstraction reaction occurred via an electron-transfer reaction. There is certainly good evidence for the

$$PhCO_2CH_2Ph \xrightarrow[Et_3N]{h\nu} Ph\dot{C}(OH)OCH_2Ph + CH_3\dot{C}HNEt_2$$
(130)

$$Ph\dot{C}(OH)OCH_2Ph \longrightarrow PhCO_2H + Ph\dot{C}H_2$$

$$Ph\dot{C}H_2 + CH_3\dot{C}H NEt_2 \longrightarrow PhCH_2CH(CH_3)NEt_2$$

formation of radical ions in the reaction of aromatic amines with esters.[340c] Thus flush photolysis of a mixture of N,N-dimethyl-β-naphthylamine and dimethyl isophthalate in polar solvents led to the production of identifiable radical-ions. If non polar solvents are utilized a fluorescent exciplex is formed.[340d]

H. Photoreactions of Heterocyclic Compounds with Amines and Amino Acids

1. Benzo(b)thiophene, methyl-3-isothiazole and oxindole

Benzo(b)thiophene has been shown to react with primary and secondary amines on irradiation and adducts of the type (131) are formed.[341] With pyrrole, (132) is produced. These reactions have been rationalized as occurring

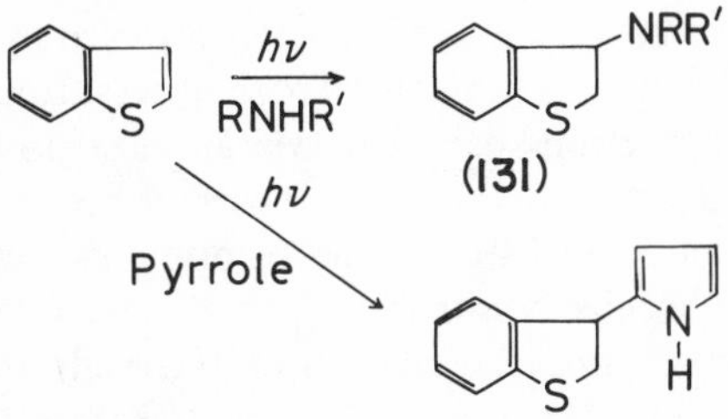

via electron transfer from the amines to the excited thiophene. The radical ions so formed undergo a proton-transfer reaction involving the N—H group of the amine. Which excited state of the thiophene is responsible for reaction has not been delineated. Formation of (132) in the reaction with pyrrole bears some analogy to the reaction of naphthalene with the same amine. Excited 3-methylisothiazoles dehydrogenate primary amines to imines.[342] The initial photoreaction leads to α-amino-alkyl radicals which can donate a hydrogen atom to a ground-state isothiazole molecule so producing the imine. The reaction of the oxindole (133) with diethylamine is reported as occurring via an electron-transfer process.[343]

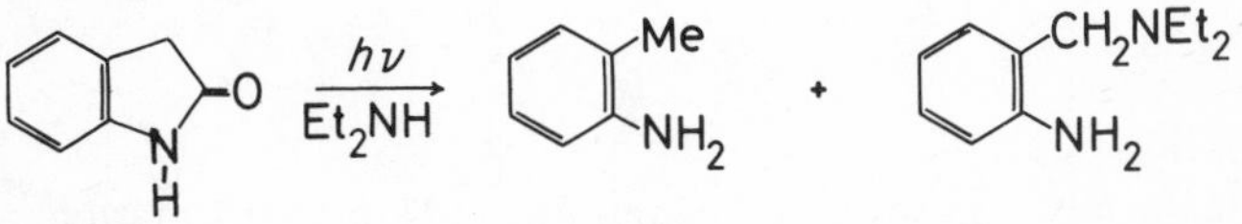

2. Acridine, quinoxalines and cinnolines

Acridine, various quinoxalines and cinnolines are reduced to dehydro compounds by amines. Acridine is reduced by dihydroacridan and several radical intermediates have been identified by esr spectroscopy.[344] It is also reduced by triethylamine to give 9,9',10,10'-tetrahydro-9,9'-bisacridine.[332] A detailed study has been made of the reduction of phenazine by triethylamine which gives 5,10-dihydrophenazine.[345] The reactive state was identified as the $n\pi^*$-singlet state of the heterocycle. A similar study was also made of the reduction of 9,10-diazaphenanthrene and once again the $n\pi^*$-singlet state of the heterocycle was identified as the reactive species.[346] In contrast, the reduction of 2-phenylquinoxaline by triethylamine is proposed as occurring via the $\pi\pi^*$-triplet state.[347] Detailed kinetic studies were made of the reduction of this quinoxaline by triethylamine, di-n-butylamine and n-butylamine. The rate constant for reaction decreases as the ionization potential of the amine increases, and increases for each amine as the solvent is changed from benzene to acetonitrile. These observations were taken to be consistent with the reductions occurring via an electron-transfer process, a mechanism proposed for the reduction of the other heterocyclic compounds.

If in fact the reduction of heterocyclic compounds, such as acridine, by amines does involve an electron-transfer process, we might anticipate that they should sensitize the decarboxylation of N-arylglycines. This has been shown to be the case.[348] Acridine, phenazine and various quinoxalines act as efficient sensitizers for decarboxylation of these acids, and for other acids such as phenylthioacetic acid and phenoxyacetic acid. However, there is the possibility that in these reactions, the acids, initially protonate the singlet $\pi\pi^*$-state of the heterocycle and that this is followed by an electron-transfer process. Fluorescence from the protonated heterocycles was not detected in these reactions and therefore it was presumed that reaction occurred primarily by electron transfer from the amino-acid to the excited heterocycle. This primary reaction is relatively efficient as judges by the observation that the decarboxylation reactions readily occur in the presence of oxygen.

Heterocyclic compounds which efficiently sensitize the decarboxylation of N-arylglycines also sensitize the fragmentation of the amino-alcohols (96).[259] Once again there is the possibility that the primary reaction is one of protonation of the excited heterocycle. However, as with the amino-acids, no emission from the protonated heterocycles could be detected in these systems.

3. Purines

The reduction of purines by alkyl amines leads to the formation of alkylated purines.[349, 350] By use of Cu^{2+} and Ni^{2+} ions as quenchers,[350] it was shown that the triplet state of the purine is the reactive species. The reaction is thought to involve hydrogen abstraction from the amines, possibly by an electron-transfer mechanism, followed by combination of the radicals to give an adduct such as (134). Elimination of ammonia gives the observed

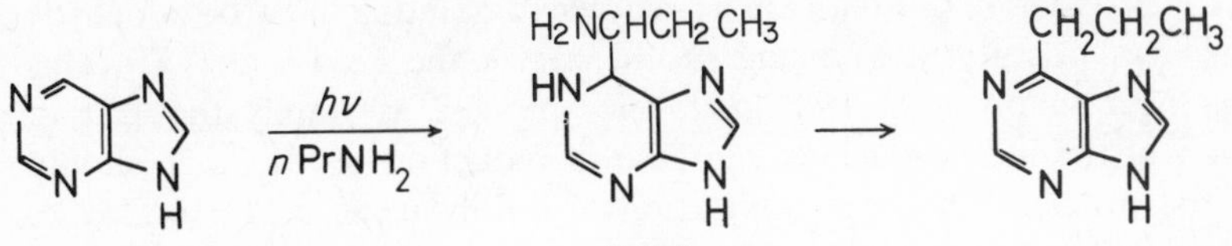

(134)

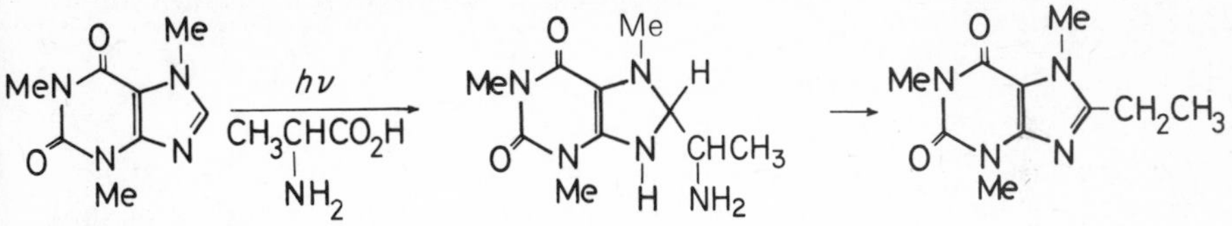

(135)

product. The position of alkylation is dependent upon the purine used. Purines, e.g. (135) also sensitize the decarboxylation of α-amino-acids.[351] The products of these reactions are alkylated purines.

4. Flavines

Flavines, compounds which are derivatives of 7,8-dimethylisoalloxazine, (136) are reduced by amines[352] and sensitize[352, 353a, b)] the decarboxylation of α-amino-acids. From a detailed kinetic study of these reactions, Penzer and Radda concluded that they initially involved electron transfer from the

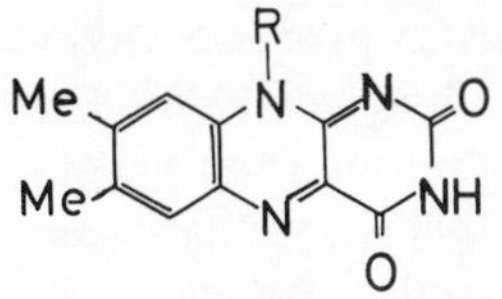

(136) 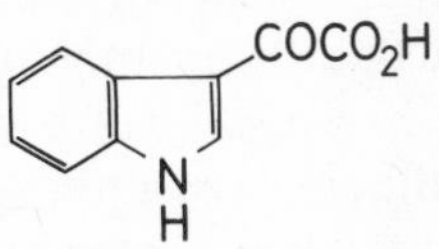   (137)

amino-compound to the triplet flavin. This conclusion seems very reasonable when one considers that flavins in the ground state have low reduction potentials and these are probably even lower for the triplet flavins. Some amino-acids, e.g. phenylalanine have been shown to react with the triplet flavin to give α-keto acids probably via intermediate α-imino-acid.[353a] Thus in these reactions, the primary chemical reaction does not appear to be decarboxylation but rather hydrogen abstraction from the C—H bond α to the carboxyl group by the excited flavin.

$$RCH(NH_2)CO_2H \rightarrow R\dot{C}(NH_2)CO_2H \rightarrow RC{=}NHCO_2H$$

Riboflavin also sensitizes the decarboxylation of indole-3-acetic acid.[354] One of the products of this reaction was shown to be indole-3-carboxaldehyde and this was postulated as being produced via the keto-acid (137). This reaction may involve an electron-transfer process. Certainly the ease of oxidation of the indole nucleus and ease of reduction of the flavin should favour the process. The decarboxylation of indolylacetic acid can also be sensitized by carbonyl compounds and quinoxalines; a point which does rather suggest that the acid is susceptible to undergoing electron-transfer reactions.[355]

5. Chlorophyll, porphyrins and various dyes

Chlorophyll,[356] magnesium and copper phthalocyanine[357a, b] and porphyrins[358] have all been shown to be reduced by amines to dihydro compounds. Electron-transfer mechanisms have been proposed for these reactions. All these compounds have low-energy triplet states but also they all have very low reduction potentials. These points together with the recorded observations strongly support the suggested mechanism. Extensive investigations of the photoreduction of thionine[359a] and thiopyronine[359b] by allylthiourea have been made, utilizing the technique of flash photolysis. These indicated that the reactions, not surprisingly, involve an electron-transfer process. Methylene blue and rose bengal are photoreduced by amines, the efficiency being tertiary > secondary > primary amines.[360] These dyes are reduced by N-arylglycines[361] and in the process the acids are decarboxylated.[348] The effects of various substituents placed in the aryl ring, upon the rate of the reductions can be best rationalized in terms of the reactions occurring by an electron-transfer process. Methylene blue also sensitizes the decarboxylation of N,N-diethylglycine.[362] The mechanism of this reaction may also involve electron donation from the amino-acid to the excited dye although the authors suggest that the primary reaction is electron transfer from the hydroxyl anion to the excited dye.

I. Photoreactions of Carbonyl and Heterocyclic Compounds with Hydrocarbons, Ethers and Sulphides

1. Carbonyl compounds

The physical and chemical processes which attend the interaction of aromatic carbonyl compounds with aromatic hydrocarbons have been extensively investigated because of the common usage of hydrocarbons, such as benzene and its derivatives, as solvents for photochemical reactions. The triplet lifetime of benzophenone is very much shorter in benzene than in the very inert solvent, perfluoromethylcyclohexane.[363a, b, 364a–c] Deactivation of the triplet has been shown to involve in reaction with solvent[363b, 364c, 365] and its reaction with ground-state ketone molecules.[364c, 366] Reaction with the solvent ultimately leads to phenyl radical formation by a hydrogen-abstraction process. However, from the fact that the logarithm of the quenching rate-constant for quenching of triplet ketone by a variety of substituted benzenes increases linearly as the ionization potential of hydrocarbon increases suggests that the initial interaction involves a charge-transfer process[364b] in which transfer-of-charge from the carbonyl group to the hydrocarbon occurs, i.e. transfer occurs in the opposite direction to that normally observed. However, in a related study on the quenching of triplet acetone by aromatic hydrocarbons it was concluded that the quenching resulted from enhanced coupling between the locally-excited triplet ketone and the triplet quencher due to interaction with a triplet charge-transfer state.[366] The quenching of benzophenone triplets is relatively efficient compared with the hydrogen-abstraction process.[363b, 364c] Therefore one concludes that the initially formed complex may collapse to products (radicals) or regenerate starting materials in their ground state. Reaction of triplet benzophenone with ground-state benzophenone molecules is relatively efficient and it is proposed that this interaction gives a triplet excimer.[364c] This may then either collapse into ground-state molecules or give radicals, e.g. the diphenylhydroxymethyl radicals. Rate constants for all the processes have been determined.[364c] An illustration of the effectiveness of the quenching of benzophenone triplets by ground-state ketone molecules compared with reaction with the solvent is the finding that irradiation of benzophenone in hexadeuteriobenzene produces diphenylhydroxymethyl as the only identifiable radical.[367] If reaction had occurred with the solvent the deuterated radical so produced would have been readily identified.

Irradiation of α,α,α-trifluoroacetophenone in benzene containing a proton donor, e.g. methanol or trifluoracetic acid, produces the pinacol of the ketone as well as trifluoromethyldiphenylcarbinol (**138**).[368] It was proposed that in this reaction charge-transfer from the hydrocarbon to the excited ketone occurred which seems very reasonable since the trifluoroacetophenone should be a much better electron-acceptor than benzophenone.

α,α,α-Trifluoroacetophenone also reacts with a number of alkylbenzenes to give reduction products, e.g. (**139**).[136] The kinetics of this reaction have been the subject of a detailed study. Quenching of the triplet ketone by the hydrocarbons increases as the ionization potential of the hydrocarbon decreases. Thus, this reaction is a further example of the trifluoro ketone acting as an electron acceptor. The reduction reactions were suggested as occurring via two simultaneous processes—a charge-transfer and a radical process.

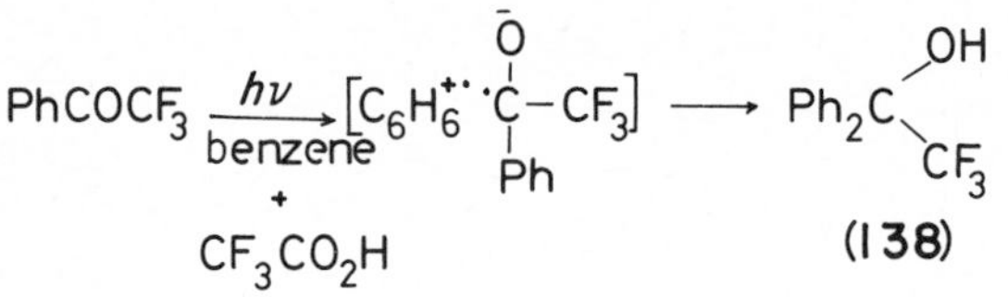

$$PhCOCF_3 \xrightarrow[\text{benzene}]{h\nu} \left[C_6H_6^{+\cdot} \cdot \overset{\overset{\displaystyle O}{|}}{\underset{\underset{\displaystyle Ph}{|}}{C}}-CF_3\right] \longrightarrow Ph_2C\overset{\displaystyle OH}{\underset{\displaystyle CF_3}{\diagdown}}$$

$$+$$

$$CF_3CO_2H \qquad\qquad (138)$$

$$PhCOCF_3 \xrightarrow{h\nu} \left[Ph\overset{\overset{\displaystyle O^-}{|}}{\underset{\displaystyle \cdot}{C}}-CF_3 \quad PhCH_3^{+\cdot}\right]$$

$$+$$

$$PhCH_3 \qquad\qquad \downarrow \text{ proton transfer}$$

$$Ph\overset{\overset{\displaystyle OH}{|}}{\underset{\displaystyle \cdot}{C}}-CF_3 + Ph\overset{\cdot}{C}H_2 \longrightarrow Ph\overset{\overset{\displaystyle OH}{|}}{\underset{\underset{\displaystyle CF_3}{|}}{C}}-CH_2Ph$$

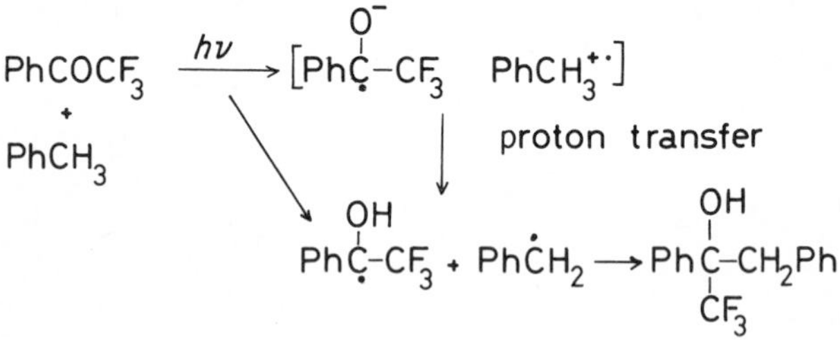

$$(139)$$

Alkylbenzenes and alkoxybenzenes also appear to act as electron donors for excited benzophenone and biacetyl.[292] Alkoxybenzenes quench biacetyl phosphorescence, but not fluorescence and the quenching is relatively inefficient. The triplet energies for the ethers are very much higher than the triplet energy of the ketone and it was for this reason that quenching was proposed as occurring via an electron-transfer mechanism. Phenols quench both the fluorescence and phosphorescence of biacetyl.[292] This quenching is very much more efficient than that by the alkoxybenzenes and therefore it was suggested that most of the quenching is due to a reversible hydrogen-abstraction mechanism.

Baum and Norman have reported the very interesting finding that biacetyl can sensitize the decarboxylation of phenylacetic and phenoxyacetic acids.[370] By the use of esr spectroscopy they identified the monoprotonated biacetyl semidione radical as an intermediate. Since this report, benzophenone and several other aryl ketones and quinones have been found to sensitize the decarboxylation of these acids.[171, 172] From kinetic studies it was concluded that the reactions involved the excited ketone reacting with the acid in an electron-transfer reaction the mechanism being analogous to that for the decarboxylation of N-arylglycines. The occurrence of these decarboxylation reactions is further evidence in support of the suggestion that alkoxybenzenes react with excited ketones by an electron-transfer process.

Aryl ethers, e.g. 2-ethoxynaphthalene, have been shown to quench the fluorescence of aromatic esters and when non-polar solvents are employed exciplex emission can be observed.[340d] The stability of this exciplex was shown to increase as the solvent was changed from benzene through ether to ethyl acetate. Reference has already been made to the ability of aromatic esters to sensitize the dimerization of aryl vinyl ethers via radical-cation intermediates.[47] Aromatic hydrocarbons, e.g. toluene, p-xylene, diphenylmethane, act as reductants for excited aromatic esters and these reactions may well involve an exciplex or radical-ion intermediates.[371] The type of products formed in these reactions appears to be dependent upon the bulkiness of the reducing hydrocarbon. Thus the reduction of dimethyl terephthalate can take two possible reaction pathways which give the products (140)–(143).

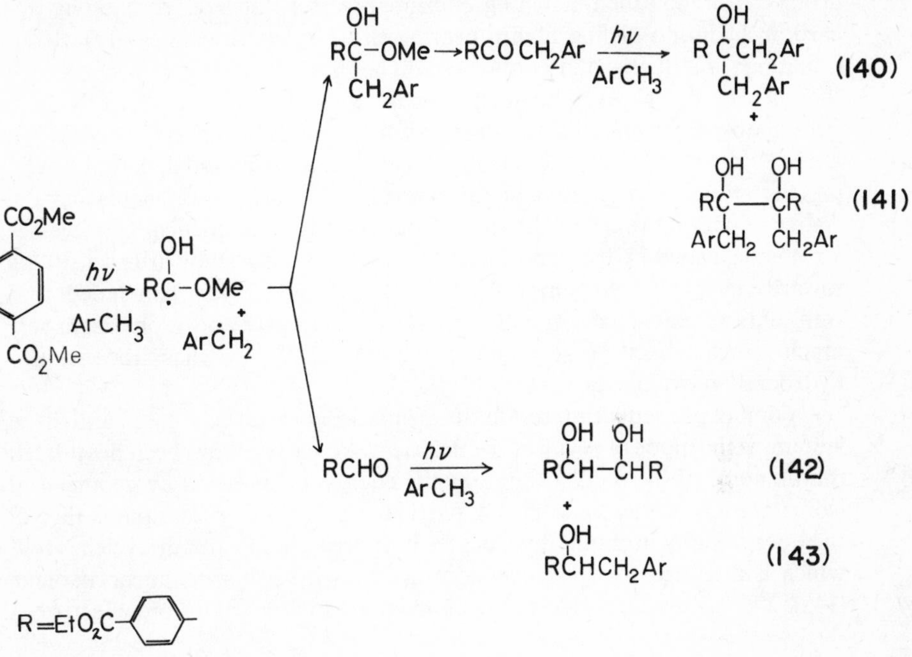

2. Heterocyclic compounds

Acridine is photoreduced by aromatic hydrocarbons such as toluene and 4-methylanisole.[372] These reactions were shown to involve the $^1n\pi^*$-state of the heterocycle. It was suggested that the hydrocarbons reacted by an electron-transfer mechanism because of the small primary isotope effect for the hydrogen-abstraction reaction.

Acridine, like many other heterocyclic compounds, sensitizes the decarboxylation of a wide variety of carboxylic acids, e.g. phenylacetic acid. benzylic acid, and phenoxyacetic acid.[348] Riboflavin reacts in a similar way.[373] It seems likely that these reactions also involve an electron-transfer process akin to that proposed for the decarboxylation of N-arylglycines sensitized by heterocyclic compounds.

J. Photoreactions of Aromatic Hydrocarbons with Mono-Olefins and Conjugated Dienes

1. Some general mechanistic considerations

We have previously noted that some mono-olefins and conjugated dienes form fluorescent exciplexes with aromatic hydrocarbons.[31c, 45] Usually these hydrocarbons contained electron-withdrawing groups, e.g. cyano-groups. Although the quenching of excited-singlet states of aromatic hydro-carbons by mono-olefins, olefins bearing electron-withdrawing and electron-donating substituents and conjugated dienes is a relatively common process, that of fluorescent exciplex-formation in these systems is relatively rare. Presumably electron-withdrawing substituents in the aromatic hydrocarbon aids exciplex formation with unsubstituted mono-olefins and dienes by lower-ing the reduction potential of the hydrocarbon and hence increasing the degree of charge-transfer character of the exciplex. The quenching processes for unsubstituted hydrocarbon–olefin or diene systems show little sensitivity towards change in solvent polarity and therefore it was concluded that transient exciplexes had little charge-transfer character and probably have a much greater exciton resonance contribution[74, 374] than the amine–hydrocarbon exciplexes.

A point of particular interest in the aromatic hydrocarbon–olefin and diene systems is the mode of decay of the transient exciplexes. It has been postulated that in some of the diene exciplexes the energy is dissipated by geometrical isomerization of the diene.[375] A particularly intriguing example is that of the quenching of aromatic hydrocarbon fluorescence by quadricyclene (144) which is attended by cycloreversion of the hydrocarbon to norbornadiene (145).[376a, b, c] Quite a wide range of aromatic hydrocarbons are effective as

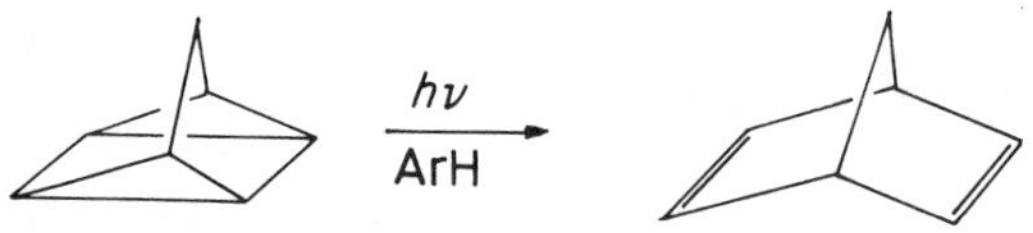

(144)   (145)

sensitizers[376b] and it was proposed that the isomerization resulted from break up of the hydrocarbon–quadricyclene exciplex, i.e. the electronic energy of the exciplex was being dissipated as vibrational energy.[376a, b] This view has been challenged and an alternative mechanism proposed.[376c] It was pointed out that in the exciplex the quadricyclene will have lost some electronic density and consequently its structure approaches that of the quadricyclene radical-cation. Neutralization of this charge may then produce either quadricyclene or norbornadiene. Before considering a most important mode of dissipation of electronic energy of exciplex, namely chemical reaction, it should be pointed out that dissipation of electronic energy via decay of the singlet exciplex to the triplet state of one of the reaction partners as a route for chemical reaction has been totally neglected. (Section III.B.2). It has been found that the quenching of naphthalene fluorescence by *cis-trans*-cyclo-octa-1,3-diene is attended by isomerization of the diene to the all-*cis* diene.[375] This isomerization could well have occurred via the triplet state of the diene which was in turn produced by decay of the singlet exciplex. Similarly, the dimerization of cyclopentene which is observed in the reactions of benzene in the presence of cyclopentene may well have occurred via the triplet cyclopentene which is populated[377] by decay of singlet benzene–cyclopentene exciplex.[377]

Many aromatic–olefin and aromatic–diene exciplexes decay by undergoing chemical reaction—usually an addition reaction. The stereospecificity of these reactions is often controlled by the geometry of the exciplexes. Thus the 1,3-photo-addition of cyclobutenes to benzene and toluene has been shown to be remarkably stereospecific.[378a, b] Addition of 1,2-dimethylcyclobutene to toluene gave (146) as the sole product.[378a] It was proposed that such

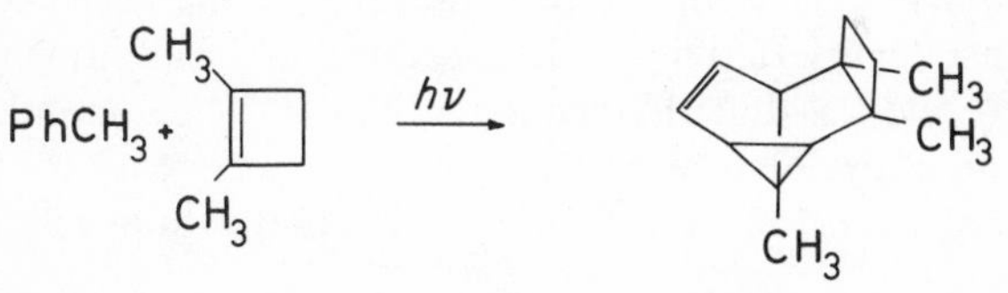

(146)

(147)   (148)

reactions occur via exciplexes in which there is maximum overlap of the olefin with the hydrocarbon, e.g. as in (147) as opposed to (148). A configuration such as (147) for the exciplex explains the preferential formation of the endo-

cyclo-addition product. Increasing the ring size of cyclo-olefins decreases
the degree of overlap that can occur in the exciplex and consequently the
amount of exo-isomer formed increases.[377, 378] However, an alternative
theory has been put forward which suggests that the 1,3-cyclo-addition reactions of olefins with the ^1B_2u state are concerted whereas the 1,2- and 1,4-
addition reactions involve exciplex intermediates.[121d] This theory was
developed from the observation that the amount of 1,2- and 1,4-cyclo-
addition increases as the ionization potential of the olefin is decreased,
whereas that of 1,3-cyclo-addition decreases. Addition of the butadiene
(149) to trans-stilbene was shown to give (150) in very much higher yield than
(151).[379] Formation of (150) can be rationalized in terms of an intermediate

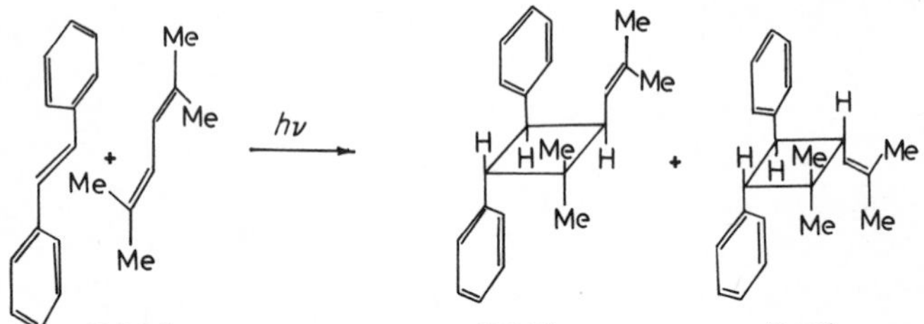

(149)   (150)   (151)

exciplex in which there is maximum overlap of the diene systems and the
stilbene. A salient fact is that direct addition of the diene to cis-stilbene
does not occur. Regioselectivity is also observed in the addition of cyclohexa-
1,3-diene[380a] and 2,4-dimethylpenta-1,3-diene[380b] to naphthalene. Thus,
in the addition of cyclohexadiene there is a 9:1 preference for formation of
(153) compared with (152). The formation of (154) with acyclic diene is also

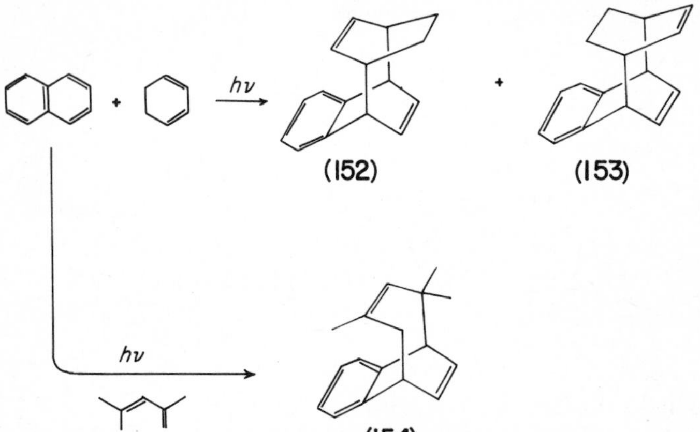

(152)   (153)

(154)

consistent with the reaction occurring via an exciplex in which there is maximum overlap of the π-orbitals of the diene with those of the naphthalene.

Since the formation of cyclo-addition products from reaction of aromatic hydrocarbons with dienes demands the intermediacy of an exciplex with fairly fixed configurational requirements we may anticipate that the geometry of the diene, i.e. whether it is *s-cis* or *s-trans* will be important. Also, the degree of substitution in either the diene or hydrocarbon may be important because this will bring into play steric factors. As might be anticipated the *s-cis* dienes are better quenchers than *s-trans* dienes since one can get better overlap of the π-system of these dienes with the π-system of the hydrocarbon.[374] However, the efficiency of quenching of various alkylated benzenes by *cis*-piperylene is relatively insensitive to the number and type of alkyl substituents.[381] Only when a number of *t*-butyl substituents are present does the quenching efficiency fall off. This has led to an estimate of the interplanar distance of the alkylbenzene–*cis*-piperylene exciplexes as being between 3·7 to 5·2 Å.

Another observation which suggests that the cyclo-addition of olefins to aromatic compounds involves an exciplex relates to the finding that the quantum yield for addition of tetramethylethylene to *trans*-stilbene to give a cyclobutane increases as the temperature of the system is decreased.[382] Presumably the lowering of the temperature increases the stability of the exciplex and consequently gives a greater chance for the cyclo-addition reaction to occur.

As noted earlier, the introduction of strongly electron-withdrawing groups into the aromatic hydrocarbon or into the olefin increases the charge-transfer character of the exciplex and enhances the possibility of radical-ion formation when polar solvents are employed. There is in fact good evidence for radical-ion formation in the reaction of acrylonitrile with naphthalene in which the substituted naphthalene (155) is formed.[383a, b] Formation of (155) is favoured by the use of polar solvents and when either CH_3CO_2D or MeOD are used as solvents, deuterium is incorporated into the methyl group of (155).[383a] There is also evidence that the addition of tetramethylethylene to benzene to give (156) involves radical ions.[384] Thus addition of MeOD to the reaction mixture leads to deuterium incorporation at the 4-position of the cyclohexadiene ring. Formation of (156) competes with the formation of the 1,2-cyclo-addition product, (157). As yet, attempts to trap zwitterionic intermediates in 1,3-cyclo-addition reactions of olefins with benzene have been unsuccessful. Tetramethylethylene also reacts with naphthalene in a similar manner to which it reacts with benzene.

2. Particular examples of addition

Besides 1,3-addition[377a, b, 385, 386] benzene undergoes 1,2- and 1,4-cyclo-

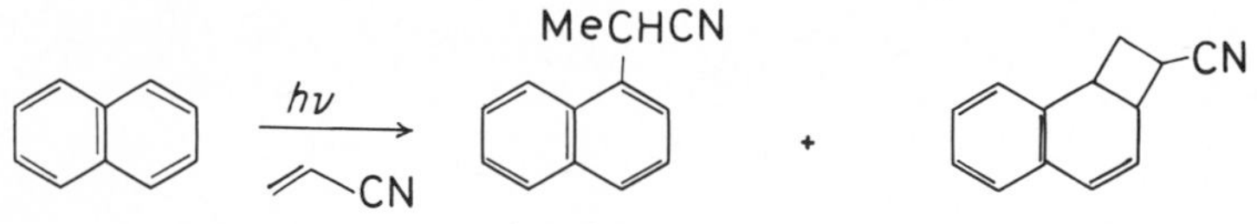

MeCHCN

(155)

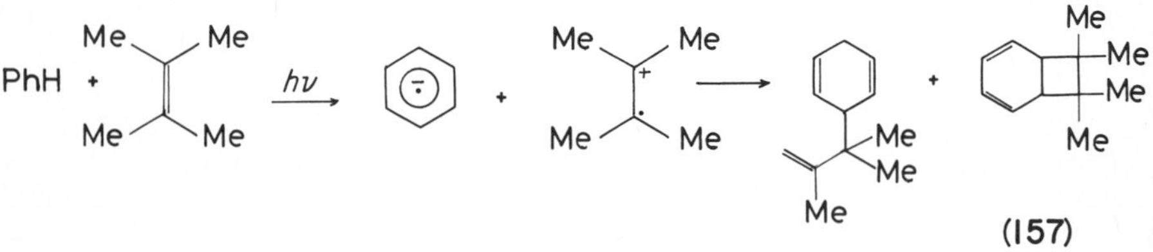

(157)

(156)

addition reactions with olefins.[121d, 384a] It also forms a 1,2-cyclo-addition product with dimethyl acetylenedicarboxylate which can be trapped with tetracyanoethylene to give (158).[387a] When the addition reaction is carried

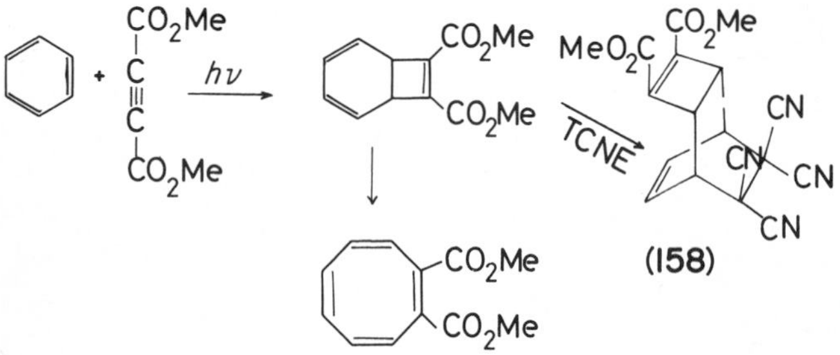

(158)

(159)

out in the presence of acids dimethyl 1-phenylmaleate and the corresponding furmarate are formed.[387b] It would appear that the acid has intercepted an exciplex or radical-ion pair formed by reaction of the acetylene with the singlet state of benzene. Cycloreversion of the initially formed adduct gives (159). Since the ester is a relatively good electron-acceptor it is likely that the reaction occurs via an exciplex. The photoreactions of 1,2-dichloroethylene with benzene are rather interesting in that they appear to involve (161) as an intermediate.[388] Formation of this compound may be rationalized as occurring via the biradical (160) which in turn may be formed via an exciplex.

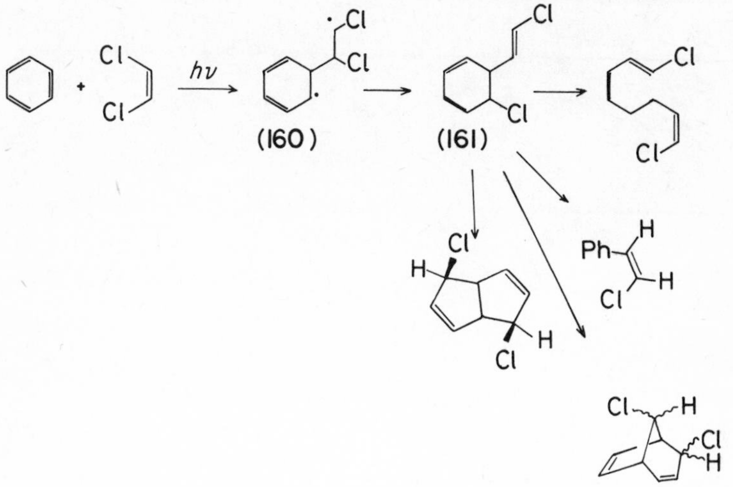

At present it is pure speculation as to whether these reactions and those of dichlorovinylene carbonate[389] and norbornadiene[390] involve exciplex intermediates. The intramolecular cyclo-addition reactions of 6-phenylhex-2-ene (162)[391] and 6-phenylhex-2-yne (163)[392] have been interpreted as occurring via singlet intramolecular exciplexes. The olefinic and the acetylenic

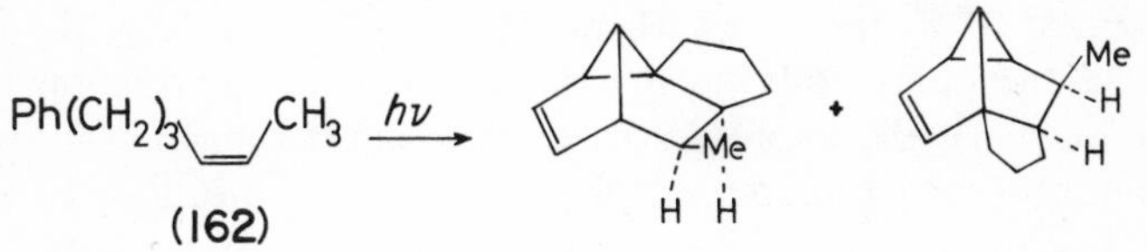

(162)

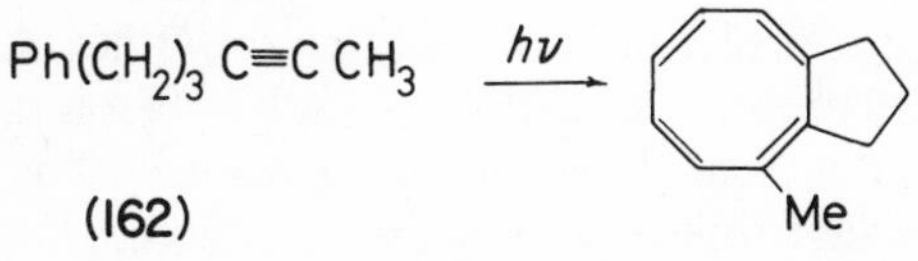

(162)

systems lead to quenching of the excited-singlet state of the toluene nucleus and in the case of (162) it was suggested that quenching of vibrationally excited-singlet states occurred. The fact that the quantum yield for cyclo-adduct formation is wavelength dependent is consistent with this suggestion. In the case of 1-phenyl-but-2-ene, intramolecular triplet–triplet energy transfer leads to olefin isomerization but no products from reaction of the excited-singlet state of the hydrocarbon were isolated.[393]

Diphenylacetylene has been shown to give (2 + 2) cyclo-addition products such as (164) with a variety of substituted naphthalenes.[394] These adducts

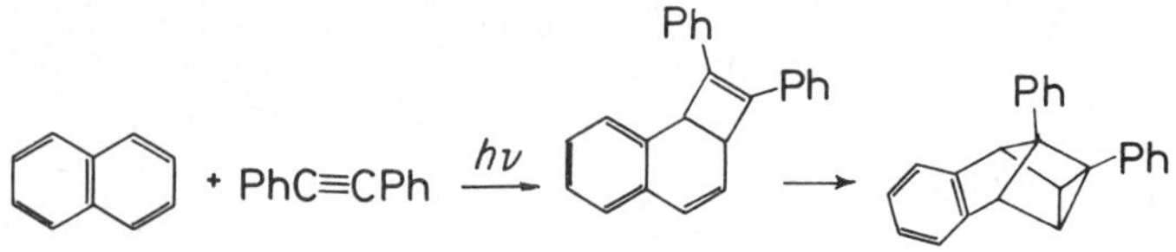

(164)   (165)

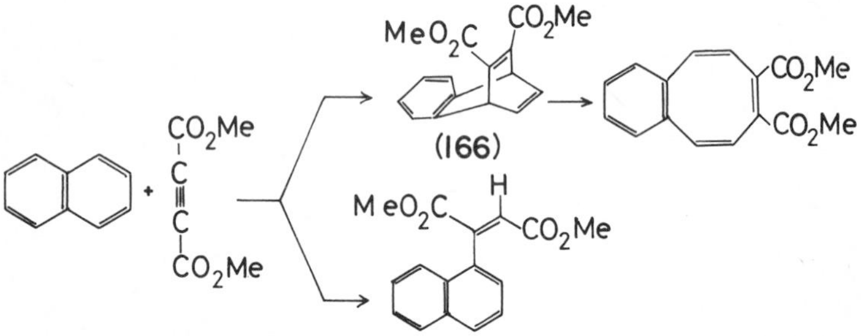

(166)

(167)

are isomerized to compounds such as **(165)** which are usually the isolated products of the photo-addition reactions. From the observation that the acetylene quenches the fluorescence of the naphthalenes it was postulated that addition occurred via an exciplex intermediate. Addition of dimethyl acetylenedicarboxylate to naphthalene takes a different course.[395] One of the main primary products appears to be the (2 + 4) cyclo-addition product **(166)** which readily undergoes further photo-isomerization. Other interesting products include the formation of dimethyl (1-naphthyl)fumarate **(167)** which may well be produced in analogous manner to **(156)** which is isolated from the reaction of tetramethylethylene with benzene. Reference has already been made to the reaction of acrylonitrile with naphthalene for which there is good evidence for radical-ion intermediates.[383a, b] Reaction of acrylonitrile with indene also gives products, e.g. **(168)** which can be rationalized

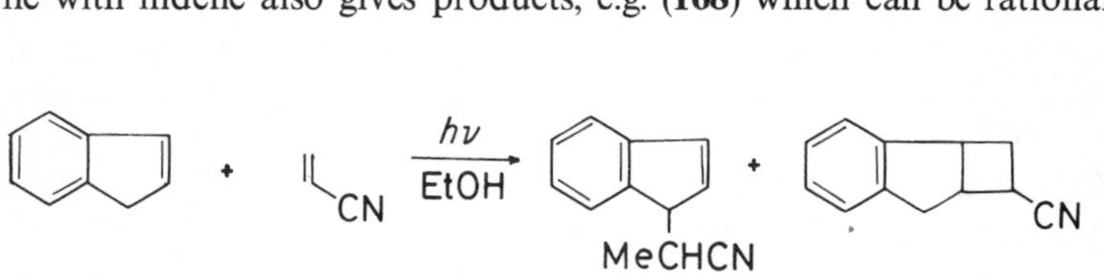

MeCHCN

(168)

as being produced by a similar mechanism.[396] (2 + 2)-Cyclo-addition products have been isolated from reactions of methoxynaphthalenes with acrylonitrile[397] and from cyanonaphthalenes with alkylvinyl ether.[398] Rather surprisingly a (2 + 4)-cyclo-addition product has been isolated from the reaction of 2-acetylnaphthalene with methyl cinnamate.[389] Furan has been shown to react with 1-cyanonaphthalene to give a (4 + 4)-cyclo-addition product.[400a] The heterocycle quenched the fluorescence of the hydrocarbon and the efficiency of quenching displayed a negative-temperature dependency.[400b] The efficiency of adduct formation displayed a similar negative-temperature dependency. On the basis of these facts, it was postulated that the addition reaction occurs via an exciplex intermediate. Addition of furan to 2-cyanonaphthalene occurs in a rather unusual manner to give (169).[400c] Presumably this reaction also occurs via an exciplex intermediate.

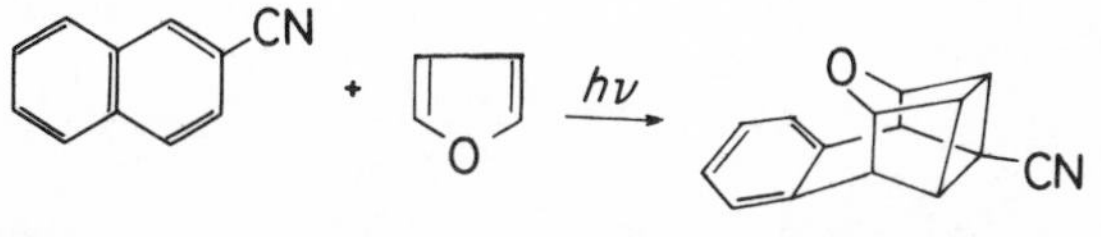

(169)

Other cyclo-addition reactions which may involve exciplex intermediates include reaction of phenanthrene with dimethyl maleate[401a] and reaction of anthracene with cycloheptatriene.[401b] Anthracene also undergoes addition reactions with alkenes[402] and dienes.[403, 404a, b] Addition of cyclopentadiene gives both (2 + 4) and (4 + 4) cyclo-addition products and reaction via a biradical intermediate (170) was suggested. Cyclohexa-1,3-diene also reacts to give a (4 + 4)-cyclo-addition product.[404a] Addition of 2,5-dimethylhexa-2,4-diene gives (171). By contrast, 9-cyanoanthracene reacts with the diene to give a (2 + 4)-cyclo-addition product.[404b] This behaviour was thought to result from the transoid diene having one of its double bonds aligned with cyano-group in an intermediate such as (172). In all cases the dienes were shown to quench the fluorescence of anthracene efficiently and exciplex intermediates are implicated.

VIII. PHOTOREACTIONS POSTULATED AS OCCURRING VIA EXCIPLEX INTERMEDIATES FOR WHICH SPECTROSCOPIC EVIDENCE IS NOT AVAILABLE

A. Photoreactions of Carbonyl Compounds with Olefins and Dienes

1. Photoreactions of ketones with mono-olefins
Reference has already been made to the reaction of dialkyl ketones with

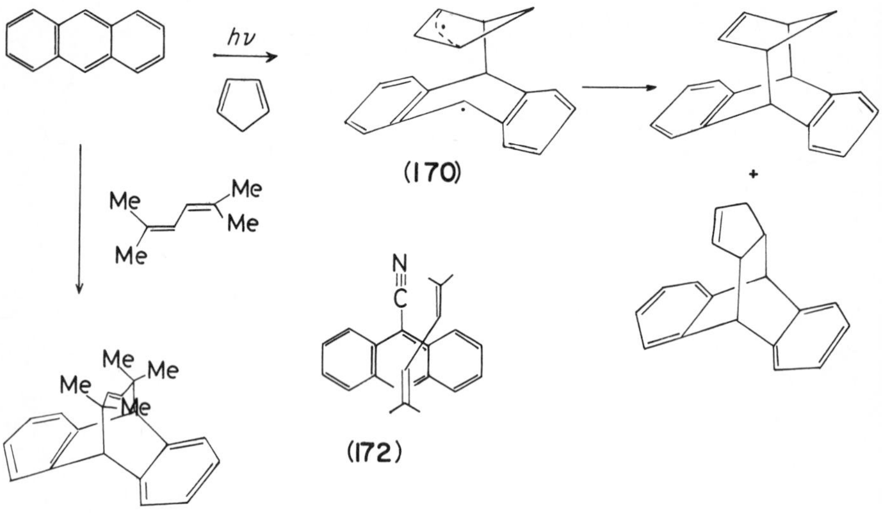

(170)

(172)

(171)

cyanoethylenes and alkoxyethylenes (II.B.2c). Several other papers, not referred to in this previous section deal in more detail with some of the aspects of these reactions.[405]

Carbonyl compounds such as benzophenone have been shown to react with a wide variety of olefins to yield oxetanes.[406] In many cases oxetane formation is accompanied by cis–trans isomerization of the olefin. Since there is insufficient triplet energy available from the ketone to sensitize this process it was initially postulated that isomerization occurred by break-up of a biradical which had been formed by attack of the carbonyl group on the olefin (II.B.2d). However, detailed kinetic analysis of these reactions revealed another complicating factor. Initial reaction of the excited ketone with the olefin is extremely efficient. In many cases the rate constants for this process exceed those for attack by alkoxyl radicals upon such olefins.[41b] It therefore seemed unlikely that the primary reaction led directly to biradical formation and for this reason exciplex formation between the excited carbonyl-group and the olefin in which the olefin is acting as an electron donor was postulated.[41b, c] This intermediate may then proceed to give a biradical (173) or lead to regeneration of carbonyl compound and olefin. For a quite a number of systems it has been shown that substantial energy wastage occurs and this can in principle occur from the exciplex or the biradical.[407] Another very telling observation which supports the exciplex postulate is that the efficiency of quenching of aromatic carbonyl compound triplets by cis-but-2-ene is not determined so much by the triplet energies of the ketones but rather by their reduction potentials.[41c] For example, the rate constants for

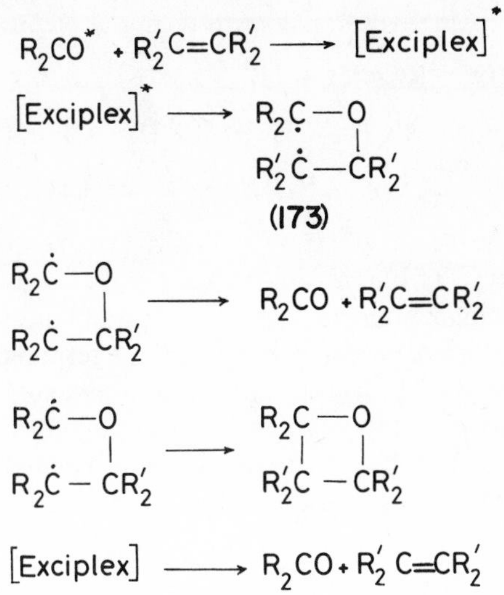

$$R_2CO^* + R_2'C{=}CR_2' \longrightarrow [\text{Exciplex}]^*$$

$$[\text{Exciplex}]^* \longrightarrow \begin{array}{c} R_2\dot{C}-O \\ | \\ R_2'\dot{C}-CR_2' \end{array}$$

(173)

$$\begin{array}{c} R_2\dot{C}-O \\ | \\ R_2\dot{C}-CR_2' \end{array} \longrightarrow R_2CO + R_2'C{=}CR_2'$$

$$\begin{array}{c} R_2\dot{C}-O \\ | \\ R_2\dot{C}-CR_2' \end{array} \longrightarrow \begin{array}{cc} R_2C-O \\ | \quad | \\ R_2'C-CR_2' \end{array}$$

$$[\text{Exciplex}] \longrightarrow R_2CO + R_2'C{=}CR_2'$$

quenching triplet 4,4'-dimethoxybenzophenone and 4-trifluormethylbenzo-phenone were determined as 4×10^6 and $2\cdot2 \times 10^8 \, \text{mol}^{-1} \, \text{sec}^{-1}$. Further detailed kinetic analysis of these reactions, in particular, secondary deuterium isotope effects, corroborate with these findings.[41c, 407] Energy wastage has also been observed in the intramolecular attack of carbonyl groups upon olefins, e.g. in hept-2-en-5-one, and it has been attributed to an exciplex intermediate.[408a] In contrast energy wastage was not observed in the cis-styrylalkyl phenyl ketones. A rather interesting case of energy wastage is that of the triplet sensitized isomerization of stilbene and other arylalkenes by fluorenone[409] and 2-acetylnaphthalene.[410] In these systems, triplet energy-transfer should occur efficiently because the triplet energy of the carbonyl compound exceeds that of the olefin. With fluorenone as sensitizer the isomerization of trans to cis-α-cyanostilbene was found to be 15–22 per cent inefficient. Similarly the quantum yields for cis → trans and trans→ cis isomerization of stilbene sensitized by 2-acetylnaphthalene do not add up to unity and in fact isomerization of the cis compounds appears to be only about 50 per cent efficient. Exciplex formation in these systems may be particularly favoured by the fact that the triplet energies of the ketones lie relatively close to those of the olefins and hence increasing stabilization by exciton resonance.

2. Photoreactions of aldehydes with mono-olefins

Exciplex intermediates have been postulated as occurring in reactions of aldehydes with alkenes to explain non-linearity of Stern–Volmer plots for the quenching process, energy wastage[412] and the stereoselectivity of the reactions.[412, 413] Aldehydes react with terminal olefins to give ketones, e.g.

$$RCHO \rightarrow R\dot{C}{=}O + H^{\cdot}$$

$$R\dot{C}{=}O + R'CH{=}CH_2 \rightarrow R'\dot{C}H_2{-}CH_2COR$$

$$R'\dot{C}HCH_2COR + RCHO \rightarrow R'CH_2CH_2COR + R\dot{C}{=}O$$

whereas with other more substituted olefins oxetane formation takes place. Since the latter type of olefins are better donors it was suggested that they reacted via an exciplex intermediate.[412] Certainly these olefins should react with the excited aldehyde more readily than the terminal olefins and thereby suppress production of acyl radicals. 1-Naphthaldehyde reacts with olefins via its excited singlet state ($^1n\pi^*$) and oxetane formation is reasonably stereoselective.[413] The stereoselectivity was attributed to the intermediate exciplex controlling the mode of addition.

3. Photoreactions of ketones with dienes

Conjugated dienes quench the fluorescence of a number of ketones and the efficiency of quenching increases as the degree of substitution in the diene is increased.[414] From the stereochemical requirements for oxetane formation in these reactions it was shown that the dienes act as nucleophiles. Benzophenone also gives oxetanes with conjugated dienes although this process is very much less efficient that triplet–triplet energy transfer.[415]

4. Photoreactions of α,β-unsaturated ketones

Exciplex formation in the (2 + 2)-cyclo-addition reactions of enones has been postulated in order to account for the stereoselectivity of the reactions[25, 416] (Section II.B.1) and energy wastage.[417]

5. Photoreactions of ortho-quinones with olefins and acetylenes

Ortho-quinones, e.g. phenanthra-9,10-quinone, react with olefins to give (2 + 2)-[418] and (2 + 4)-[418, 419] cyclo-addition products. Other quinones react in a similar way.[37]

Reactions were postulated as occurring via either biradical or zwitterionic intermediates, e.g. (12). Addition products are also formed on reaction of ortho-quinones with acetylenes.[420] Presumably these reactions may occur via exciplex intermediates although no evidence on this point is presently available.

6. Photoreactions of esters and carboxylic acids with olefins

Several aromatic esters, e.g. dimethyl terephthalate,[421a, b] tetramethyl pyromellitoate[421c] and methyl benzoate[422a] form oxetanes on reaction with olefins. In many cases these oxetanes, e.g. (174), suffer further decompo-

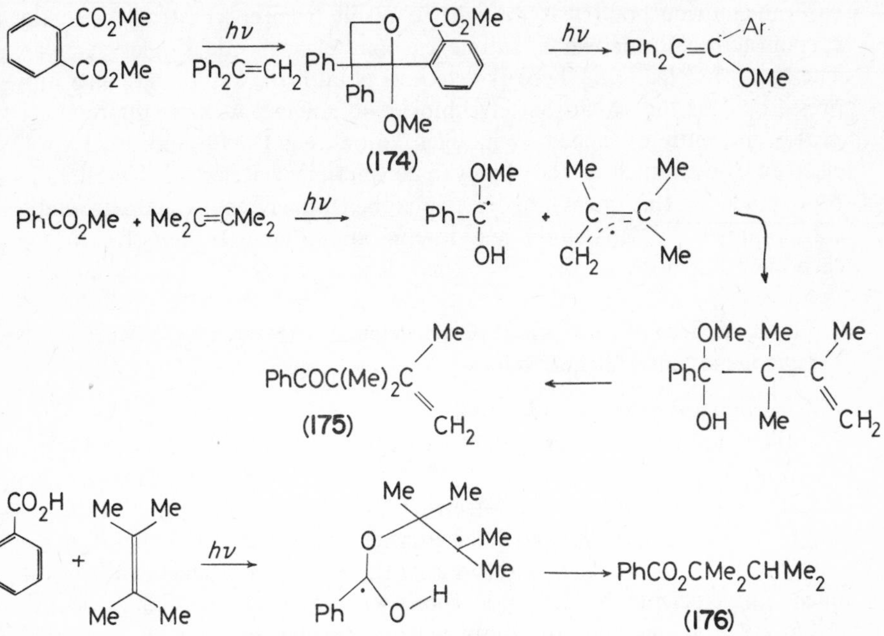

(174)

$PhCO_2Me$ + $Me_2C=CMe_2$ $\xrightarrow{h\nu}$

(175)

$\xrightarrow{}$ $PhCO_2CMe_2CHMe_2$

(176)

sition.[421a, c] Tetramethylethylene reacts with methyl benzoate and benzoic acid to give oxetanes together with other products (175) formed by the excited ester abstracting hydrogen from the olefin.[422a] Formation of (176) from benzoic acid is particularly interesting because it is derived by a "retro-Norrish Type II" reaction.[422b] Oxetane formation has also been observed in the reaction of esters with conjugated dienes.[421c] From the observation that the olefins efficiently quench the fluorescence of some of the esters it was proposed that the reaction involved a singlet exciplex which could decay to give the oxetane or the triplet ester.[421b] The latter could then give the oxetane by reaction with the olefin. It would be interesting to know if the reactions of the ester group which lead to hydrogen abstraction and oxetane formation occur from the same excited state.

B. Photoreactions of Carbonyl Compounds with Halogen Containing Compounds and Halogen Anions

Although carbon tetrachloride is often considered as an inert solvent for photochemical reactions it has been shown to quench the fluorescence of aliphatic ketones quite efficiently.[423] This process is accompanied by homolysis of the carbon–chlorine bond and products are obtained which are derived by attack of a chlorine radical upon the ketone. Homolysis of the carbon–chlorine bond was thought to result from energy transfer to the carbon tetrachloride via a ketone–carbon tetrachloride singlet exciplex. The Norrish Type I and Type II reactions of aliphatic ketones are also suppressed by the use of carbon tetrachloride as solvent. Acetone fluorescence is also efficiently quenched by halogen anions, e.g. I⁻, Br̄, and other good electron-donors such as the thiocyanate anion.[424] Reference has already been made to the quenching of triplet benzophenone by halo-aromatic compounds.[367] When aliphatic iodo-compounds are used homolysis of the carbon–iodine bond occurs.[425]

C. Photoreactions of Aromatic Hydrocarbons with Halogen Containing Compounds and Halogen Anions

1. General mechanistic considerations

The quenching of the fluorescence of aromatic hydrocarbons, e.g. benzene,[426] naphthalene[427] and anthracene[428a, b, c] by carbon tetrachloride and chloroform is well documented. Charge-transfer absorption bands between benzene and the halocarbons have been detected when solvents at low temperature are employed.[426] In the case of anthracene the wavelength dependency of the quenching efficiency by the halocarbons was interpreted as being evidence for more than one type of complex being formed in these systems.[428c] The quenching by carbon tetrachloride always appears to be accompanied by homolysis of a carbon–chlorine bond (cf. the carbonyl sensitized reaction) and this was suggested as being the mode of dissipation of the electronic energy of the hydrocarbon–halocarbon excited complex. The fluorescence of several benzenoid systems, e.g. 1,4-dimethoxybenzene is quenched by chloro-esters such as methyl chloro-acetate.[429] In these systems quenching via exciplex formation was proposed. The energy of this species may be dissipated by vibrational stretching of the carbon–halogen bond.

2. Product studies

Benzene reacts with chloroform and carbon tetrachloride to give hexatrienes, e.g. (178). Presumably reaction occurs via initial addition of the halocarbon

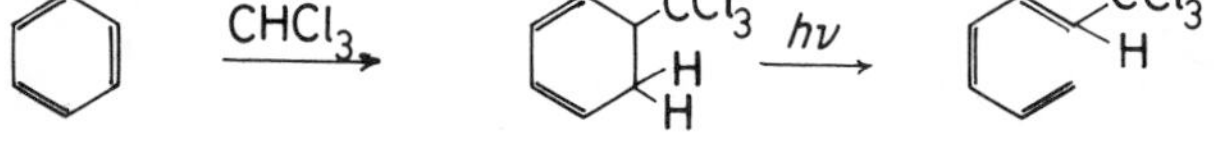

(177)   (178)

+

isomers

to the hydrocarbon to give a diene (177) which then undergoes electrocyclic ring opening.[426] Naphthalene has been shown to act as a photo-sensitizer for the chain reaction of carbon tetrachloride with methanol which leads to the formation of formaldehyde, chloroform, and hydrogen chloride.[427] The anthracene-sensitized decomposition of carbon tetrachloride appears to give substituted anthracenes.[428a] A particularly interesting reaction is that of compound (179), since the type of products formed are dependent on the

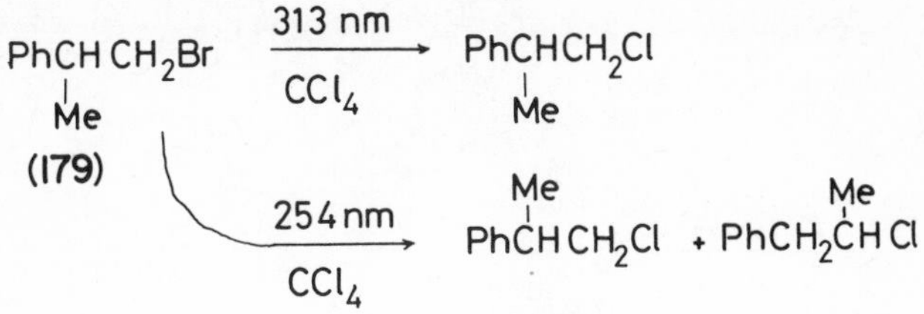

$$
\begin{array}{c}
\text{PhCH CH}_2\text{Br} \xrightarrow[\text{CCl}_4]{313 \text{ nm}} \text{PhCHCH}_2\text{Cl} \\
\underset{\text{Me}}{|} \qquad\qquad \underset{\text{Me}}{|} \\
(179)
\end{array}
$$

$$
\xrightarrow[\text{CCl}_4]{254\,\text{nm}} \underset{\text{Me}}{\overset{\text{Me}}{\text{PhCH CH}_2\text{Cl}}} + \underset{}{\overset{\text{Me}}{\text{PhCH}_2\text{CH Cl}}}
$$

PhCHMe ĊH₂

PhCH₂ĊHMe

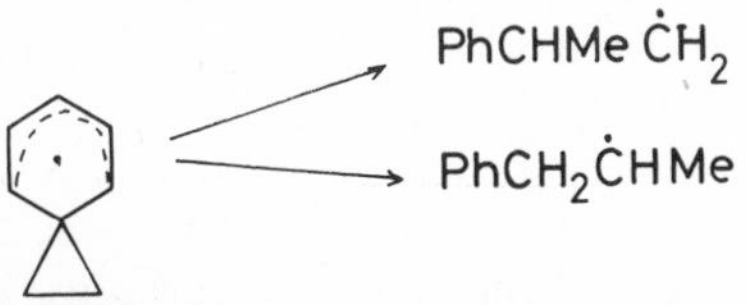

(180)

excitation wavelength.[430] It was proposed that the higher energy excitation leads to intramolecular interaction between the excited benzenoid system and the halogen group. This interaction is thought to give a radical of the type (180).

Intramolecular interaction of excited-singlet aromatic hydrocarbons with halogeno-groups has been been put to good use in the synthesis of a wide variety of compounds [431a, b] Reactions in the benzenoid series are well illustrated by (181).[431a] Several similar derivatives of phenol have also been cyclized.[432a, b] In the case of (182), investigation of the mechanism of the reaction by flash photolysis revealed the intermediacy of phenoxyl radicals and the tautomer (183).[432a] Similar reactions have been reported for substituted naphthalenes.[433] Aromatic hydrocarbons also intermolecularly sensitize the homolysis of halo-acetic.[425] Thus irradiation of anthracene in the presence of iodoacetic acid gives 9-anthracenylacetic acid.

D. Photoreactions Involving Azo-compounds

Polycyclic aromatic hydrocarbons, e.g. 9,10-diphenylanthracene are effective photosensitizers for the decomposition of acyclic azo-compounds.[434] The rate constants for quenching the hydrocarbon fluorescence by the azo-compound are all very high, $\sim 1\text{--}10 \times 10^9 \text{ mol}^{-1} \text{ sec}^{-1}$ and yet the efficiency of decomposition of the azo-compound is relatively low. In many cases the

L*

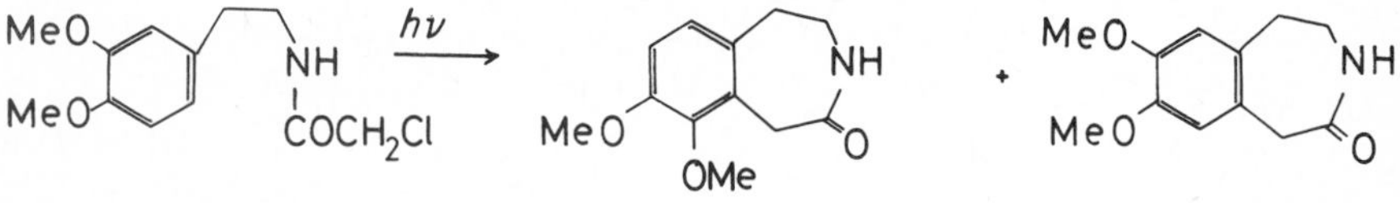

MeO (hv) ... + MeO

(181)

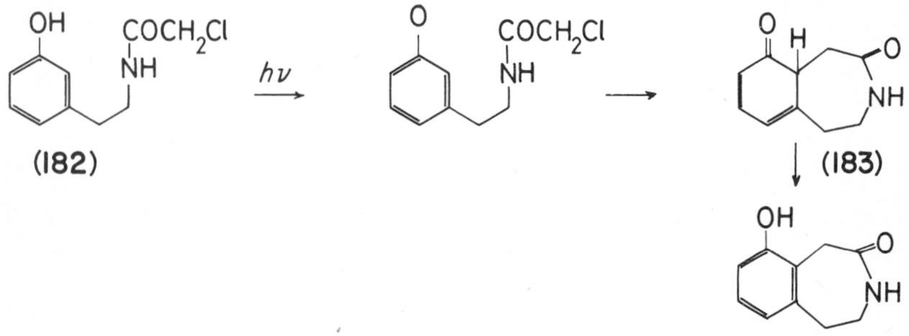

(182)   hv   →   (183)

singlet energy of the azo-compound was in excess of that of the hydrocarbon. On the basis of the fact that the quenching rate-constants are relatively insensitive to solvent polarity it was proposed that the quenching could not be due to exciplex formation. However, since the quenching efficiency is so high in non-polar solvents one would not expect much of an increase on going to the more polar solvent. Furthermore, quenching via an exciplex would account for the apparent ability for the hydrocarbons to act as sensitizers in an endothermic energy-transfer process.

The fluorescence of azo-compounds such as 2,3-diazabicyclo[2,2,2]oct-2-ene is quenched by a number of dienes and olefins.[435a, b, c] The efficiency of quenching has been shown to be related to the ionization potential of the olefin and therefore the conclusion was reached that exciplex intermediates were involved.

E. Photoreactions Involving Anions

Halogen anions[69, 85, 436] and other easily oxidized anions quench the excited-singlet states of polycyclic aromatic hydrocarbons and this has been shown to lead to production of hydrocarbon triplets.[85] Recently it was shown that use of cetylpyridinium halides as micelles for aromatic hydrocarbons in aqueous and alcoholic solutions led to quenching of the hydrocarbon fluorescence.[437] This was attributed to the pyridinium nucleus acting as an electron acceptor. However, if the pyridinium salts are not fully dissociated the quenching may well be due to the anions. Halogen anions also appear to be able to undergo electron-transfer reactions with quinones.[438] Hydroxyl anions react with anthraquinones in a similar manner[438, 439a, b] and hydroxyl-ated quinones are produced.[439a] Excited aromatic nitro-compounds are also quenched by chloride and bromide ions.[440, 441a, b] Irradiation of nitro-benzene in the presence of hydrochloric acid ultimately yields 2,4-dichloro-aniline.[441b] The chloride ions presumably react with the excited nitro-compound to give chlorine atoms which then attack aniline produced by reduction of the nitro-group. Triplet methylene blue apparently undergoes a redox reaction with benzenesulphinate anions to give radicals which are capable of initiating the polymerization of acrylamide.[442]

IX. CONCLUSION

Whilst the contents of Sections IV–VIII give many examples of reactions which probably involve an electron-transfer process, it will not have escaped the reader's attention that unequivocal experimental proof in favour of such a mechanism is often hard to come by. At present our armoury of methods for tackling the problem is insufficient although no doubt further physical methods methods will be developed which will help to overcome this deficiency.

Although chemists may think they have a difficult problem in investigat-ing photo-induced electron-transfer reactions, the photobiologists have often a much more difficult task. In spite of all these difficulties they have made a significant progress in the understanding of such processes as photosyn-thesis.[443, 444] For example many of the details of the electron-transfer processes of excited chlorophyll in vitro[445, 446] and in vivo[447, 448, 449] have been elucidated. The in vitro work has included a demonstration of the excited chlorophyll molecules, e.g. bacteriochlorophyll, act as electron-donors towards quinones.[445] The radical-ion species in these reactions were detected and identified by esr spectroscopy. For the examination of the electron-transfer processes of excited chlorophyll in vivo, a method was developed for monitoring the production of chlorophyll radical cations by esr spectroscopy and simultaneously monitoring the electronic absorption spectrum of the chlorophyll. Using this method it was possible to show that the light-induced bleaching of the chlorophyll pigments P700 and P870 is due to production of chlorophyll radical-cations. A rather interesting facet of the photosynthetic process is that wastage of electronic energy by back electron-transfer from the acceptor radical anion, e.g. $\dot{Q}^-$ to the chlorophyll radical cation is obviated by the placement of suitable acceptor:

$$Chl^* + Q \rightarrow \dot{Chl}^+ + \dot{Q}^-$$

$$\dot{Q}^- + A \rightarrow Q + \dot{A}^-$$

Chlorophyll is regenerated from its radical cation by accepting an electron from a donor other than either Q or A.

No doubt the next few years will see a much deeper probing of the mechanism of photo-induced electron-transfer mechanisms and of examples of reactions which occur by such a process. If nothing else is accomplished, this chapter demonstrates the explosion of interest in this subject which has developed over the past decade.

REFERENCES

1. R. Foster. "Organic Charge-Transfer Complexes", Academic Press London and New York (1969); R. Foster. "Molecular Complexes Vol. 1", Elek Science, London; Crane Russak, New York (1973); G. Briegleb, "Elektronen-Donator-Acceptor-Komplexe", Springer-Verlag, Berlin (1961); R. S. Mulliken and W. B. Pearson, in "Molecular Complexes: A Lecture and Reprint Volume", Wiley-Interscience, New York (1969).
2. M. J. S. Dewar and C. C. Thompson. *Tetrahedron Supplement No. 7*, 97, (1966).
3. C. A. Parker, in "Photoluminescence of Solutions", Elsevier, Amsterdam (1968), p. 73.
4a. A. Weller in "Progress in Reaction Kinetics", 1, 187 (1961).
4b. J. B. Birks, "Photophysics of Aromatic Molecules", Wiley-Interscience, London (1970), ch. 9.
5. E. J. Bowen and W. S. Metcalf. *Proc. Roy. Soc., Ser. A*, **206**, 437 (1951).
6. T. G. Beaumont and K. M. C. Davis. *J. Chem. Soc.* (B), 456 (1970).
7. C. Lewis and W. R. Ware. *Chem. Phys. Lett.*, **15**, 290 (1972).
8. M. Kasha. *Radiation Research*, **20**, 55 (1963).
9. E. G. McRae and M. Kasha. *J. Chem. Phys.* **28**, 721 (1958).
10a. P. J. Wagner and D. J. Bucheck. *J. Amer. Chem. Soc.*, **92**, 181 (1970).
10b. G. J. Fisher and H. E. Johns. *Photochem. Photobiol.*, **11**, 429 (1970).
11a. R. Lisewski and K. L. Wierzchowski. *Chem. Commun.*, 348 (1969).
11b. R. Kleopfer and H. Morrison. *J. Amer. Chem. Soc.*, **94**, 255 (1972).
12. T. N. Solle and J. A. Schellman. *J. Mol Biol.*, **33**, 61 (1968).
13. R. Breslow and P. C. Scholl. *J. Amer. Chem. Soc..* **93**, 2331 (1971).
14. M. A. El-Bayoumi and M. Kasha. *J. Chem. Phys.*, **34**, 2181 (1961).
15. T. S. Spencer and C. M. O'Donnell. *J. Amer. Chem. Soc.*, **94**, 4846 (1972).
16. R. S. Davidson and M. Santhanam. *J. Chem. Soc.* (B), 1151 (1971); *Chem. Commun.*, 1114 (1971).
17. R. O. Loutfy, D. F. Williams and R. W. Yip. *Can. J. Chem.*, **51**, 2502 (1973).
18. K. C. Ingham, M. Abu-Elgheit and M. A. El-Bayoumi. *J. Amer. Chem. Soc.*, **93**, 5023 (1971).
19. P. J. Wagner and I. Kochevar. *J. Amer. Chem. Soc.*, **90**, 2232 (1968).
20. J. B. Birks and J. C. Conte. *Proc. Roy. Soc., Ser. A*, **303**, 85 (1968).
21. D. J. Cram and H. Steinberg. *J. Amer. Chem. Soc.*, **73**, 5691 (1951).
22. M. T. Vala, J. Haebig and S. A. Rice. *J. Chem. Phys.*, **43**, 886 (1965).
23. W. Rebafka and H. A. Staab. *Angew Chem. Int. Ed. Engl.*, **12**, 776 (1973).
24. L. G. Schroff, A. J. A. van der Weerdt, D. J. H. Staalman, J. W. Verhoeven and Th. J. de Boer. *Tetrahedron Lett.*, 1649 (1973).

25. B. D. Challand and P. de Mayo. *Chem. Commun.*, 982 (1968).
26. P. E. Eaton. *Accounts Chem. Res.*, **1**, 50 (1968); P. de Mayo. *Accounts Chem Res.*, **4**, 41 (1971); W. C. Herndon. *Mol. Photochem.*, **5**, 253 (1973).
27. Th. Förster and K. Kasper. *Z. Phys. Chem.* (Frankfurt am Main), **1**, 275 (1954).
28. Th. Förster. *Pure Appl. Chem.*, **34**, 225 (1973).
29. J. Menter and Th. Förster. *Photochem. and Photobiol.*, **15**, 289 (1972).
30. H. Beens and A. Weller. *Acta Physica Polonica*, **34**, 593 (1968).
31a. H. Knibbe, D. Rehm and A. Weller. *Z. Phys. Chem.* (Frankfurt am Main), **56**, 95 (1967); D. Rehm and A. Weller. *Z. Phys. Chem.* (Frankfurt am Main), **69**, 183 (1970).
31b. E. A. Chandross and J. Ferguson. *J. Chem. Phys.*, **47**, 2557 (1967).
31c. G. N. Taylor. *Chem. Phys. Lett.*, **10**, 355 (1971).
32a. J. B. Guttenplan and S. G. Cohen. *J. Amer. Chem. Soc.*, **94**, 4040 (1972).
32b. S. G. Cohen and G. Parsons. *J. Amer. Chem. Soc.*, **92**, 7603 (1970).
33. R. W. Yip, R. O. Loutfy, Y. L. Chow and L. K. Magdzinski. *Can. J. Chem.*, **50**, 3426 (1972).
34. J. B. Guttenplan and S. G. Cohen. *Tetrahedron Lett.*, 2163 (1972).
35a. H. Beens, H. Knibbe and A. Weller. *J. Chem. Phys.*, **47**, 1183 (1967); H. Knibbe, D. Rehm and A. Weller. *Ber. Bunsenges. Phys. Chem.*, **72**, 257 (1968).
35b. J. H. Borkent, J. W. Verhoeven and Th. J. de Boer. *Tetrahedron Lett.*, 3363 (1972).
36. J. B. Birks, "Photophysics of Aromatic Molecules", Wiley-Interscience, London (1970), ch. 7, p. 327.
37. D. Bryce-Smith and A. Gilbert. *Chem. Commun.*, 1701, 1702 (1968).
38. N. J. Turro, J. C. Dalton, K. Dawes, G. Farrington, R. Hautala, D. Morton, M. Niemczyk and N. Schore. *Accounts Chem. Res.*, **5**, 92 (1972).
39. N. J. Turro, C. Lee, N. Schore, J. Barltrop and H. A. J. Carless. *J. Amer. Chem. Soc.*, **93**, 3079 (1971).
40. J. C. Dalton, D. M. Pond and N. J. Turro. *J. Amer. Chem. Soc.*, **92**, 2173 (1970).
41a. N. C. Yang, J. I. Cohen and A. Shani. *J. Amer. Chem. Soc.*, **90**, 3264 (1968); J. Saltiel, K. R. Nueberger and M. Wrighton. *J. Amer. Chem. Soc.*, **91**, 3658 (1969); N. J. Turro. *Photochem. and Photobiol.*, **9**, 555 (1969).
41b. I. H. Kochevar and P. J. Wagner. *J. Amer. Chem. Soc.*, **92**, 5742 (1970); **94**, 3859 (1972).
41c. R. A. Caldwell and R. P. Gajewski. *J. Amer. Chem. Soc.*, **93**, 532 (1971); R. A. Caldwell, G. W. Sovocool, and R. P. Gajewski. *J. Amer. Chem. Soc.*, **95**, 2549 (1973).
42a. C. DeBoer. *J. Amer. Chem. Soc.*, **91**, 1855 (1969).
42b. H. Morrison, H. Curtis and T. McDowell. *J. Amer. Chem. Soc.*, **88**, 5415 (1966).
42c. R. W. Yip, A. G. Szabo and P. K. Tolg. *J. Amer. Chem. Soc.*, **95**, 4471 (1973); C. D. DeBoer and R. H. Schlessinger. *J. Amer. Chem. Soc.*, **94**, 655 (1972); O. L. Chapman and G. Wampfler. *J. Amer. Chem. Soc.*, **91**, 5390 (1969).
42d. E. A. Ogryzlo and C. W. Tang. *J. Amer. Chem. Soc.*, **92**, 5034 (1970); I. B. C. Matheson and J. Lee. *J. Amer. Chem. Soc.*, **94**, 3310 (1972); R. H. Young, R. L. Martin, D. Feriozi, D. Brewer and R. Kayser. *Photochem. and Photobiol.*, **17**, 233 (1973).
42e. R. A. Ackerman, I. Rosenthal and J. N. Pitts. *J. Chem. Phys.*, **54**, 4960 (1971).
43. R. S. Davidson, P. F. Lambeth and M. Santhanam. *J. Chem. Soc., Perkin II*, 2351 (1972).
44. G. S. Hammond and J. Saltiel. *J. Amer. Chem. Soc.*, **85**, 2516 (1963); W. G. Herk-

stroeter and G. S. Hammond. *J. Amer. Chem. Soc.*, **88**, 4769 (1966); G. S. Hammond, J. Saltiel, A. A. Lamola, N. J. Turro, J. S. Bradshaw, D. O. Cowan, R. C. Counsell, V. Vogt and C. Dalton. *J. Amer. Chem. Soc.*, **86**, 3197 (1964).
45. J. Saltiel and D. E. Townsend. *J. Amer. Chem. Soc.*, **95**, 6140 (1973).
46. R. O. Campbell and R. S. H. Liu. *Chem. Commun.*, 1191 (1970); *Mol. Photochem.*, **6**, 207 (1974).
47. S. Kuwata, Y. Shigemitsu and Y. Odaira. *J. Org. Chem.*, **38**, 3803 (1973).
48. S. Kaneta and M. Koizumi. *Bull. Chem. Soc. Jap.*, **40**, 2254 (1967), H. Yamashita, H. Kokubun and M. Koizumi, *Bull. Chem. Soc. Jap.*, **41**, 2312 (1968).
49. H. Beens and A. Weller. *Chem. Phys. Lett.*, **2**, 140 (1968).
50. G. Briegleb, H. Schuster and W. Herre. *Chem. Phys. Lett.*, **4**, 53, (1969).
51. R. A. Curruthers, R. A. Crellin and A. Ledwith. *Chem. Commun.*, 252 (1969).
52. S. Farid and S. E. Shealer. *Chem. Commun.*, 677 (1973).
53. R. A. Neunteufel and D. R. Arnold. *J. Amer. Chem. Soc.*, **95**, 4080 (1973).
54a. J. Czekalla, G. Briegleb, W. Herre and R. Glier. *Z. Elektrochem.*, **61**, 537 (1957); J. Czekalla, G. Briegleb and W. Herre. *Z. Elektrochem.*, **63**, 712 (1959); J. Czekalla, A. Schmillen and K. J. Mager. *Z. Elektrochem.*, **61**, 1053 (1957); **63**, 623 (1959); S. P. McGlynn and J. D. Boggus. *J. Amer. Chem. Soc.*, **80**, 5096 (1958); K. B. Eisenthal and M. A. El-Sayed. *J. Chem. Phys.*, **42**, 794 (1965); M. S. Walker, T. W. Bednar and R. Lumry. *J. Chem. Phys.*, **45**, 3455 (1966); **47**, 1020 (1967); J. B. Birks. *Nature*, **214**, 1187 (1967); J. B. Birks and L. G. Christophorou, *Nature*, **196**, 33 (1962); H. Leonhardt and A. Weller. *Z. Elektrochem.*, **67**, 791 (1963); E. A. Chandross and J. Ferguson. *J. Chem. Phys.*, **47**, 2557 (1967); N. Mataga, T. Okada and K. Ezumi. *Mol. Phys.*, **10**, 203 (1966); K. Mutai. *Tetrahedron Lett.*, 1125 (1971); K. Mutai. *Bull. Chem. Soc. Jap.*, **45**, 2635 (1972); G. Briegleb, G. Betz and W. Herre. *Z. Phys. Chem.* (Frankfurt am Main), **64**, 85 (1969); G. Briegleb, J. Trencséni and W. Herre. *Chem. Phys. Letters*, **3**, 146 (1969).
54b. J. H. Borkent, J. W. Verhoeven and Th. J. de Boer. *Tetrahedron Lett.*, 3363 (1972); R. S. Davidson and A. Lewis. *Tetrahedon Lett.*, 611 (1974).
54c. J. Prochorow and R. Siegoczynski. *Chem. Phys. Lett.*, **3**, 635 (1969); J. Czekalla and K.-O. Meyer. *Z. Phys. Chem.* (Frankfurt am Main), **27**, 185 (1961).
55a. C. Reid. *J. Chem. Phys.*, **20**, 1212 (1952); M. M. Moodie and C. Reid. *J. Chem. Phys.*, **22**, 252 (1954); N. Mataga and Y. Murata. *J. Amer. Chem. Soc.*, **91**, 3144 (1969); J. Czekalla, G. Briegleb, W. Herre and H. J. Vahlensieck. *Z. Elektrochem*, **63**, 715 (1959).
55b. S. Iwata, J. Tanaka and S. Nagakura. *J. Chem. Phys.*, **47**, 2203 (1967).
56a. G. D. Short. *Chem. Commun.*, 1500 (1968).
56b. G. Briegleb and D. Wolf. *Angew. Chem. Int. Ed. Engl.*, **9**, 171 (1970). H. Beens and A. Weller in "Molecular Luminescence", ed. E. C. Lim, Benjamin, New York (1969).
57a. R. Potashnik and M. Ottolenghi. *Chem. Phys. Lett.*, **6**, 525 (1970).
57b. H. Masuhara and N. Mataga. *Chem. Phys. Lett.*, **6**, 608 (1970); H. Masuhara and N. Mataga. *Z. Phys. Chem.* (Frankfurt am Main), **80**, 113 (1972); *Bull. Chem. Soc. Jap.*, **45**, 43 (1972).
58a. S. Matsumoto, S. Nagakura, S. Iwata and J. Nakamura. *Chem. Phys. Lett.*, **13**, 463 (1972).
58b. N. Tsujino, H. Masuhara and N. Mataga. *Chem. Phys. Lett.*, **15**, 360 (1972).
59. H. Hayashi, S. Iwata and S. Nagakura. *J. Chem. Phys.*, **50**, 993 (1969).
60. K. H. Grellmann, A. R. Watkins and A. Weller. *J. Phys. Chem.*, **76**, 3132 (1972).

61. H. Masuhara, N. Tsujino and N. Mataga. *Chem. Phys. Lett.*, **12**, 481 (1972); T. Kobayashi, K. Yoshihara and S. Nagakura. *Bull. Chem. Soc. Jap.*, **44**, 2603 (1971).

62. G. Briegleb, H. Schuster and W. Herre. *Chem. Phys. Lett.*, **4**, 53 (1969); G. Briegleb, G. Betz and W. Herre. *Z. Phys. Chem.* (Frankfurt am Main), **64**, 85 (1969); G. Briegleb and H. Schuster. *Angew. Chem. Int. Ed. Engl.*, **8**, 771 (1969); S. Briegleb, W. Herre and D. Wolf. *Spectrochim. Acta*, **25**, 39 (1969); G. Briegleb and H. Schuster. *Z. Phys. Chem.* (Frankfurt am Main), **77**, 269 (1972); T. Kobayashi and S. Nagakura. *Bull. Chem. Soc. Jap.*, **45**, 987 (1972); M. Irie, S. Tomimoto and K. Hayashi. *J. Phys. Chem.*, **76**, 1419 (1972).

63. N. S. Isaacs and J. Paxton. *Photochem. and Photobiol.*, **11**, 137 (1970); M. Soma. *J. Amer. Chem. Soc.*, **92**, 3289 (1970); H. Masuhara, M. Shimada and N. Mataga. *Bull. Chem. Soc. Jap.*, **43**, 3316 (1970); K. Egawa, N. Nakashima, N. Mataga and C. Yamanaka. *Bull. Chem. Soc. Jap.* **44**, 3287 (1971); M. Shimada, H. Masuhara and N. Mataga. *Chem. Phys. Lett.*, **15**, 364 (1972); Y. Taniguchi, Y. Nishina and N. Mataga. *Bull. Chem. Soc. Jap.*, **45**, 764 (1972); Y. P. Pilette and K. Weiss. *J. Phys. Chem.*, **75**, 3805 (1971); S. Arimitsu and H. Tsubomura. *Bull. Chem. Soc. Jap.*, **45**, 2433 (1972); Y. Shirota, K. Kawai, N. Yamamoto, K. Tada, T. Shida, H. Mikawa and H. Tsubomura. *Bull. Chem. Soc. Jap.*, **45**, 2683 (1972); Y. Achiba, S. Katsumata and K. Kimura. *Chem. Phys. Lett.*, **13**, 213 (1972).

64. N. Christodouleas and S. P. McGlynn. *J. Chef. Phys.*, **40**, 166 (1964).

65. R. Potashnik, C. R. Goldschmidt and M. Ottolenghi. *J. Phys. Chem.*, **73**, 3170 (1969).

66. K. Kawai, N. Yamamoto and H. Tsubomura. *Bull. Chem. Soc. Jap.*, **42**, 369 (1969).

67a. A. Weller. *Pure and Appl. Chem.*, **16**, 115 (1968).

67b. H. Knibbe, D. Rehm and A. Weller. *Ber. Bunsenges. Phys. Chem.*, **73**, 839 (1969); D. Rehm and A. Weller. *Z. Phys. Chem.* (Frankfurt am Main), **69**, 183 (1970).

67c. H. Beens and A. Weller. *Acta Physica Polonica*, **34**, 593 (1968); H. Beens, H. Knibbe and A. Weller. *J. Chem. Phys.*, **47**, 1183 (1967).

67d. H. Knibbe, D. Rehm and A. Weller. *Ber. Bunsenges. Phys. Chem.*, **72**, 257 (1968).

67e. H. Knibbe, K. Röllig, F. P. Schäfer and A. Weller. *J. Chem. Phys.*, **47**, 1184 (1967).

68a. E. A. Chandross and H. T. Thomas. *Chem. Phys. Lett.*, **9**, 393, 397 (1971).

68b. D. R. G. Brimage and R. S. Davidson. *Chem. Commun.*, 1385 (1971).

68c. G. S. Beddard, R. S. Davidson and A. Lewis. *J. Photochem.*, **1**, 491 (1972/3).

68d. R. Ide, Y. Sakata, S. Misumi, T. Okada and N. Mataga. *Chem. Commun.*, 1009 (1972).

69. R. S. Davidson and A. Lewis. *Chem. Commun.*, 262 (1973).

70. A. E. W. Knight and B. K. Selinger. *Chem. Phys. Lett.*, **10**, 43 (1971).

71. G. S. Beddard and R. S. Davidson, unpublished results.

72. M. G. Kuzmin and L. N. Guseva. *Chem. Phys. Lett.*, **3**, 71 (1969).

73. M. Ottolenghi. *Accounts Chem. Res.*, **6**, 153 (1973); C. R. Goldschmidt and M. Ottolenghi, *Chem. Phys. Lett.* **4**, 570 (1970); R. Potashnik, C. R. Goldschmidt, M. Ottolenghi and A. Weller. *J. Chem. Phys.*, **55**, 5344 (1971).

74. G. N. Taylor and G. S. Hammond. *J. Amer. Chem. Soc.*, **94**, 3684 (1972).

75a. Th. Förster. *Angew. Chem. Int. Ed. Engl.*, **8**, 333 (1969).

75b. J. B. Birks, "Photophysics of Aromatic Molecules," Wiley-Interscience, London (1970), ch. 7.

75c. J. B. Birks, Progress in Reaction Kinetics, **5**, 273 (1970); W. Klöpffer in "Organic Molecular Photophysics", ed. J. B. Birks, Wiley, London (1973), ch. 7.

75d. B. Stevens. *Advan Photochem.,* **8**, 161 (1971).

75e. E. A. Chandross and C. J. Dempster. *J. Amer. Chem. Soc.,* **92**, 703, 704 (1970).

76. M. F. M. Post, J. Langelaar and J. D. W. Van Voorst. *Chem. Phys. Letters,* **10**, 468 (1971); J. T. Richards and J. K. Thomas. *Chem. Phys. Letters,* **5**, 527 (1970).

77. G. Castro and R. M. Hochstrasser. *J. Chem. Phys.,* **45**, 4352 (1966).

78. S. G. Cohen, A. Parola and G. H. Parsons. *Chem. Rev.,* **73**, 141 (1973).

79. K. A. Zachariasse, Ph.D. Thesis, University of Amsterdam (1972).
See also R. E. Hemingway, S.-M. Park and A. J. Bard. *J. Amer. Chem. Soc.,* **97**, 200 (1975).

80. C. Lewis and W. R. Ware. *Mol. Photochem.,* **5**, 261 (1973).

81a. D. Rehm and A. Weller. *Ber. Bunsenges. Phys. Chem.,* **73**, 834 (1969); D. Rehm and A. Weller. *Israel J. Chem.,* **8**, 259 (1970).

81b. K. H. Grellmann, A. R. Watkins and A. Weller. *J. Phys. Chem.,* **76**, 469 (1972).

82. Y. Taniguchi, Y. Nishina and N. Mataga. *Bull. Chem. Soc. Jap.,* **45**, 764 (1972); Y. Taniguchi and N. Mataga. *Chem. Phys. Lett.,* **13**, 596 (1972).

83a. N. Orbach, R. Potashnik and M. Ottolenghi. *J. Phys. Chem.,* **76**, 1133 (1972).

83b. N. Nakashima, N. Mataga, F. Ushio and C. Yamanaka. *Z. Phys. Chem.* (Frankfurt am Main), **79**, 150 (1972).

84. H. Schomburg, H. Staerk and A. Weller. *Chem. Phys. Lett.,* **21**, 433 (1973).

85. A. R. Watkins. *J. Phys. Chem.,* **77**, 1207 (1973).

86. R. S. Davidson, P. F. Lambeth, J. F. McKellar, P. H. Turner and R. R. Wilson. *Chem. Comm.,* 732 (1969); R. F. Bartholomew, R. S. Davidson, P. F. Lambeth, J. F. McKellar and P. H. Turner. *J. Chem. Soc., Perkin II,* 577 (1972).

87. T. Medinger and F. Wilkinson. *Trans. Faraday Soc.,* **62**, 1785 (1966).

88. W. Heinzelmann and H. Labhart. *Chem. Phys. Lett.,* **4**, 20 (1969).

89. R. B. Cundall, L. C. Pereira and D. A. Robinson. *Chem. Phys. Lett.,* **13**, 253 (1972); R. R. Hentz and R. M. Thibault. *J. Phys. Chem.,* **77**, 1105 (1973).

90. J. Jortner and U. Sokolov. *Nature,* **190**, 1003 (1961); *J. Phys. Chem.,* **65**, 1633 (1961).

91. K. S. Wei and A. H. Adelman. *Tetrahedron Lett.,* 3297 (1969).

92. A. U. Munck and J. F. Scott. *Nature,* **177**, 587 (1956).

93. L. E. Orgel and R. S. Mulliken. *J. Amer. Chem. Soc.,* **79**, 4839 (1957).

94. J. B. Birks, "Photophysics of Aromatic Molecules", Wiley-Interscience, London (1970), ch. 10.

95. G. J. Hoytink. *Accounts. Chem. Res.,* **2**, 114 (1969).

96a. E. J. Bowen and W. S. Metcalf. *Proc. Roy. Soc., Ser. A,* **206**, 437 (1951); W. R. Ware. *J. Phys. Chem.,* **66**, 455 (1962).

96b. T. Brewer. *J. Amer. Chem. Soc.,* **93**, 775 (1971).

96c. L. K. Patterson, G. Porter and M. R. Topp. *Chem. Phys. Lett.,* **7**, 612 (1970).

96d. R. Potashnik, C. R. Goldschmidt and M. Ottolenghi. *Chem. Phys. Lett.,* **9**, 424 (1971).

96e. H. Ishida, H. Takahashi and H. Tsubomura. *Bull. Chem. Soc. Jap.,* **43**, 3130 (1970); B. Stevens and B. E. Algar. *Ann. N.Y. Acad. Sci.,* **171**, 50 (1970).

96f. H. Ishida, H. Takahashi, H. Sato and H. Tsubomura. *J. Amer. Chem. Soc.,* **92**, 275 (1970).

96g. J. Yguerabide. *J. Chem. Phys.,* **49**, 1018, 1026 (1968).

96h. I. B. Berlman, C. R. Goldschmidt, G. Stein, Y. Tomkiewicz and A. Weinreb. *Chem. Phys. Lett.,* **4**, 338 (1969).

97a. C. S. Foote. *Accounts Chem. Res.*, **1**, 104 (1968); F. Gollnick. *Advan. Photochem.*, **6**, 2 (1968); D. R. Kearns. *Chem. Rev.*, **71**, 395 (1971).

97b. C. S. Foote in "Free Radicals and Biological Systems", ed. W. A. Pryor, Academic Press, New York (1975).

98. A. U. Khan and M. Kasha. *Ann. N.Y. Acad. Sci.*, **171**, 24 (1970).

99. O. L. J. Gijzeman, F. Kaufman and G. Porter, *J. Chem. Soc.*, *Faraday II*, **69**, 708 (1973); O. L. J. Gijzeman and F. Kaufman, *J. Chem. Soc.*, *Faraday II*, **69**, 721 (1973).

100. E. J. Bowen. *Advan. Photochem.*, **1**, 23, (1963).

101. J. Rigaudy, J. Guillaume and D. Maurette. *Bull. Chem. Soc. Fr.*, 144 (1971).

102. H. Brockmann and F. Dicke. *Chem. Ber.*, **103**, 7 (1970).

103. B. Stevens and B. E. Algar. *J. Phys. Chem.*, **72**, 3468 (1968).

104. N. Sugiyama, M. Iwata, M. Yoshioka, K. Yamada and H. Aoyama. *Chem. Commun.*, 1563 (1968); N. Sugiyama, M. Iwata, M. Yoshioka, K. Yamada and and H. Aoyama. *Bull. Chem. Soc. Jap.*, **42**, 1377 (1969).

105. H. Hart and A. Oku. *Chem. Commun.*, 254 (1972).

106. J. Rigaudy and M. Maumy. *Bull. Chem. Soc. Fr.*, 3936 (1972).

107. H. Shizuka, K. Sorimachi, T. Morita, K. Nishiyama and T. Sato. *Bull. Chem. Soc. Jap.*, **44**, 1983 (1971).

108. M. Chubachi and M. Hamada. *Tetrahedron Lett.*, 3537 (1971); J. Betts and J. C. Robb. *Trans. Faraday Soc.*, **65**, 2144 (1969).
     M. Höfert, *Photochem. and Photobiol.*, **9**, 427 (1969).

109. J. C. W. Chien. *J. Phys. Chem.*, **69**, 4317 (1965).

110. H. H. Wasserman, P. S. Mariano and P. M. Keehn. *J. Org. Chem.*, **36**, 1765 (1971).

111. T. Matsuura, A. Nishinaga, N. Yoshimura, T. Arai, K. Omura, H. Matsushima, S. Kato and I. Saito. *Tetrahedron Lett.*, 1673 (1969).

112. A. Miyake and M. Tomoeda. *J. Chem. Soc.*, *Perkin 1*, 663 (1972).

113. R. P. Seiber and H. L. Needles. *Chem. Commun.*, **209** (1972).

114. N. Kulevsky, P. V. Sneeringer, L. D. Grina and V. I. Stenberg. *Photochem. and Photobiol.*, **12**, 395 (1970).

115. V. I. Stenberg, P. V. Sneeringer, C. Niu and N. Kulevsky. *Photochem. and Photobiol.*, **16**, 81 (1972); N. Kulevsky, P. V. Sneeringer and V. I. Stenberg. *J. Org. Chem.*, **37**, 438 (1972); N. Friedman, M. Gorodetsky and Y. Mazur. *Chem. Commun.*, 874 (1971).

116. Y. P. Pilette and K. Weiss. *J. Phys. Chem.*, **75**, 3805 (1971).

117. M. Shimada, H. Masuhara and N. Mataga. *Bull. Chem. Soc. Jap.*, **46**, 1903 (1973).

118. D. E. Breen and R. A. Keller. *J. Amer. Chem. Soc.*, **90**, 1935 (1968).

119. R. S. Davidson and A. Lewis, unpublished results.

120. D. Bryce-Smith, B. E. Connett and A. Gilbert. *J. Chem. Soc.* (*B*), 816 (1968); D. Bryce-Smith and M. A. Hems. *J. Chem. Soc.* (*B*), 812 (1968).

121a. H. J. F. Angus and D. Bryce-Smith. *J. Chem. Soc.*, 4791 (1960).

121b. D. Bryce-Smith and A. Gilbert. *J. Chem. Soc.*, 918 (1965).

121c. G. O. Schenck and R. Steinmetz. *Tetrahedron Lett.*, **21**, 1 (1960); E. Grovenstein, D. V. Rao and J. W. Taylor. *J. Amer. Chem. Soc.*, **83**, 1705 (1961).

121d. D. Bryce-Smith. *Pure Appl. Chem.*, **34**, 193 (1973).

122. D. Bryce-Smith and A. Gilbert. *Chem. Commun.*, **19**, 1968.

123. D. Bryce-Smith, R. Deshpande, A. Gilbert and J. Grzonka. *Chem. Commun.*, 561 (1970).

124. Z. Raciszewski. *J. Chem. Soc.* (*B*), 1147 (1966).

125. Z. Raciszewski. *J. Chem. Soc.* (*B*), 1142 (1966).

126. G. B. Vermont, P. X. Riccobono and J. Blake. *J. Amer. Chem. Soc.*, **87**, 4024 (1965).
127. T. Matsuo, Y. Tanoue, T. Matsunaga and K. Nagaleski. *Chem. Letters (Tokyo)*, 709 (1972).
128. M. Verbeek, H.-D. Scharf and F. Korte. *Chem. Ber.*, **102**, 2471 (1969).
129. D. Bryce-Smith, A. Gilbert and B. H. Orger. *Chem. Commun.*, 512 (1966).
130. W. Hartmann and H. G. Heine. *Angew. Chem. Int. Ed. Engl.*, **10**, 272 (1971).
131. J. Meinwald, G. E. Samuelson and M. Ikeda. *J. Amer. Chem. Soc.*, **92**, 7604 (1970).
132. S. Farid and S. E. Shealer. *Chem. Commun.*, 296 (1973).
133. D. Bryce-Smith and M. A. Hems. *Tetrahedron Lett.*, 1895 (1966); D. Bryce-Smith and M. A. Hems, *Tetrahedron*, **25**, 247 (1969).
134a. K. H. Grellmann, A. R. Watkins and A. Weller. *J. Phys. Chem.*, **76**, 469 (1972).
134b. K. H. Grellmann, A. R. Watkins and A. Weller. *J. Phys. Chem.*, **76**, 3132 (1972).
134c. T. Okada, H. Oohari and N. Mataga. *Bull. Chem. Soc. Jap.*, **43**, 2750 (1970).
135. A. Yoshino, M. Ohashi and T. Yonezawa. *Chem. Commun.*, 97 (1971).
136. K. Yamasaki, A. Yoshino, T. Yonezawa and M. Ohashi. *Chem. Commun.*, 9 (1973).
137a. M. B. Rubin and Z. Neuwirth-Weiss. *Chem. Commun.*, 1607 (1968).
137b. K. Maruyuma, K. Ono and J. Osugi. *Bull. Chem. Soc. Jap.*, **42**, 3357 (1969).
137c. P. A. Carapellucci, H. P. Wolf and K. Weiss. *J. Amer. Chem. Soc.*, **91**, 4635 (1969).
138. J. A. Barltrop and R. Robson. *Tetrahedron Lett.*, 597 (1963).
139. H. Tsubomura and R. S. Mulliken. *J. Amer. Chem. Soc.*, **82**, 5966 (1960).
140. R. A. Ackerman, I. Rosenthal and J. N. Pitts. *J. Chem. Phys.*, **54**, 4960 (1971).
141a. E. A. Ogryzlo and C. W. Tang. *J. Amer. Chem. Soc.*, **92**, 5034 (1970); I. B. C. Matheson and J. Lee. *J. Amer. Chem. Soc.*, **94**, 3310 (1972); R. H. Young, R. L. Martin, D. Feriozi, D. Brewer and R. Kayser. *Photochem. and Photobiol.*, **17**, 233 (1973).
141b. R. H. Young and R. L. Martin. *J. Amer. Chem. Soc.*, **94**, 5183 (1972).
142. B. Pouyet and R. Chapelon. *C.R. Acad. Sci. Ser. C.* **272**, 1753 (1971).
143. E. A. Mayeda and A. J. Bard. *J. Amer. Chem. Soc.*, **95**, 6223 (1973).
144. V. I. Stenberg, R. D. Olson, C. T. Wang and N. Kulevsky. *J. Org. Chem.*, **32**, 3227 (1967).
145. C. S. Foote and J. W. Peters. *J. Amer. Chem. Soc.*, **93**, 3795 (1971).
146. K. Gollnick and J. H. E. Lindner. *Tetrahedron Lett.*, 1903 (1973).
147. E. G. E. Hawkins. *J. Chem. Soc., Perkin 1*, 13 (1972).
148a. R. F. Bartholomew and R. S. Davidson. *J. Chem. Soc. (C)*, 2342 (1971).
148b. R. F. Bartholomew, R. S. Davidson and M. J. Howell. *J. Chem. Soc. (C)*, 2804 (1971).
148c. R. F. Bartholomew, D. R. G. Brimage and R. S. Davidson. *J. Chem. Soc. (C)*, 3482 (1971).
148d. R. F. Bartholomew and R. S. Davidson. *J. Chem. Soc. (C)*, 2347 (1971).
148e. F. C. Schaefer and W. D. Zimmermann. *J. Org. Chem.*, **35**, 2165 (1970).
148f. D. Herlem, Y. Hubert-Brierre, F. Khuong-Huu and R. Goutarel. *Tetrahedron*, **29**, 2195 (1973); F. Khuong-Huu and D. Herlem. *Tetrahedron Lett.*, 3649 (1970).
148g. J. H. E. Lindner, H. J. Kuhn and K. Gollnick. *Tetrahedron Lett.*, 1705 (1972).
148h. M. H. Fisch, J. C. Gramain and J. A. Oleson. *Chem. Commun.*, 13 (1970); M. H. Fisch, J. C. Gramain and J. A. Olesen. *Chem. Commun.*, 663 (1971).
149. M. Kawana and S. Emoto. *Bull. Chem. Soc. Jap.*, **41**, 2552 (1968).
150. R. S. Atkinson, D. R. G. Brimage, R. S. Davidson and E. Gray. *J. Chem. Soc., Perkin 1*, 960 (1973).

151. W. F. Smith. *J. Amer. Chem. Soc.*, **94**, 186 (1972).
152. R. Chapelon, G. Perichet and B. Pouyet. *Mol. Photochem.*, **5**, 77 (1973).
153. N. Kulevsky, C. Niu and V. I. Stenberg. *J. Org. Chem.*, **38**, 1154 (1973).
154. I. R. C. Bick, J. B. Bremner and P. Wiriyachitra. *Tetrahedron Lett.*, 4795 (1971);
     I. R. C. Bick, J. B. Bremner, H. M. Leow and P. Wiriyachitra. *Tetrahedron Lett.*,
     33 (1972).
155. M. Hori, H. Itoi and H. Tsubomura. *Bull. Chem. Soc. Jap.*, **43**, 3765 (1970).
156. T. Iwaoka, H. Kokubun and M. Koizumi. *Bull. Chem. Soc. Jap.*, **44**, 341 (1971);
     T. Iwaoka, H. Kokubun and M. Koizumi. *Bull. Chem. Soc. Jap.*, **44**, 3466 (1971).
157. D. P. Stevenson and G. M. Coppinger. *J. Amer. Chem. Soc.*, **84**, 149 (1962).
158. K. M. C. Davis and M. F. Farmer. *J. Chem. Soc. (B)*, 28 (1967).
159. W. C. Meyer. *J. Phys. Chem.*, **74**, 2118 (1970); S. Matsuda, H. Kokado and
     E. Inoue. *Bull. Chem. Soc. Jap.*, **43**, 2994 (1970).
160. T. Tosa, C. Pac and H. Sakurai. *Tetrahedron Lett.*, 3635 (1969); C. Pac, T. Tosa
     and H. Sakurai. *Bull. Chem. Soc. Jap.*, **45**, 1169 (1972).
161. T. Latowski. *Z. Naturforsch*, **23a**, 1127 (1968).
162. M. Ohashi, K. Tsujimoto and K. Seki. *Chem. Commun.*, 384 (1973).
163a. E. M. Kosower in "Progress in Physical Organic Chemistry, vol. 3", *eds.*
     S. G. Cohen, A. Streitwieser, Jr. and R. W. Taft, Interscience, New York (1965),
     p. 81.
163b. W. J. Lautenberger, E. N. Jones and J. G. Miller. *J. Amer. Chem. Soc.*, **90**, 1110
     (1968).
164. K. G. Hancock and D. A. Dickinson. *Chem. Commun.*, 783 (1973).
165. L. L. Miller, R. S. Narang and G. D. Nordblom. *J. Org. Chem.*, **38**, 340 (1973).
166. L. L. Miller and R. S. Narang. *Science*, **169**, 368 (1970).
167. K.-I. Hirao and O. Yonemitsu. *Chem. Commun.*, 812 (1972).
168. E. A. Fitzgerald, P. Wuelfing and H. H. Richtol. *J. Phys. Chem.*, **75**, 2737 (1971).
169. J. Nasielski, A. Kirsch-Demesmaeker, P. Kirsch and R. Nasielski-Hinkens.
     Chem. Commun., 302 (1970); J. Nasielski and A. Kirsch-Demesmaeker, *Tetra-
     hedron*, **29**, 3153 (1973).
170. S. Naruto and O. Yonemitsu. *Tetrahedron Lett.*, 2297 (1971).
171. R. S. Davidson and P. R. Steiner. *J. Chem. Soc. (C)*, 1682 (1971).
172. R. S. Davidson and P. R. Steiner. *J. Chem. Soc.*, *Perkin II*, 1357 (1972).
173. W. N. White. *J. Amer. Chem. Soc.*, **81**, 2912 (1959).
174. J. A. Barltrop and N. J. Bunce. *J. Chem. Soc. (C)*, 1467 (1968).
175. M. Takami, T. Matsuura and I. Saito. *Tetrahedron Lett.*, 661 (1974).
176. R. S. Davidson, S. Korkut and P. R. Steiner. *Chem. Commun.*, 1052 (1971).
177a. D. J. Neadle and R. J. Pollitt. *J. Chem. Soc. (C)*, 2127 (1969); P. H. MacFarlane
     and D. W. Russell. *Tetrahedron Lett.*, 725 (1971).
177b. R. Fielden, O. Meth-Cohn and H. Suschitzky. *Tetrahedron Lett.*, 1229 (1970);
     O. Meth-Cohn. *Tetrahedron Lett.*, 1235 (1970).
178. D. Doepp. *Chem. Ber.*, **104**, 1058 (1971).
179. R. S. Goudie and P. N. Preston. *J. Chem. Soc. (C)*, 3081 (1971).
180. K. Dimroth, C. Reichardt, T. Siepmann and F. Bohlmann. *Justus Liebigs. Ann.
     Chem.*, **661**, 1 (1963).
181. E. M. Kosower and L. Lindqvist. *Tetrahedron Lett.*, 4481 (1965).
182. R. F. Cozzens and T. A. Gover. *J. Phys. Chem.*, **74**, 3003 (1970).
183a. K. D. Legg and D. M. Hercules. *J. Phys. Chem.*, **74**, 2114 (1970).
183b. D. G. Whitten, J. W. Happ, G. L. B. Carlson and M. T. McCall. *J. Amer. Chem.
     Soc.*, **92**, 3499 (1970).

184. A. Ledwith. *Accounts Chem. Res.*, **5**, 133 (1972).
185. A. Ledwith, P. J. Russell and L. H. Sutcliffe. *Chem. Commun.*, 964 (1971).
186. N. M. D. Brown, D. J. Cowley and W. J. Murphy. *Chem. Commun.*, 592 (1973).
187. J. F. McKellar and P. H. Turner. *Photochem. and Photobiol.*, **13**, 437 (1971).
188. J. R. Barnett, A. S. Hopkins and A. Ledwith. *J. Chem. Soc., Perkin II*, 80 (1973).
189. T. D. Walsh and R. C. Long. *J. Amer. Chem. Soc.*, **89**, 3943 (1967).
190. C. Pac and H. Sakurai. *Tetrahedron Lett.*, 1865 (1968); C. Pac and H. Sakurai. *Kogyo Kagaku Zasshi*. **72**, 230 (1969); *Chem. Abstr.*, **70**, 114373 (1969).
191. C. Pac and H. Sakurai. *Chem. Commun.*, 20 (1969).
192. J. F. McKellar. *Proc. Roy. Soc., Ser. A.*, **287**, 363 (1965).
193. S. L. Nickol and J. A. Kampmeier. *J. Amer. Chem. Soc.*, **95**, 1908 (1973).
194. C. E. Griffin and M. L. Kaufman. *Tetrahedron Lett.*, 773 (1965).
195. Y. Nagao, K. Shima and H. Sakurai. *Tetrahedron Lett.*, 1101 (1971).
196. Z. Yoshida and T. Kobayashi. *Tetrahedron*. **26**, 267 (1970).
197. G. O. Schenck, W. Hartmann, S.-P. Mannsfeld, W. Metzner and C. H. Krauch. *Chem. Ber.*, **95**, 1642 (1962).
198. G. O. Schenck, W. Hartmann and R. Steinmetz. *Chem. Ber.*, **96**, 498 (1963).
199. R. S. Davidson and A. Lewis. *Tetrahedron Lett.*, 611 (1974).
200. R. S. Davidson and A. Lewis, unpublished results.
201. J. M. Fayadh and G. A. Swan. *J. Chem. Soc. (C)*, 1781 (1969).
202. I. Saito, M. Takami and T. Matsuura. *Tetrahedron Lett.*, 659 (1974).
203. J. L. Charlton and P. de Mayo. *Can. J. Chem.*, **46**, 1041 (1968).
204. C. A. Mudry and A. R. Frasca. *Chem. Ind. (London)*, 1038 (1971); H. B. Land and A. R. Frasca. *Tetrahedron*, **26**, 5793 (1970); C. A. Mudry and A. R. Frasca, *Tetrahedron*, **29**, 603 (1973).
205. D. H. Iles and A. Ledwith. *Chem. Commun.*, 364 (1969).
206. R. Calas, R. Lalande and P. Mauret. *Bull. Chem. Soc. Fr.*, 140 (1960).
207. A. S. Cherkasov and T. M. Vember. *Opt. Spectrosk.*, **6**, 319 (1959).
208. B. Stevens, T. Dickinson and R. R. Sharpe. *Nature*, **204**, 876 (1964); B. Stevens, R. R. Sharpe and S. A. Emmons, *Photochem. and Photobiol.*, **4**, 603 (1965).
209a. R. B. Cundall and D. A. Robinson. *J. Chem. Soc., Faraday II*, 1133 (1972); R. B. Cundall and D. A. Robinson. *Chem. Phys. Lett.*, **13**, 257 (1972); R. B. Cundall and D. A. Robinson. *J. Chem. Soc., Faraday II*, 1323 (1972).
209b. R. L. Barnes and J. B. Birks. *Proc. Roy. Soc., Ser A*, **291**, 570 (1966).
210. E. A. Chandross. *J. Chem. Phys.*, **43**, 4175 (1965).
211. R. Calas and R. Lalande. *Bull. Chem. Soc. Fr.*, 763, 766, 770 (1959).
212. R. Calas, R. Lalande, J. G. Gangere and F. Moulines. *Bull. Chem. Soc. Fr.*, 119 (1965).
213. R. Calas, P. Mauret and R. Lalande. *C.R. Acad. Sci. Ser.*, **247**, 2146 (1958); R. Lalande and R. Calas, *Bull. Chem. Soc. Fr.*, 144, 148 (1960).
214. F. D. Greene, S. L. Misrock and J. R. Wolfe. *J. Amer. Chem. Soc.*, **77**, 3852 (1955).
215. R. Lapouyade, A. Castellan and H. Bouas-Laurent. *C.R. Acad. Sci. Ser. C*, **268**, 217 (1969).
216. H. Bouas-Laurent and R. Lapouyade. *Chem. Commun.*, 817 (1969).
217. J. B. Birks and T. A. King. *Proc. Roy. Soc., Ser. A*, **291**, 244 (1966).
218a. P. Wilairat and B. Selinger. *Aust. J. Chem.*, **21**, 733 (1968).
218b. T. Teitei, D. Wells and W. H. F. Sasse. *Tetrahedron Lett.*, 367 (1974).
219. J. S. Bradshaw, N. B. Neilson and D. P. Rees, *J. Org. Chem.*, **33**, 259 (1968).
220a. D. O. Cowen and R. L. E. Drisko. *J. Amer. Chem. Soc.*, **92**, 6281 (1970); D. O. Cowan and R. L. E. Drisko. *J. Amer. Chem. Soc.*, **92**, 6286 (1970).

220b. I. M. Hartmann, W. Hartmann and G. O. Schenck. *Chem. Ber.,* **100,** 3146 (1967).
221. N. Y. C. Chu and D. R. Kearns. *J. Phys. Chem.,* **74,** 1255 (1970).
222. E. A. Chandross and H. T. Thomas. *J. Amer. Chem. Soc.,* **94,** 2421 (1972).
223. M. D. Cohen, A. Elgavi, B. S. Green, Z. Ludmer and G. M. J. Schmidt. *J. Amer. Chem. Soc.,* **94,** 6776 (1972).
224. M. D. Cohen, B. S. Green, Z. Ludmer and G. M. J. Schmidt. *Chem. Phys. Lett.,* **7,** 486 (1970).
225. H. Morrison, H. Curtis and T. McDowell. *J. Amer. Chem. Soc.,* **88,** 5415 (1966).
226. R. Hoffman, P. Wells and H. Morrison. *J. Org. Chem.,* **36,** 102 (1971)
227. D. Weinblum and H. E. Johns. *Biochim. Biophys. Acta,* **114,** 450 (1966).
228. H. D. Roth and A. A. Lamola. *J. Amer. Chem. Soc.,* **94,** 1013 (1972).
229. A. A. Lamola. *Pure Appl. Chem.,* **24,** 599 (1970).
230. J. Eisinger and A. A. Lamola. *Biochim. Biophys. Res. Commun.,* **28,** 558 (1967); J. Eisinger. *Photochem. and Photobiol.,* **7,** 597 (1968); A. A. Lamola and J. Eisinger, *Proc. Nat. Acad. Sci. U.S.,* **59,** 46 (1968).
231. W. Klöpffer, in "Organic Molecular Photophysics", ed. J. B. Birks. Wiley, London (1973), ch. 7.
232. F. Hirayama. *J. Chem. Phys.* **42,** 3163 (1965).
233. E. A. Chandross and C. J. Dempster. *J. Amer. Chem. Soc.,* **92,** 3586 (1970).
234. E. A. Chandross and C. J. Dempster. *J. Amer. Chem. Soc.,* **92,** 704 (1970).
235. E. A. Chandross and A. H. Schiebel. *J. Amer. Chem. Soc.,* **95,** 611 (1973).
236a. E. A. Chandross and A. H. Schiebel. *J. Amer. Chem. Soc.,* **95,** 1671 (1973).
236b. M. Itoh, T. Mimura, H. Usui and T. Okamoto. *J. Amer. Chem. Soc.,* **95,** 4388 (1973).
237a. J. H. Golden. *J. Chem. Soc.,* 3741 (1961).
237b. N. M. Weinshenker and F. D. Greene. *J. Amer. Chem. Soc.,* **90,** 506 (1968).
237c. R. Livingston and K. S. Wei. *J. Amer. Chem. Soc.,* **89,** 3098 (1967).
237d. D. E. Applequist, M. A. Lintner and R. Searle. *J. Org. Chem.,* **33,** 254 (1968).
238. F. C. DeSchryver, M. De Brackeleire, S. Toppet and M. Van Schoor. *Tetrahedron Lett.,* 1253 (1973).
239. L. Leenders and F. C. DeSchryver. *Angew. Chem. Int. Ed. Engl.,* **10,** 338 (1971); L. H. Leenders, E. Schouteden and F. C. De Schryver. *J. Org. Chem.,* **38,** 957 (1973).
240. M. Van Meerbeck, S. Toppet and F. C. DeSchryver. *Tetrahedron Lett.,* 2247 (1972).
241. F. C. DeSchryver, I. Bhardwaj and J. Put. *Angew. Chem. Int. Ed. Engl.,* **8,** 213 (1969); J. Put and F. C. DeSchryver. *J. Amer. Chem. Soc.,* **95,** 137 (1973); F. C. DeSchryver. *Pure. Appl. Chem.,* **34,** 213 (1973).
242. J. Eisinger, M. Guéron, R. G. Shulman and T. Yamane. *Proc. Nat. Acad. Sci. U.S.,* **55,** 1015 (1966); M. Gueron, R. G. Shulman and J. Eisinger. *Proc. Nat. Acad. Sci. U.S.,* **56,** 814 (1966); J. Eisinger and R. G. Shulman. *Science,* **161,** 1311 (1968).
243. D. T. Browne, J. Eisinger and N. J. Leonard. *J. Amer. Chem. Soc.,* **90,** 7302 (1968).
244a. N. J. Leonard, R. S. McCredie, M. W. Logue and R. L. Cundall. *J. Amer. Chem. Soc.,* **95,** 2320 (1973).
244b. N. J. Leonard, K. Golankiewicz, R. S. McCredie, S. M. Johnson and I. C. Paul. *J. Amer. Chem. Soc.,* **91,** 5855 (1969).
245. K. Golankiewicz and L. Strekowski. *Bull. Acad. Pol. Sci. Ser. Sci. Chem.,* **19,** 171 (1971); *Chem. Abstr.,* **77,** 87454 (1972).
246. A. Lablache-Combier. *Bull. Chem. Soc. Fr.,* 4791 (1972).
247. R. S. Davidson. *Chem. Commun.,* 1450 (1969).

248. C. Pac and H. Sakurai. *Tetrahedron Lett.*, 3829 (1969).
249. J. A. Barltrop and R. J. Owers. *Chem. Commun.*, 1462 (1970).
250. V. R. Rao and V. Ramakrishnan. *Chem. Commun.*, 971 (1971).
251. R. F. Bartholomew, D. R. G. Brimage and R. S. Davidson. *J. Chem. Soc. (C)*, 3482 (1971).
252. R. C. Cookson, S. M. de B. Costa and J. Hudec. *Chem. Commun.*, 753 (1969).
253. M. Bellas, D. Bryce-Smith and A. Gilbert. *Chem. Commun.*, 263 (1967).
254. D. Bryce-Smith, M. T. Clarke, A. Gilbert, G. Klunklin and C. Manning. *Chem. Commun.*, 916 (1971).
255. M. Bellas, D. Bryce-Smith and A. Gilbert. *Chem. Commun.*, 862 (1967).
256. J. J. McCullough, C. W. Huang and W. S. Wu. *Chem. Commun.*, 1368 (1970); J. J. McCullough and W. S. Wu. *Tetrahedron Lett.*, 3951 (1971); J. J. McCullough, W. S. Wu and C. W. Huang. *J. Chem. Soc., Perkin II*, 370 (1972).
257. N. C. Yang and J. Libman. *J. Amer. Chem. Soc.*, **95**, 5783 (1973).
258. D. R. G. Brimage and R. S. Davidson. *J. Chem. Soc., Perkin I*, 496 (1973).
259. R. S. Davidson and S. P. Orton. *Chem. Commun.*, 209 (1974).
260. D. Bryce-Smith and G. B. Cox. *Chem. Commun.*, 915 (1971).
261. R. S. Davidson and P. F. Lambeth. *Chem. Commun.*, 1098 (1969).
262. E. A. Chandross and H. T. Thomas. *Chem. Phys. Lett.*, **9**, 393 (1971).
263. D. R. G. Brimage and R. S. Davidson. *Chem. Commun.*, 1385 (1971).
264. G. S. Beddard, R. S. Davidson and A. Lewis. *J. Photochem.*, **1**, 491 (1972/73).
265. R. Ide, Y. Sakata, S. Misumi, T. Okada and N. Mataga. *Chem. Commun.*, 1009 (1972).
266. D. Bryce-Smith, A. Gilbert and G. Klunklin. *Chem. Commun.*, 330 (1973).
267. A. Matsuyama and C. Nagata. *Photochem. and Photobiol.*, **15**, 319 (1972).
268. C. Antonello, F. Carlassare and L. Musajo. *Gazz. Chim. Ital.*, **98**, 30 (1968).
269. J. M. Rice. *J. Amer. Chem. Soc.*, **86**, 1444 (1964).
270. G. M. Blackburn, R. G. Fenwick and M. H. Thompson. *Tetrahedron Lett.*, 589 (1972).
271. E. Cavalieri and M. Calvin. *Photochem. and Photobiol.*, **14**, 641 (1971).
272. C. Antonello and F. Carlassare. *Z. Naturforsch*, **25B**, 1269 (1970).
273a. G. R. Seely. *J. Phys. Chem.*, **73**, 125 (1969).
273b. G. R. Seely. *J. Phys. Chem.*, **73**, 117 (1969).
274. D. G. Whitten, I. G. Lopp and P. D. Wildes. *J. Amer. Chem. Soc.*, **90**, 7196 (1968); I. G. Lopp, R. W. Hendren, P. D. Wildes and D. G. Whitten. *J. Amer. Chem. Soc.*, **92**, 6440 (1970).
275. J. K. Roy and D. G. Whitten. *J. Amer. Chem. Soc.*, **93**, 7093 (1971).
276. J. K. Roy and D. G. Whitten. *J. Amer. Chem. Soc.*, **94**, 7162 (1972).
277. J. K. Roy, F. A. Carroll and D. G. Whitten, submitted for publication to J. Amer. Chem. Soc.
278. R. S. Davidson and P. F. Lambeth. *Chem. Commun.*, 1265 (1967).
279. S. G. Cohen and J. I. Cohen. *J. Amer. Chem. Soc.*, **89**, 164 (1967).
280. S. G. Cohen and J. B. Guttenplan. *Tetrahedron Lett.*, 5353 (1968).
281. S. G. Cohen and R. J. Baumgarten. *J. Amer. Chem. Soc.*, **87**, 2996 (1965); **89**, 3471 (1967).
282. R. S. Davidson. *Chem. Comm.*, 575 (1966).
283. S. G. Cohen and N. M. Stein. *J. Amer. Chem. Soc.*, **93**, 6542 (1971).
284. R. S. Davidson, P. F. Lambeth, F. A. Younis and R. Wilson. *J. Chem. Soc. (C)*, 2203 (1969).
285. A. L. Buley and R. O. C. Norman. *Proc. Chem. Soc. London*, 225 (1964).

286. S. Arimitsu and H. Tsubomura. *Bull. Chem. Soc. Jap.,* **45,** 1357 (1972); S. Arimitsu and H. Tsubomura. *Bull. Chem. Soc. Jap.,* **44,** 2288 (1971).
287. R. S. Davidson and R. Wilson. *J. Chem. Soc. (B),* 71 (1970).
288. R. S. Davidson and M. Santhanam. *J. Chem. Soc., Perkin II,* 2355 (1972).
289. H. D. Roth. *Mol. Photochem.,* **5,** 91 (1973).
290. P. W. Atkins, K. A. McLauchlan and P. W. Percival. *Chem. Commun.,* 121 (1973).
291. H. L. J. Backstrom and K. Sandros. *Acta Chem. Scand.,* **12,** 823 (1958).
292. N. J. Turro and R. Engel. *Mol. Photochem.,* **1,** 143 (1969); N. J. Turro and R. Engel. *J. Amer. Chem. Soc.,* **91,** 7113 (1969).
293. G. A. Davis and S. G. Cohen. *Chem. Commun.,* 622 (1970).
294. L. A. Singer. *Tetrahedron Lett.,* 923 (1969).
295. R. A. Caldwell. *Tetrahedron Lett.,* 2121 (1969).
296. B. M. Monroe and R. P. Groff. *Tetrahedron Lett.,* 3955 (1973).
297. R. S. Davidson, Unpublished results.
298. A. K. Davies, G. A. Gee, J. F. McKellar and G. O. Phillips. *J. Chem. Soc., Perkin II,* 1742 (1973).
299. S. G. Cohen and A. D. Litt. *Tetrahedron Lett.,* 837 (1970).
300. S. G. Cohen, G. A. Davis and W. D. K. Clark. *J. Amer. Chem. Soc.,* **94,** 869 (1972).
301. S. G. Cohen and N. Stein. *J. Amer. Chem. Soc.,* **91,** 3690 (1969).
302. C. Pac, H. Sakurai and T. Tosa. *Chem. Commun.,* 1311 (1970).
303. L. A. Singer, R. E. Brown and G. A. Davis. *J. Amer. Chem. Soc.,* **95,** 8638 (1973); A. A. Baum and L. Karnischky. *J. Amer. Chem. Soc.,* **95,** 3072 (1973).
304. G. Irick and G. C. Newland. *Tetrahedron Lett.,* 4151 (1970).
305. S. G. Cohen and B. Green. *J. Amer. Chem. Soc.,* **91,** 6824 (1969).
306. D. Seebach and H. Daum. *J. Amer. Chem. Soc.,* **93,** 2795 (1971).
307. S. G. Cohen, M. D. Saltzman and J. B. Guttenplan. *Tetrahedron Lett.,* 4321 (1969).
308. S. G. Cohen and J. I. Cohen. *J. Phys. Chem.,* **72,** 3782 (1968).
309. T. H. Koch and A. H. Jones. *J. Amer. Chem. Soc.,* **92,** 7503 (1970).
310. C. C. Wamser, G. S. Hammond, C. T. Chang and C. Baylor. *J. Amer. Chem. Soc.,* **92,** 6362 (1970).
311. L. A. Singer, G. A. Davis and V. P. Muralidharan. *J. Amer. Chem. Soc.,* **91,** 897 (1969).
312. J. B. Guttenplan and S. G. Cohen. *Tetrahedron Lett.,* 2125 (1969).
313. G. A. Davis, P. A. Carapellucci, K. Szoc and J. D. Gresser. *J. Amer. Chem. Soc.,* **91,** 2264 (1969).
314. R. S. Davidson, K. Harrison and P. R. Steiner. *J. Chem. Soc. (C),* 3480 (1971).
315. G. Porter, S. K. Dogra, R. O. Loutfy, S. E. Sugamori and R. W. Yip. *J. Chem. Soc., Faraday I,* 1462 (1973).
316. H.-S. Ryang and H. Sakurai. *Chem. Commun.,* 824 (1973).
317. D. L. Bunbury and T. M. Chan. *Can. J. Chem.,* **50,** 2499 (1972).
318. R. A. Clasen and S. Searles. *Chem. Commun.,* 289 (1966).
319. J. Hill and J. Townend. *Tetrahedron Lett.,* 4607 (1970); J. Hill and J. Townend. *J. Chem. Soc., Perkin I,* 1210 (1972).
320. E. H. Gold. *J. Amer. Chem. Soc.,* **93,** 2793 (1971).
321. H. J. Roth and M. H. El Raie. *Tetrahedron Lett.,* 2445 (1970).
322a. A. Padwa and L. Hamilton. *J. Amer. Chem. Soc.,* **89,** 102 (1967).
322b. A. Padwa and A. Battisti. *J. Org. Chem.,* **36,** 230 (1971).
323. A. Padwa. *Accounts Chem. Res.,* **4,** 48 (1971).
324. A. Padwa and W. Eisenhardt. *J. Amer. Chem. Soc.,* **93,** 1400 (1971).

325. A. Padwa and R. Gruber. *J. Amer. Chem. Soc.*, **92**, 107 (1970); T.-Y. Chen. *Bull. Chem. Soc. Jap.*, **41**, 2540 (1968).
326. A. Padwa, F. Albrecht, P. Singh and E. Vega. *J. Amer. Chem. Soc.*, **93**, 2928 (1971).
327. P. J. Wagner, A. E. Kemppainen and T. Jellinek. *J. Amer. Chem. Soc.*, **94**, 7512 (1972).
328. P. J. Wagner and A. E. Kemppainen. *J. Amer. Chem. Soc.*, **91**, 3085 (1969).
329. P. J. Wagner and T. Jellinek. *J. Amer. Chem. Soc.*, **93**, 7328 (1971).
330. A. Padwa, W. Eisenhardt, R. Gruber and D. Pashayan. *J. Amer. Chem. Soc.*, **93**, 6998 (1971).
331. P. J. Wagner. *Accounts Chem. Res.*, **4**, 168 (1971).
332. R. S. Davidson and P. F. Lambeth, unpublished results.
333. R. S. Cooke and G. S. Hammond. *J. Amer. Chem. Soc.*, **90**, 2958 (1968); R. S. Cooke and G. S. Hammond. *J. Amer. Chem. Soc.*, **92**, 2739 (1970).
334. K. Mislow, M. Axelrod, D. R. Rayner, H. Gotthardt, L. M. Coyne and G. S. Hammond. *J. Amer. Chem. Soc.*, **87**, 4958 (1965); G. S. Hammond, H. Gotthardt, L. M. Coyne, M. Axelrod, D. R. Rayner and K. Mislow. *J. Amer. Chem. Soc.*, **87**, 4959 (1965).
335. G. Balavoine, S. Jugé and H. B. Kagan. *Tetrahedron Lett.*, 4159 (1973).
336a. J. Guttenplan and S. G. Cohen. *Chem. Commun.*, 247 (1969).
336b. J. B. Guttenplan and S. G. Cohen. *J. Org. Chem.*, **38**, 2001 (1973).
336c. W. Ando, J. Suzuki and T. Migita. *Bull. Chem. Soc. Jap.*, **44**, 1987 (1971).
337. A. Padwa and D. Pashayan. *J. Org. Chem.*, **36**, 3550 (1971).
338. R. G. Zepp and P. J. Wagner. *Chem. Commun.*, 167 (1972).
339a. H. Gruen, H. N. Schott, G. W. Byers, H. G. Giles and J. A. Kampmeier. *Tetrahedron Lett.*, 3925 (1972).
339b. G. W. Byers, H. Gruen, H. G. Giles, H. N. Schott and J. A. Kampmeier. *J. Amer. Chem. Soc.*, **94**, 1016 (1972).
340a. R. C. Cookson, J. Hudec and N. A. Mirza. *Chem. Commun.*, 180 (1968).
340b. R. C. Cookson, J. Hudec and N. A. Mirza. *Chem. Commun.*, 824 (1967).
340c. H. Yamashita, H. Kokubun and M. Koizumi. *Bull. Chem. Soc. Jap.*, **41**, 2312 (1968).
340d. S. Murata, H. Kokubun and M. Koizumi. *Z. Phys. Chem.*, (*Frankfurt am Main*), **70**, 47 (1970).
341. P. Grandclaudon and A. Lablache-Combier. *Chem. Commun.*, 892 (1971); P. Grandclaudon, A. Lablache-Combier and C. Parkanyi. *Tetrahedron.* **29**, 651 (1973).
342. A. Lablache-Combier and A. Pollet. *Tetrahedron*, **28**, 3141 (1972).
343. J. Nasielski, A. Kirsch-Demesmaeker, P. Kirsch and R. Nasielski-Hinkens. *Chem. Commun.*, 302 (1970).
344. S. Niizuma, H. Kokubun and M. Koizumi. *Chem. Phys. Lett.*, **7**, 279 (1970); S. Niizume, H. Kokubun and M. Koizumi, *Bull. Chem. Soc. Jap.*, **44**, 335 (1971).
345. G. A. Davis, J. D. Gresser and P. A. Carapellucci. *J. Amer. Chem. Soc.*, **93**, 2179 (1971).
346. G. A. Davis and S. G. Cohen. *Chem. Commun.*, 675 (1971).
347. N. Baumann. *Helv. Chim. Acta*, **56**, 2227 (1973).
348. D. R. G. Brimage, R. S. Davidson and P. R. Steiner. *J. Chem. Soc.*, *Perkin I*, 526 (1973).
349. N. C. Yang, L. S. Gorelic and B. Kim. *Photochem. and Photobiol.*, **13**, 275 (1971).
350. A. Stankunas, I. Rosenthal and J. N. Pitts. *Tetrahedron Lett.*, 4779 (1971).
351. D. Elad and I. Rosenthal. *Chem. Commun.*, 905 (1969).

352. G. R. Penzer and G. K. Radda. *Biochem. J.*, **109**, 259 (1968).

353a. P. Byrom and J. H. Turnbull. *Photochem. and Photobiol.*, **6**, 125 (1967).

353b. P. Byrom and J. H. Turnbull. *Photochem. and Photobiol.*, **8**, 243 (1968).

354. B. Nathanson, M. Brody, S. Brody and S. B. Broyde. *Photochem. and Photobiol.*, **6**, 177 (1967).

355. R. S. Davidson and P. R. Steiner, unpublished results.

356. W. Hendrich. *Biochim. Biophys. Acta*, **162**, 265 (1968).

357a. Y. M. Stolovitskii and V. B. Evstigneev. *Mol. Biol.*, **3**, 176 (1969); *Chem. Abs.*, **71**, 2735 (1969).

357b. G. Eigenmann. *Helv. Chim. Acta*, **46**, 855 (1963).

358. G. R. Seely and K. Talmadge. *Photochem. and Photobiol.* **3**, 195 (1964).

359a. H. Fischer, H. E. A. Kramer and A. Maute. *Z. Phys. Chem.* (*Frankfurt am Main*), **69**, 113 (1970); H. E. A. Kramer and A. Maute. *Photochem. and Photobiol.*, **15**, 7 (1972); M. Zügel, Th. Förster and H. E. A. Kramer. *Photochem. and Photobiol.*, **15**, 33 (1972).

359b. M. Morita and S. Kato. *Bull. Chem. Soc. Jap.*, **42**, 25 (1969).

360. R. F. Bartholomew and R. S. Davidson. *J. Chem. Soc. C*, 2347 (1971).

361. S. Matsumoto. *Bull. Chem. Soc. Jap.*, **35**, 1860, 1866 (1962); **37**, 491, 499 (1964).

362. F. C. Goodspeed, B. L. Scott and J. G. Burr. *J. Phys. Chem.*, **69**, 1149 (1965).

363a. J. Saltiel, H. C. Curtis, L. Metts, J. W. Miley, J. Winterle and M. Wrighton. *J. Amer. Chem. Soc.*, **92**, 410 (1970).

363b. J. Saltiel, H. C. Curtis and B. Jones. *Mol. Photochem.*, **2**, 331 (1970).

364a. D. I. Schuster, T. M. Weil and M. R. Topp. *Chem. Commun.*, 1212 (1971).

364b. D. I. Schuster, T. M. Weil and A. M. Halpern. *J. Amer. Chem. Soc.*, **94**, 8248 (1972).

364c. D. I. Schuster and T. M. Weil. *J. Amer. Chem. Soc.*, **95**, 4091 (1973).

365. J. Dedinas, *J. Phys. Chem.*, **75**, 181 (1971); A. V. Buettner and J. Dedinas. *J. Phys. Chem.*, **75**, 187 (1971).

366. R. O. Loutfy and R. W. Yip. *Can. J. Chem.*, **51**, 1881 (1973).

367. R. S. Davidson and R. Wilson. *Mol. Photochem.*, **6**, 231 (1974).

368. D. Bryce-Smith, G. B. Cox and A. Gilbert. *Chem. Commun.*, 914 (1971).

369. P. J. Wagner and R. A. Leavitt. *J. Amer. Chem. Soc.*, **95**, 3669 (1973).

370. E. J. Baum and R. O. C. Norman. *J. Chem. Soc.* (*B*), 227 (1968).

371 K. Fukui and Y. Odaira. *Tetrahedron Lett.*, 5255 (1969); K. Fukui, K. Senda, Y. Shigemitsu and Y. Odaira. *J. Org. Chem.*, **37**, 3176 (1972).

372. D. G. Whitten and Y. J. Lee. *J. Amer. Chem. Soc.*, **93**, 961 (1971).

373. W. H. Walker, P. Hemmerich and V. Massey. *Eur. J. Biochim.*, **13**, 258 (1970).

374. G. N. Taylor and G. S. Hammond. *J. Amer. Chem. Soc.*, **94**, 3687 (1972).

375. S. L. Murov, L.-S. Yu and L. P. Giering. *J. Amer. Chem. Soc.*, **95**, 4329 (1973).

376a. S. L. Murov, R. S. Cole and G. S. Hammond. *J. Amer. Chem. Soc.*, **90**, 2957 (1968); G. S. Hammond, N. J. Turro and A. Fischer. *J. Amer. Chem. Soc.*, **83**, 4674 (1961); G. S. Hammond, P. Wyatt, C. D. DeBoer and N. J. Turro. *J. Amer. Chem. Soc.*, **86**, 2532 (1964).

376b. S. Murov and G. S. Hammond. *J. Phys. Chem.*, **72**, 3797 (1968).

376c. B. S. Solomon, C. Steel and A. Weller. *Chem. Commun.*, 927 (1969).

377. V. Y. Merritt, J. Cornelisse and R. Srinivasan. *J. Amer. Chem. Soc.*, **95**, 8250 (1973).

378a. J. Cornelisse, V. Y. Merritt and R. Srinivasan. *J. Amer. Chem. Soc.*, **95**, 6197 (1973).

378b. R. Srinivasan. *J. Amer. Chem. Soc.*, **93**, 3555 (1971).

379. F. D. Lewis and R. H. Hirsch. *Tetrahedron Lett.*, 4947 (1973).
380a. N. C. Yang and J. Libman. *J. Amer. Chem. Soc.*, **94**, 9228 (1972).
380b. N. C. Yang, J. Libman and M. F. Savitzky. *J. Amer. Chem. Soc.*, **94**, 9226 (1972).
381. P. M. Froehlich and H. A. Morrison. *J. Amer. Chem. Soc.*, **96**, 332 (1974).
382. O. L. Chapman and R. D. Lura. *J. Amer. Chem. Soc.*, **92**, 6352 (1970).
383a. R. M. Bowman, T. R. Chamberlain, C. W. Huang and J. J. McCullough. *J. Amer. Chem. Soc.*, **92**, 4106 (1970).
383b. R. M. Bowman and J. J. McCullough. *Chem. Commun.*, 948 (1970); R. M. Bowman, C. Calvo, J. J. McCullough, R. C. Miller and I. Singh. *Can. J. Chem.*, **51**, 1060 (1973).
384a. D. Bryce-Smith, B. E. Foulger, A. Gilbert and P. J. Twitchett. *Chem. Commun.*, 794 (1971).
384b. J. J. McCullough and W. S. Wu. *Chem. Commun.*, 1136 (1972).
385. C. S. Angadiyavar, J. Cornelisse, V. Y. Merritt and R. Srinivasan. *Tetrahedron Lett.*, 4407 (1973).
386. D. Bryce-Smith, A. Gilbert and B. H. Orger. *Chem. Commun.*, 512 (1966).
387a. D. Bryce-Smith, A. Gilbert and J. Grzonka. *Chem. Commun.*, 498 (1970).
387b. D. Bryce-Smith, R. Deshpande, A. Gilbert and J. Grzonka. *Chem. Commun.*, 561 (1970).
388. D. Bryce-Smith, B. E. Foulger and A. Gilbert. *Chem. Commun.*, 769 (1972).
389. P. Lechtken and G. Hesse. *Annalen,* **754**, 1, 8 (1971); H.-D. Scharf and R. Klar. *Chem. Ber.*, **105**, 575 (1972).
390. R. Srinivasan. *J. Phys. Chem.*, **76**, 15 (1972).
391. H. Morrison and W. I. Ferree. *Chem. Commun.*, 268 (1969); W. Ferree, J. B. Grutzner and H. Morrison. *J. Amer. Chem. Soc.*, **93**, 5502 (1971).
392. W. Lippke, W. Ferree and H. Morrison. *J. Amer. Chem. Soc.*, **96**, 2134 (1974).
393. H. Morrison, J. Pajak and R. Peiffer. *J. Amer. Chem. Soc.*, **93**, 3978 (1971).
394. R. J. McDonald and B. K. Selinger. *Z. Phys. Chem.* (Frankfurt am Main), **69**, 132 (1970); W. H. F. Sasse, P. J. Collin, D. B. Roberts and G. Sugowdz. *Aust. J. Chem.*, **24**, 2151 (1971); G. Sugowdz, P. J. Collin and W. H. F. Sasse. *Aust. J. Chem.*, **26**, 147 (1973).
395. E. Grovenstein, T. C. Campbell and T. Shibata. *J. Org. Chem.*, **34**, 2418 (1969).
396. J. J. McCullough and C. W. Huang. *Can. J. Chem.*, **47**, 757 (1969).
397. T. R. Chamberlain and J. J. McCullough. *Can. J. Chem.*, **51**, 2578 (1973).
398. C. Pac, T. Sugioka, K. Mizumo and H. Sakurai. *Bull. Chem. Soc. Jap.*, **46**, 238 (1973); K. Mizuno, C. Pac and H. Sakurai. *Chem. Commun.*, 219 (1973).
399. D. R. Arnold, L. B. Gillis and E. B. Whipple. *Chem. Commun.*, 918 (1964).
400a. C. Pac, T. Sugioka and H. Sakurai. *Chem. Letters* (*Tokyo*), 39 (1972).
400b. T. Sugioka, C. Pac and H. Sakurai. *Chem. Letters* (*Tokyo*), 667 (1972).
400c. T. Sugioka, C. Pac and H. Sakurai. *Chem. Letters* (*Tokyo*), 791 (1972).
401a. R. A. Caldwell. *J. Amer. Chem. Soc.*, **95**, 1690 (1973); G. Kaupp. *Angew. Chem. Int. Ed.*, **12**, 765 (1973).
401b. T. Sasaki, K. Kanematsu and K. Hayakawa. *J. J. Amer. Chem. Soc.*, **95**, 5632 (1973).
402. G. Kaupp. *Angew. Chem. Int. Ed. Engl.,* **11**, 313 (1972).
403. G. Kaupp. *Angew. Chem. Int. Ed. Engl.*, **11**, 718 (1972).
404a. N. C. Yang and J. Libman. *J. Amer. Chem. Soc.*, **94**, 1405 (1972).
404b. N. C. Yang, J. Libman, L. W. Barrett, M. H. Hui and R. L. Loeschen. *J. Amer. Chem. Soc.*, **94**, 1406 (1972).
405. N. J. Turro and P. A. Wriede. *J. Org. Chem.*, **34**, 3562 (1969); N. J. Turro, M.

Niemczyk and D. M. Pond. *Mol. Photochem.*, **2**, 345 (1970); N. J. Turro, C. Lee, N. Schore, J. Barltrop and H. A. J. Carless. *J. Amer. Chem. Soc.*, **93**, 3079 (1971); J. A. Barltrop and H. A. J. Carless. *J. Amer. Chem. Soc.*, **94**, 1951 (1972); M. P. Niemczyk, N. E. Schore and N. J. Turro. *Mol. Photochem.*, **5**, 69 (1973).

406. D. R. Arnold. *Adv. Photochem.*, **6**, 301 (1968).
407. R. A. Caldwell and S. P. James. *J. Amer. Chem. Soc.*, **91**, 5184 (1969).
408a. S. R. Kurowsky and H. Morrison. *J. Amer. Chem. Soc.*, **94**, 507 (1972).
408b. D. O. Cowan and A. A. Baum. *J. Amer. Chem. Soc.*, **93**, 1153 (1971).
409. R. A. Caldwell and R. P. Gajewski. *J. Amer. Chem. Soc.*, **93**, 532 (1971).
410. G. F. Vesley and B. A. Prichard. *Mol. Photochem.*, **5**, 355 (1973).
411. J. T. Yardley. *J. Amer. Chem. Soc.*, **94**, 7283 (1972).
412. N. C. Yang and W. Eisenhardt. *J. Amer. Chem. Soc.*, **93**, 1277 (1971).
413. N. C. Yang, M. Kimura and W. Eisenhardt. *J. Amer. Chem. Soc.*, **95**, 5058 (1973).
414. R. R. Hautala and N. J. Turro. *J. Amer. Chem. Soc.*, **93**, 5595 (1971).
415. J. A. Barltrop and H. A. J. Carless. *J. Amer. Chem. Soc.*, **93**, 4794 (1971).
416. E. J. Corey, J. D. Bass, R. Le Mahieu and R. B. Mitra. *J. Amer. Chem. Soc.*, **86**, 5570 (1964).
417. P. J. Wagner and D. J. Buchek. *J. Amer. Chem. Soc.*, **91**, 5090 (1969); P. J. Wagner and D. J. Bucheck. *Can. J. Chem.*, **47**, 713 (1969).
418. S. Farid and D. Hess. *Chem. Ber.*, **102**, 3747 (1969).
419. Y. L. Chow, T. C. Joseph, H. H. Quon and J. N. S. Tam. *Can. J. Chem.*, **48**, 3045 (1970).
420. H. J. T. Bos, H. Polman and P. F. E. van Montfort. *Chem. Commun.*, 188 (1973).
421a. Y. Shigemitsu, H. Nakai and Y. Odaira. *Tetrahedron*, **25**, 3039 (1969).
421b. Y. Shigemitsu, Y. Katsuhara and Y. Odaira. *Tetrahedron Lett.*, 2887 (1971).
421c. Y. Katsuhara, Y. Shigemitsu and Y. Odaira. *Bull. Chem. Soc. Jap.*, **44**, 1169 (1971).
422a. T. S. Cantrell. *Chem. Commun.*, 468 (1973).
422b. T. S. Cantrell. *J. Amer. Chem. Soc.*, **95**, 2714 (1973).
423. J. O. Pavlik, P. I. Plooard, A. C. Somersall and J. E. Guillet. *Can. J. Chem.*, **51**, 1435 (1973).
424. A. Harriman and B. W. Rockett. *J. Chem. Soc.*, *Perkin II*, 1624 (1973).
425. H. van Zwet. *Rec. Trav. chim. Pays-Bas*, **87**, 1201 (1968).
426. N. C. Perrins and J. P. Simons. *Trans. Faraday Soc.*, **65**, 390 (1969). J. D. Simons and A. L. Smith. *Chem. Phys. Letts.*, **16**, 536 (1972).
427. G. Schlicht and D. Schulte-Frohlinde. *Photochem. and Photobiol.*, **16**, 183 (1972).
428a. E. J. Bowen and K. K. Rohatgi. *Disc. Faraday Soc.*, **14**, 146 (1953).
428b. E. J. Bowen and J. Sahu. *J. Phys. Chem.*, **63**, 4 (1959); M.S.S.C. Leite and K. R. Naqvi. *Chem. Phys. Letts.*, **4**, 35 (1969).
428c. W. R. Ware and C. Lewis. *J. Chem. Phys.*, **57**, 3546 (1972); C. Lewis and W. R. Ware. *Chem. Phys. Letts.*, **15**, 290 (1972).
429. M. T. McCall, G. S. Hammond, O. Yonemitsu and B. Witkop. *J. Amer. Chem. Soc.*, **92**, 6991 (1970); F. A. Carroll, M. T. McCall and G. S. Hammond. *J. Amer. Chem. Soc.*, **95**, 315 (1973).
430. J. J. Dannenberg, K. Dill and H. P. Waits. *Chem. Commun.*, 1348 (1971).
431a. O. Yonemitsu, Y. Okuno, Y. Kanaoka and B. Witkop. *J. Amer. Chem. Soc.*, **92**, 5686 (1970).
431b. Y. Okuno, M. Kawamori and O. Yonemitsu. *Tetrahedron Lett.*, 3009 (1973); O. Yonemitsu, H. Nakai, Y. Okuno, S. Naruto, K. Hemmi and B. Witkop.

Photochem. and Photobiol., **15**, 509 (1972); O. Yonemitsu, H. Nakai, Y. Kanaoka, I. L. Karle and B. Witkop. *J. Amer. Chem. Soc.*, **92**, 5691 (1970).

432a. S. Naruto, O. Yonemitsu, N. Kanamaru and K. Kimura. *J. Amer. Chem. Soc.*, **93**, 4053 (1971).

432b. T. Iwakuma, O. Yonemitsu, N. Kanamaru, K. Kimura and B. Witkop. *Angew. Chem. Int. Ed. Engl.*, **12**, 72 (1973); T. Iwakuma, H. Nakai, O. Yonemitsu, D. S. Jones, I. L. Karle and B. Witkop. *J. Amer. Chem. Soc.*, **94**, 5136 (1972); H. H. Ong and E. L. May. *J. Org. Chem.*, **37**, 712 (1972); H. H. Ong and E. L. May. *J. Org. Chem.*, **38**, 924 (1973).

433. H. H. Ong and E. L. May. *J. Org. Chem.*, **35**, 2544 (1970); C. M. Foltz. *J. Org. Chem.*, **36**, 24 (1971).

434. P. S. Engel and P. D. Bartlett. *J. Amer. Chem. Soc.*, **92**, 5883 (1970).

435a. A. C. Day and T. R. Wright. *Tetrahedron Lett.*, 1067 (1969).

435b. B. S. Solomon, T. F. Thomas and C. Steel. *J. Amer. Chem. Soc.*, **90**, 2249 (1968).

435c. T. R. Evans. *J. Amer. Chem. Soc.*, **93**, 2081 (1971).

436. R. Beer, K. M. C. Davis and R. Hodgson. *Chem. Commun.*, 840 (1970).

437. G. A. Davis. *Chem. Commun.*, 728 (1973).

438. G. O. Phillips, N. W. Worthington, J. F. McKellar and R. R. Sharpe. *J. Chem. Soc. (A)*, 767 (1969).

439a. K. P. Clark and H. I. Stonehill. *J. Chem. Soc.*, *Faraday I*, 1676 (1972); K. Clark and H. I. Stonehill. *J. Chem. Soc.*, *Faraday I*, 577 (1972).

439b. M. Ahmed, A. K. Davies, G. O. Phillips and J. T. Richards. *J. Chem. Soc.*, *Perkin I*, 1386 (1973); M. S. Walker, M. A. Abkowitz, R. W. Bieglow and J. H. Sharp. *J. Phys. Chem.* **77**, 987 (1973); A. K. Davies, R. Ford, G. A. Gee, J. F. McKellar and G. O. Phillips. *Chem. Commun.*, 873 (1972); A. D. Broadbent and R. P. Newton. *Can. J. Chem.*, **50**, 381 (1972).

440. J. den Heijer, T. Spee, G. P. de Gunst and J. Cornelisse. *Tetrahedron Lett.*, 1261 (1973).

441a. A. Cu and A. C. Testa. *J. Phys. Chem.*, **77**, 1487 (1973).

441b. G. G. Wubbels, J. W. Jordan and N. S. Mills. *J. Amer. Chem. Soc.*, **95**, 1281 (1973).

442. J. D. Margerum, A. M. Lackner, M. J. Little and C. T. Petrusis. *J. Phys. Chem.*, **75**, 3066 (1971).

443. A fascinating and simple account of photosynthesis is given in "Light and Living Matter, Vol. 2", R. K. Clayton, McGraw Hill Book Co., New York, 1971.

444. P. Loach and J. J. Katz. *Photochem. and Photobiol.*, **17**, 195 (1973).

445. J. R. Harbour and G. Tollin. *Photochem. and Photobiol.*, **19**, 163 (1974).

446. J. R. Harbour and G. Tollin. *Photochem. and Photobiol.*, **19**, 69 (1974); J. R. Harbour and G. Tollin. *Photochem. and Photobiol.*, **19**, 147 (1974).

447. J. T. Warden and J. R. Bolton. *J. Amer. Chem. Soc.*, **94**, 4351 (1972).

448. J. R. Bolton, R. K. Clayton and D. W. Reed. *Photochem. and Photobiol.*, **9**, 209 (1969).

449. P. A. Loach and K. Walsh. *Biochem.*, **8**, 1908 (1969); J. D. McElroy, G. Feker and D. C. Manzerall. *Biochim. Biophys. Acta*, **172**, 180 (1969).

Author Index

335

M

Subject Index

A

Absorption spectra of exciplexes, 234
Acceptor charge-density in the ground state, 35
Acenaphthylene, 268
  photodimerization of, 260
  reaction with amines, 244
  reaction with maleic anhydride, 269
Acetone, fluorescence, 314
  reaction with amines, 287
Acetylenes, reaction of, 307
2-Acetylnaphthalene, reaction of, 279, 286, 309, 311
Acid anhydride, complexes of, 183, 241
  dipole moments of, 139
Acridine, reaction with amines, 296
  reduction of, 301
Acridine, reaction with amines, 296
  reduction of, 301
Acridinium, quenching of, 255
Acridinium salts, 170, 218
Acrylonitrile, reaction with indene, 308
  reaction with naphthalene, 305
  reaction with methoxynaphthalenes, 309
Activity effects, 201
Activity method for determining association constants, 202
Adenine, complexes of, 274
Aggregation, effect on production of reactions of excited states, 219
N-alkylanilines, reaction with halides, 252
Allylthiourea, reaction of, 298
Amides, complexes with iodine, 125
Amines, complexes of, 232
  complexes of, with halogen-substituted compounds, 251
  complexes with iodine, 124, 126
  complexes of with oxygen, 246
  interaction with hydrocarbons, 272

  photoreactions of, 269
  reaction with carbonyl compounds, 275
  reaction with heterocyclic compounds, 295
  reaction with nitro-compounds, 252
  reaction with quinones, 252
  reaction with unsaturated carbonyl compounds and esters, 293
  tertiary, as quenchers, 278
Amino-acids, decarboxylation of, 270, 283, 286, 287, 295, 296, 297, 298
Aminoanthraquinones, reactions with amine, 285
Aminobenzophenones, reactions of, 279, 286
Anhydrides, complexes of, 183, 241
Aniline, reaction with benzoquinone, 227
Anilinium, photoreactions of, 256
Anions as donors, 254
Anthracenes, intramolecular excimer formation by, 264
  photodimerization of, 228
  reactions of, 238, 258, 269, 309
9-Anthracenylacetic acid, 315
Anthraquinones, reactions of, 317
Anthracene, reactions of, 309, 314
Anthraquinones, fluorescence of, 278
Antiferromagnetic exchange, 68
Aromatic hydrocarbons, photodimerization of, 258
  photoreactions of, 269
  reactions of, 301
  reactions with halogen-containing compounds, 314
  reactions with sulphides and sulph-oxides, 291
  reactions with olefins, 302
  reactions with purines and pyrimidines, 274